Juan J. Tarín · Antonio Cano (Eds.)

Fertilization in Protozoa and Metazoan Animals

Springer-Verlag Berlin Heidelberg GmbH

Juan J. Tarín · Antonio Cano (Eds.)

Fertilization in Protozoa and Metazoan Animals

Cellular and Molecular Aspects

With 43 Figures

Springer

J.J. TARÍN, PhD
Department of Animal Biology
Faculty of Biological Sciences
University of Valencia, Burjassot
46100 Valencia
Spain

A. CANO, MD
Department of Paediatrics
Obstetrics and Gynaecology
Faculty of Medicine
University of Valencia
46010 Valencia
Spain

ISBN 978-3-540-67093-3

Library of Congress Cataloging-in-Publication Data
Fertilization in protozoa and metazoan animals : cellular and molecular aspects / Juan J. Tarín, Antonio Cano (eds.).
p. cm.
Includes bibliographical references (p.).
ISBN 978-3-540-67093-3 **ISBN 978-3-642-58301-8 (eBook)**
DOI 10.1007/978-3-642-58301-8
1. Fertilization (Biology) 2. Physiology, Comparative. I. Tarín, Juan J., 1959- II. Cano, Antonio, 1954-
QH485 .F476 2000
571.8′64–dc21

Originally published by Springer-Verlag Berlin Heidelberg New York in 2000

Cover design: Design & Production GmbH, Heidelberg
Typesetting: Best-set Typesetter Ltd., Hong Kong
SPIN 106970601 31/3130 – 5 4 3 2 1 0 – Printed on acid-free paper

Preface

Reproduction is the origination of new organisms from pre-existing ones. Among more than 35 separated forms of reproduction including several types of gamogony, parthenogenesis, agamogenesis, fission and division, and plasmotomy, the bisexual mode of reproduction via fertilization provides genetic variability that allows species to adapt quickly to competitive and constantly changing environments. Several excellent reviews and books have been written in the past to analyse the mechanisms of fertilization in different eukaryotic species. During the last few years, however, renewed attention has been paid to examining the process of oocyte fertilization at the cellular/molecular level not only within a single species/group but also through different phylogenetic lineages. As a result of this effort, knowledge of the molecular pathways used by oocytes and spermatozoa at fertilization has increased, but still many questions remain to be answered. Being aware of the necessity of providing an integrated view of the process of fertilization, this book has been entirely devoted to reviewing the process of oocyte fertilization at the cellular/molecular level in two different and separated groups of eukaryotic organisms: protozoa and metazoan animals.

The book is organized into six sections dealing with oocyte fertilization in protozoa, invertebrates, teleost fishes, amphibians, birds and mammals. These sections are followed by a summary/concluding chapter that provides a comparative overview of the process of fertilization in these groups of eukaryotes. As research on fertilization at the cellular/molecular level is limited and/or fragmentary when compared to the huge diversity of protozoan and animal species, we chose a single or a reduced number of species (those from whom more data were available) as the representative(s) of each group. Although the strategy of extrapolating the mechanism(s) used at fertilization by a single or a reduced number of species to the rest of the species of the same phylogenetic group must be taken with caution because it overlooks the existent interspecies variability, it can be used for integrative purposes. Each of the six major sections begins with a general description of the sexual and asexual reproduction strategies of the species belonging to a particular group. Then, a description of the reproductive cycle as well as the design of the body and reproductive system (including gametogenesis) of the representative species is given. These form the introductory steps before focusing on the cellular/molecular aspects of fertilization. Special care was taken to analyse all the fertilization steps from sperm capacitation to karyogamy. However, in many cases, a scarce or total absence of information precluded us from following this stratagem and some groups were partially treated or even not included in this monograph.

A major inspiration for this book was the work of Prof. Juan Brines, University of Valencia, on some paediatric topics from an evolutionary perspective. His insight on infant and child feeding from a biological point of view impacted so deeply on our understanding as undergraduate students that it spurred us on to produce this book. We thank Prof. Brines, not only for his inspiring work, but also for his constant support and helpful discussions on evolutionary issues. We hope this book will be the first step in a series of additional thoughts and analyses of oocyte fertilization from a comparative and evolutionary viewpoint.

University of Valencia, Spain JUAN J. TARÍN and ANTONIO CANO

Contents

6 Fertilization in Monotreme, Marsupial and Eutherian Mammals

K.E. Mate, M.S. Harris, and J.C. Rodger

CHAPTER 1

Fertilization in Protozoa

KAZUYUKI MIKAMI[1]

1 Reproduction Strategies in Protozoa

Protozoa are unicellular eukaryotes. Most of them are free-living. Some of them are symbiotic. Some have both haploid and diploid phases in their life cycle. They show a diversity of reproduction that shares common important basic characteristics of fertilization with multicellular organisms. However, fertilization has never been found in certain groups of protozoa such as the euglenoid flagellates (Phytomastigophorea) and amoebae (Rhizopodea). (For general information on protozoa see Grell 1967, 1973; Hausmann and Hülsmann 1996.)

Fertilization is the fusion of the nuclei of the two gametes (pronuclei) to form a synkaryon (Grell 1967). Among protozoa, the gametes and the vegetative reproducing cells are usually not fundamentally different from each other. Both are unicellular and at the same time whole organisms. Fertilization occurs by the fusion of two gametes or gamete nuclei. In some cases, the gametes are sexually differentiated into usually two types (male and female or + and −). They are sometimes amoebic and sometimes flagellated. In other cases, sexual differentiation occurs in the parental cells, which are then called gamonts.

The diversity of reproduction and the differences in the fertilization process among protozoa are due to differences among gamonts, gametes and their sexual differentiation. There are three types of fertilization processes (Grell 1968): (1) fusion of gametes (gametogamy), (2) fusion of gamete nuclei produced by the same parental cell (autogamy), and (3) fusion of gamonts (gamontogamy).

1.1 Gametogamy

When gametes fuse with each other to form a zygote, we call it gametogamy. If the gametes are morphologically different, this is called anisogamy. If they are morphologically alike, then this is called isogamy.

[1] EEC, Miyagi University of Education, Aoba, Aramaki, Aoba-ku, Sendai 980-0845, Japan

Juan J. Tarín and Antonio Cano (Eds.)
Fertilization in Protozoa and Metazoan Animals

Among most of the non-colonial species, such as *Chlamydomonas*, vegetative cells differentiate directly into gametes that are of the same size and shape as the original vegetative cells (isogamy). However, a few anisogamic species, such as *C. suboogama*, produce large immotile macrogametes and small motile microgametes (anisogamy) (Tschermak-Woess 1959). In this species, a gamont divides twice to produce four daughter cells in a gelatinous mass. Three of them become large gamonts and the remaining one develops into a small gamont. The small gamont divides twice to produce four small microgametes. The large gamonts do not divide but are transformed into macrogametes. The microgametes fuse with macrogametes.

Among the colonial flagellates and ciliates, such as *Zoothamnium*, the colony is composed of somatic cells and generative cells. The isogametic genera of the flagellates include *Stephanosphaera*, *Gonium* and *Pandorina*, and the anisogametic genera include *Eudorina* and *Volvox* that are either dioecious or monoecious. In the monoecious species, each individual colony is either female or male. In the dioecious case, a clone produces female colonies or male ones.

Another example of anisogamy is the group of Apicomplexa. Infectious sporozoites invade host cells and then produce large gamonts and small ones. The large macrogamonts transform into macrogametes, and the small gamonts undergo fissions, resulting in the production of many small microgametes. Fertilization takes place by the penetration of the microgamete into the macrogamete.

1.2 Autogamy

In some heliozoa and foraminiferans, the nucleus of a gamont gives rise to gametic nuclei (Grell 1968). Two gametic nuclei or gametes derived from a nucleus fuse to form a zygote. The process is called autogamy.

Heliozoa, *Actinosphaerium eichhorni*, is a multinucleate species. Sexual reproduction takes place when cells go into starvation after active reproduction under well-fed conditions. The cytoplasm of a cell simultaneously divides many times to produce many cells that contain some surviving nuclei, though many other nuclei degenerate. Then, a cyst envelope covers each of the cells. Inside the envelope, the cell divides into two daughter cells. Each daughter cell undergoes meiosis to give rise to gametes. These gametes fuse to form a zygote. Thereafter, the young *Actinosphaerium* emerges from the zygote. In the uninucleate species *Actinophrys sol*, each individual cell behaves as in *A. eichhorni*. A cell (diploid) encysts probably under a starvation condition and then divides into two cells (gamonts) within the cyst. The nucleus of each cell undergoes meiosis. During meiosis, one of the two daughter nuclei degenerates after each of the two divisions and eventually only one haploid nucleus remains in each of the two cells within the cyst. In the cyst, one of the two cells (gametes) forms pseudopodia and then fuses with the other cell to form a zygote.

In some foraminiferans, such as *Patellina corrugata*, a gamont produces amoebic gametes (haploids), fusing to form zygotes. Autogamy also occurs in some species of ciliates such as species of the *Paramecium aurelia* species complex, and in some stocks of *P. trichium*, *Euplotes minuta* and *E. crassus* (reviewed by Miyake 1996). Two micronuclei in a cell of the *P. aurelia* species complex undergo meiosis and produce eight haploid nuclei. Seven of the eight nuclei disintegrate. The remaining haploid nucleus divides mitotically into genetically identical male- and female-gametic nuclei, which then fuse to form a completely homozygous zygote (diploid).

1.3 Gamontogamy

In gamontogamy, the fertilization process starts with the union of the gamonts. In many shelled foraminiferans, the fertilization process is initiated by the union of two or more gamonts which form an aggregate. In the aggregated gamont shells, asynchronous nuclear divisions occur. Each cytoplasm of the gamonts is subdivided into as many parts as the number of nuclei, producing many gametes. Then, the gametes fuse in pairs to form zygotes, as in *Patellina corrugata* described in Sect. 4.

In the conjugation of ciliates, the gamonts mate with each other and nuclei divide to form gametic nuclei. Conjugation occurs between individual cells of complementary mating types. If conjugation takes place between morphologically identical cells, we call it isomorphous (bizygotic) conjugation. If it occurs between two gamonts, the macrogamont and the microgamont, that are different in size, it is called anisomorphous (monozygotic) conjugation (Miyake 1996).

In most species with isomorphous conjugation, two haploid nuclei, that is, one migratory (male) nucleus and one stationary (female) nucleus, are formed after gametogenesis. After the reciprocal exchange of the migratory nuclei between mating cells, each of the migratory nuclei fuses with the stationary nucleus to form a zygote in each cell of a mating pair. Then, the conjugants separate into two exconjugants. In the conjugation of anisomorphous species, peritrichs and suctorians such as *Vorticella* and *Ephelota*, the microgamont is absorbed by the macrogamont or degenerates after transferring its pronucleus in the macrogamont. Therefore, a pair produces only one zygote. This is called "total conjugation". In this connection it should be mentioned that isomorphous monozygotic conjugation and anisomorphous bizygotic conjugation have been reported in ciliates.

In the following sections (Sects. 2 to 5), reproductive cycles and fertilization will be outlined for some representative species.

2
Fertilization in a Flagellate

2.1
Reproduction in *Chlamydomonas reinhardtii*

Each species of *Chlamydomonas* is either homothallic or heterothallic. In homothallic species such as *C. monoica*, mating can occur within a clone. *C. reinhardtii* is a heterothallic species. Each clone expresses one of two complementary mating types, either plus (mt^+) or minus (mt^-). Mating occurs always between two mating types and therefore does not occur within a clone. *C. reinhardtii* is one of the most extensively studied species at the molecular level among Protozoa. Thus, many reviews have been published (see Goodenough and Ferris 1987; Goodenough 1991; Musgrave 1993; Quarmby 1994; Snell 1993; van den Ende 1994; Goodenough et al. 1995; Beck and Haring 1996). The mating types (mt^+, mt^-) are determined by a complex locus with genes restricted to one or the other mating type (Goodenough et al. 1995). Mating reactive cells are easily induced in laboratory cultures. Moreover, they can be grown in the dark, using acetate as the sole source of carbon and energy, and gamete formation is induced by removal of the nitrogen resource. After the induction of gamete differentiation by the deprivation of nitrogen, cells initiate the sexual phase. Cells shed their cell walls and fuse to form zygotes. The zygotes undergo meiosis and produce four haploid cells. After differentiation of the haploid vegetative cells into gametes, mating can be induced again by the mixing of mt^+ and mt^- gametes.

2.2
Induction of Gametogenesis

The sexual phase is initiated by the differentiation of vegetative cells into gametes that have the ability to mate and form zygotes (Fig. 1). Gametogenesis in *Chlamydomonas* offers a good opportunity to study sexual differentiation at the molecular level in a single cell, because this differentiation is easily controlled by the two environmental signals, nitrogen starvation and light (Beck and Haring 1996).

The mt^+ and mt^- cells reproduce asexually by repeated cell divisions (Fig. 1A,B). Nitrogen deprivation is an essential signal that directs vegetative cells toward differentiation into gametes (Fig. 1C,D) by changing the intracellular ammonium concentration (Matsuda et al. 1992). When the intracellular concentration of NH_4^+ decreases, the gene expression shifts to differentiate gametes. After a deprivation of ammonium from vegetative cells, the gene expression for nitrate assimilation occurs in less than 30 min (Quesada and Fernandez 1994). Light is also an essential factor required for gamete formation (Treier et al. 1989). Cells that have reached the critical stage of the

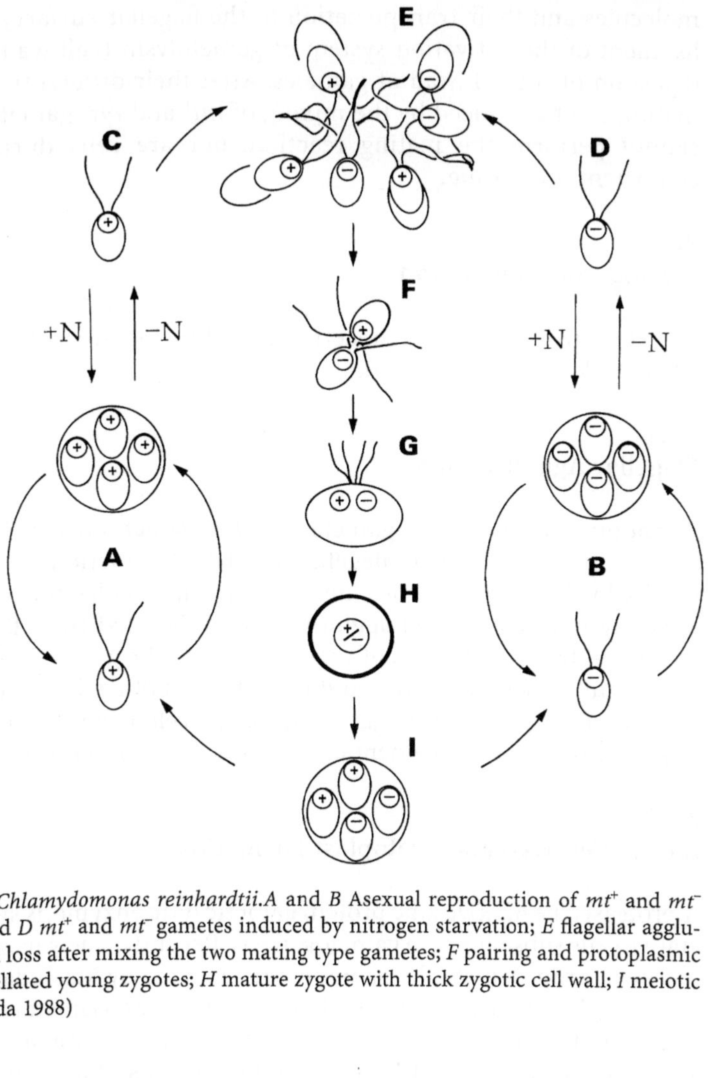

Fig. 1. Life cycle of *Chlamydomonas reinhardtii.A* and *B* Asexual reproduction of mt^{+} and mt^{-} vegetative cells; *C* and *D* mt^{+} and mt^{-} gametes induced by nitrogen starvation; *E* flagellar agglutination and cell wall loss after mixing the two mating type gametes; *F* pairing and protoplasmic fusion; *G* quadriflagellated young zygotes; *H* mature zygote with thick zygotic cell wall; *I* meiotic cell division. (Matsuda 1988)

differentiation by nitrogen starvation, i.e., pregametes, can be converted into gametes by irradiation with blue light. Ordinarily no mating-competent gametes are observed if their incubation without a nitrogen source takes place in the dark, but there are several light-independent mutants that can produce gametes in the dark (Beck and Haring 1996; Saito et al. 1998).

The differentiation of either mating type can occur even if the cells of the complementary mating type do not exist in the same culture medium. Gametes are indistinguishable from vegetative cells under the light microscope. However, the following events are involved in the gametogenesis (Matsuda 1988): (1) the construction of a mt^{+} or mt^{-} mating structure (MTS) for protoplasmic fusion, (2) the synthesis of mating type-specific agglutinin molecules and their transportation to the flagellar surface, and (3) the establishment of the activation system of gametolysin (cell wall lytic enzyme) for digestion of the cell wall of gametes. After their differentiation into gametes, mating can be induced by the mixing of mt^+ and mt^- gametes. Vegetative cells cannot perform the mating reaction, nor are cells during gametogenesis competent for mating.

2.3
Mating and Cell Fusion

The steps from the mating reaction to cell fusion are as follows (see Matsuda 1988; Beck and Haring 1996).

2.3.1
Flagellar Agglutination

A random collision of mt^+ gametes and mt^- gametes results in agglutination of cells by the tips of their flagella (Fig. 1E). Interaction of multiple gametes results in the formation of clumps. Active molecules that were identified as large glycoproteins called agglutinins were isolated from the gametes. These are absent from the flagella of vegetative cells. The gametic cells are considered to recognize each other by two different types of agglutinines. The agglutination reaction pulls the trigger leading to cell fusion, causing an elaborate sequence of events dependent on intercellular communication.

2.3.2
Losing Cell Wall and Protoplast Formation

A protease, the 62-kDa zinc protein gametic lytic enzyme, is synthesized in vegetative cells subsequent to gametogenesis. Recently, a new name, "gametolysin", has been proposed for the gametic lytic enzyme by Matsuda (1998) to denote a metabolloprotease (-lysin) that acts on gametes. Gametolysin is in an inactive form in gametes. When the mt^+ and mt^- gametes are mixed, the enzyme is activated by the signal of flagellar agglutination so that it digests the cell wall of gametes subsequent to fusion. The digestion of the cell wall changes gametes to naked protoplasts (Fig. 1E).

2.3.3
Activation of Mating Structures

Mating pairs of mt^+ and mt^- gametes which are formed by firm flagellar tip-to-tip adherence become separated from the agglutination clumps. After the flagella have become aligned along their whole length, these cells engage in the fusion reaction (Fig. 1F,G). In both cells of a pair, agglutination activates the mating structures of mt^+ and mt^- (MTS) for protoplasmic fusion. In the mt^+ gamete of a mating pair, a fertilization tubule extends somewhat (approx. 0.7 μm) from the cell surface between the two flagellar bases. The tubule reaches to the cell membrane between two flagellar bases of its mating partner. The membrane at the tip of the tubule has a surface-coating site called the "fringe", which is considered to be involved in the specific recognition of the activated mt^- mating structure (Ferris et al. 1996). In the mt^- cells, another type of fringe that is different ultrastructurally from that of mt^+ cells is present on the cell surface between two flagellar bases. The fringe of the mt^- cells is smaller than that of mt^+ cells and has electron-dense materials gathered at the central area of the cell surface. This is considered to be the specific site to recognize the activated mt^+ mating structure.

2.3.4
Formation of a Cytoplasmic Bridge

After the fertilization tubule of the mt^+ gamete contacts with the activated mt^- mating structure, a cytoplasmic bridge is formed between them.

2.3.5
Protoplasmic Fusion

In the last step, the bridge extends in diameter to fuse the two gametes of a mating pair, which transforms the mating pair into a single zygote with four flagella (Fig. 1G). These quadriflagellated cells appear within 5 min after mt^+ and mt^- gametes are mixed and swim around for a while.

2.3.6
Nuclear Fusion, Flagellar Retraction and Zygotic Wall Formation

Events following the cell fusion of male and female gametes are the regression of the flagella and the fusion of nuclei in the zygote (Fig. 1H). A thick zygospore wall is synthesized during maturation, which makes the mature zygotes resistant to adverse environmental conditions.

2.3.7
Chloroplast

The chloroplast DNA is inherited maternally, because the chloroplast DNA of the female (plus) gametes survives but that of the male (minus) gametes disappears after gametogenesis. The plus chloroplast DNA is protected from degeneration presumably due to an mt^+-specific event occurring during gametogenesis, while the unprotected minus chloroplast DNA is destroyed during early zygote development (Armbrust 1998).

2.4 Meiosis and Germination

Under appropriate environmental conditions, zygotes can undergo meiosis and germinate (Fig. 1I). This can be induced by light (Wegener and Beck 1991; Gloeckner and Beck 1995). After meiosis, four haploid cells are formed from each zygote. Then, cells break the thick zygotic cell wall. The mt^{+} and mt^{-} cells reproduce asexually by repeating a cycle of growth and division (Fig. 1A,B). Sometimes cell divisions occur inside the cell wall and then release 4–16 daughter cells by breaking down the mother cell wall.

3 Sexual Phase and Fertilization in *Plasmodium*

3.1 Reproduction in *Plasmodium falciparum*

Plasmodium, belonging to the phylum Apicomplexa, is parasitic and generally very small. This group has a unique organelle called the apical complex for the invasion of a new host cell. The human malaria parasite *P. falciparum* passes through several developmental changes during its life cycle between a mosquito and a human being. However, it is haploid for most of its life cycle. Zygote formation and meiosis occur during the part of the life cycle in the mosquito. It does not reproduce during the sexual phase but then reproduces during the asexual phase (Fig. 2).

The sexual and first asexual phases occur only in the mosquito *Anopheles*. The second asexual phase is in the human liver and the third one in the human blood. The third phase is repeated many times. There are three corresponding invasive forms, that is, the ookinete (mosquito) (Fig. 2), the hepatic merozoite (liver) and the erythrocytic merozoite (blood) (Fig. 2). In the third phase, some parasites become sexually differentiated cells called gametocytes that start a new cycle if they are taken into the mosquito. The process of transformation of the gametocytes into male and female gametes has been described at the electron microscopic level (Sinden et al. 1978).

3.2 Gametogenesis and Fertilization in the Mosquito

The sexual phase takes place in the lumen of the mid-gut (mesogaster) of the mosquito, when a female mosquito sucks the plasmodia with the blood from an infected human being. Gametocytes are of two different types, macrogametocyte (female) and microgametocyte (male). They break red blood cells and get free of them. Then, each gametocyte differentiates into male gametes from microgametocytes and female gametes from macrogametocytes (Fig. 2). The production of male gametes occurs within 10 to 20 min after feeding on a

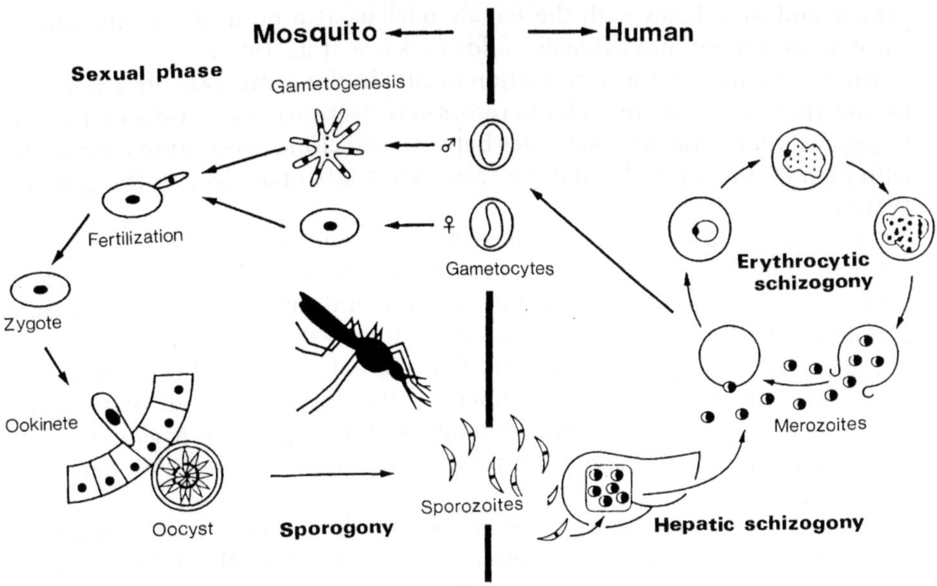

Fig. 2. Life cycle of *Plasmodium falciparum*.Sexual phase occurs in stomach of mosquito. During first asexual phase, sporogony takes place in stomach wall and body of mosquito. During second asexual phase, hepatic schizogony occurs in liver of man. During third asexual phase, erythrocytic schizogony occurs in erythrocytes of man. (Horii 1998)

malaria-infected human being. Garcia et al. (1997) isolated gamete activation factor (GAF) from the gut lumen of unfed, female *Anopheles stephensi*. It is a small, heat-stable, hydrophilic molecule. Its function in the mosquito is not yet known.

An individual male gametocyte undergoes dramatic nuclear and cytoplasmic events. It releases eight male gametes (Fig. 2, gametogenesis). During this process, cells grow eight flagella and start to beat them. As the formation of flagella is completed, their motion tears the cell apart, releasing them into the blood plasma. Some mutants that have defects in sex-specific development have been identified and characterized in *Plasmodium* (Guinet et al. 1996).

Each female gametocyte gives rise to a single, large gametic cell that appears round and is motionless. The male gametes swim around, seeking female gametes. When a male gamete encounters a female gamete, fusion occurs in the mosquito mid-gut (Fig. 2, zygotes). In the chicken malaria parasite *P. gallinaceum*, the processes of gametogenesis, fertilization and zygote transformation have been investigated at the ultrastructural level (Nijhout 1979; Rener et al. 1980; Kaushal et al. 1983). The male nucleus following the fertilization appears to travel through a channel of the endoplasmic reticulum in the female gamete and then fuses with the female nucleus at a point where the nuclear envelope is thrown into extensive folds (Aikawa et al. 1984).

Nuclear fusion, that is, fertilization occurs in the same parasite clone (self-fertilization) or in different clones (cross-fertilization), and produces a zygote. When cross-fertilization takes place between two clones, recombination occurs with high frequency and extensive rearrangements take place in the genomes (Walliker et al. 1987).

The zygote matures over the next few hours into a motile cell called an ookinete (Fig. 2). Eventually the ookinete reaches the mid-gut wall and passes between or through the epithelial cells of the mid-gut to reach the basement membrane, where it rounds up to become an oocyst (Fig. 2). In the oocyst, 14 mitotic spindles have been observed. This possibly suggests 14 chromosomes in the genome, though electron microscopic studies have failed to detect condensed mitotic chromosomes (Sinden and Strong 1978). A gametic linkage map of the chromosomes of plasmodium was reported (Walker-Jonah et al. 1992).

Within 2.5 h of fertilization, meiosis occurs in the zygote in mosquitoes. Therefore, cells in the human blood are already haploid. Ultrastructural studies indicate that the formation of the synaptonemal complex occurs at pachytene of the prophase in meiosis I. In *P. berghei*, meiotic figures with synaptonemal complexes were also detected in the nuclei of the zygotes (Sinden and Hartley 1985). Metaphase is rapidly followed by anaphase. According to the observation of Sinden and Hartley (1985), recombination of the chromosomes must be anticipated. Subsequent divisions occur within the young oocyst.

3.3 Asexual Development in the Mosquito and in the Human Body

The first asexual phase appearing in the mosquito is called sporogony. As mentioned above, the diploid oocyst grows at the outer side of the mid-gut wall in the body of the mosquito. After meiosis of the oocyst, the cells divide to produce thousands of cells called sporozoites. The mature oocyst bursts and then releases many invasive sporozoites. The free sporozoites migrate through the body of the mosquito and come together in the salivary glands. The sporozoites are formed 10 to 20 days after the parasite first invaded the mosquito.

The second asexual phase occurs in the human liver. When a mosquito carrying sporozoites in the salivary glands bites a human being, the sporozoites (haploid) are injected into the human blood. The sporozoites are transported to the liver in the bloodstream. Soon they invade liver cells and produce many cells (hepatic trophozoites). These cells grow, then divide to produce thousands of invasive forms. The infected liver cells burst, releasing the plasmodia (merozoites) into the blood (Fig. 2).

The third asexual phase occurs in the human blood. Merozoites invade red blood cells (erythrocytes) and form ring-shaped cells or amoebic cells (erythrocytic trophozoites) (Fig. 2). Each cell grows and then divides into 8 to 24 new cells in a red blood cell. When the red cell bursts, plasmodia (merozoites) are released. They invade other erythrocytes. The cycle of infection–division–burst repeats at this phase. However, some of them deviate from the cycle and develop into the male microgametocytes or the female macrogametocytes (Fig. 2). They can no longer divide in the human blood. They can go into sexual phase only when they are sucked up by the mosquito *Anopheles*. The mosquito species seem to vary in their vector potential because of environmental conditions. The interactions between the *Plasmodium* and the parasite-vector have recently been reviewed (Beier 1998).

3.4 The Sexual Phase

As mentioned above, the gametocytes that form in the human blood only develop further if they are taken up by the mosquito. The merozoites in the human red blood cells stop replication and start to differentiate into gametocytes. However, it is not well understood how the sexual differentiation is triggered. A clonal population of haploid parasites can generate both male and female gametocytes in the human blood. There should be some mechanism of the commitment to male or female determination. The male and female pathways are considered to involve differential expression of sex-specific genes. Guinet et al. (1996) reported a sex-specific developmental mutant in this species. The mutant has a defect that affects the development of competent male gametocytes. After 1 to 2 weeks in the wild type, micro- (male) and macro- (female) gametocytes mature. When they are sucked up by the mosquito, they can transform into male or female gametes free from the red blood cells. The molecular processes that cover the differentiation and development of the sexual forms are largely unknown.

4 Gamogony and Agamogony in Foraminiferans

4.1 Life Cycle of Foraminiferans

The life cycle of foraminiferans, recorded in the excellent works by Grell (1968), is composed of regular alternations of haploid and diploid phases. The zygotes (Fig. 3, stage 4) produce cells (agamonts) that cannot give rise to gametes. In the agamonts, nuclei divide usually twice and in this way the agamonts become multinucleate. Then, the agamonts hatch out from the shell (Fig. 3, stage 6). When they have reached a certain size (Fig. 3, stage 7), meiosis

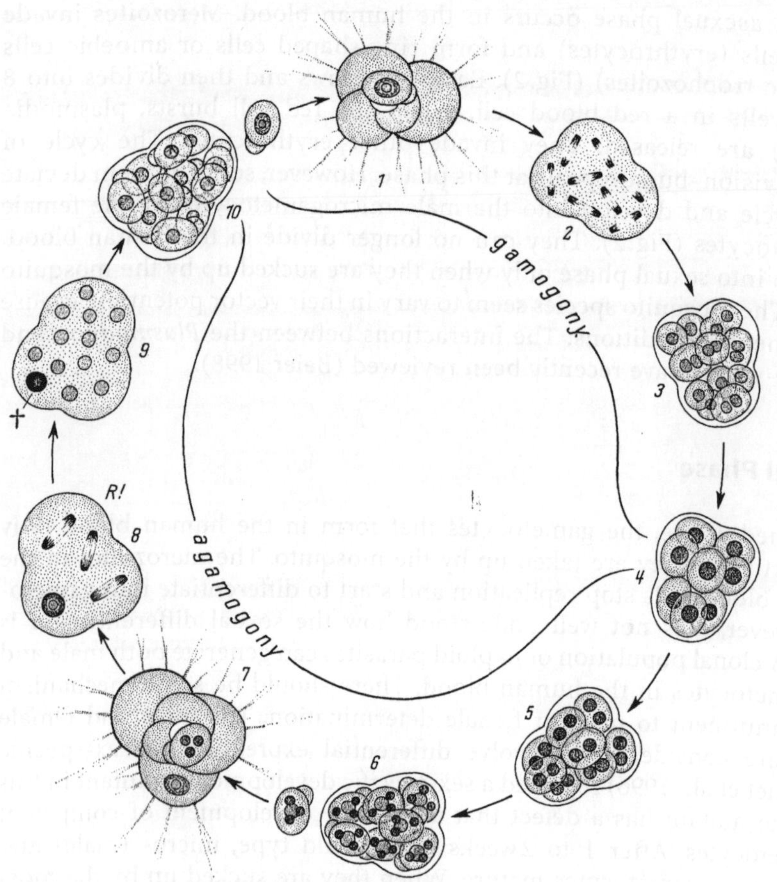

Fig. 3. Life cycle of *Rotaliella heterocaryotica.1* Adult gamont; *2* last mitosis of gamogony; *3* autogamic fusion of gametes; *4* zygotes; *5* binucleated agamonts; *6* agamonts with four nuclei; *7* adult agamont; *8* first meiotic division; *9* end of second meiotic division; *10* agametes (young gamonts). *R!* First meiotic division (reduction division). (Grell 1973)

occurs in the nuclei, producing haploid nuclei which give rise to gamonts (Fig. 3, stage 1). The gamonts are sexually differentiated in some species such as *Patellina corrugata*, but not in other species such as *Rotaliella heterocaryotica*.

In the foraminiferans, there are two types of species, homokaryotic and heterokaryotic ones. *P. corrugata* belongs to the homokaryotic species and *R. heterocaryotica* belongs to the heterokaryotic. The nuclei of homokaryotic foraminifera are all alike. However, *R. heterocaryotica* has two types of nuclei, germ nuclei that undergo meiosis and become nuclei of adult gamonts and somatic nuclei that are incapable of division and are retained only for a limited period of time.

Unfortunately, the fertilization process has never been studied at the molecular level, but the process of the sexual phase and fertilization will be briefly mentioned with regard to the two species (for particulars, see Grell 1968).

4.2 Heterokaryotic Species *Rotaliella heterocaryotica*

In the adult gamont of *R. heterocaryotica*, (Fig. 3, stage 1), the nucleus divides and produces amoebic gametes. The gametes unite in pairs in the shell of the gamont to form the zygote (diploid) (Fig. 3, stages 3 and 4). This process is called autogamy. Within the shell of the gamont, each nucleus of the zygotes undergoes two nuclear divisions and produces four nuclei (Fig. 3, stage 6). Soon after the nuclear division, one of the nuclei differentiates into a somatic nucleus that increases in size, and the other three differentiate into generative nuclei that retain their original size. Both of the nuclear types have the same content of DNA (Zech 1964). In the shell, several such diploid agamonts are formed. The agamonts leave the shell soon after nuclear differentiation. As the agamonts grow and produce several chambers, the somatic nucleus moves into a younger (outer) chamber (Fig. 3, stage 7). All three gamete nuclei are thought to undergo meiosis and produce 12 nuclei, which results in the formation of 12 agametes. The somatic nucleus disintegrates before agamete formation. They then emerge from the shell and become young gamonts.

4.3 Homokaryotic Species *Patellina corrugata*

P. corrugata is a dioecious species that may have genetic sex determination. In this species, the gamonts (haploid) are sexually differentiated into two mating types of + and – (Grell 1959). When the two types of gamonts are mixed, they move toward each other and form aggregates (Fig. 4, stage 2). In each gamont of an aggregate, the nucleus divides and produces a certain number of nuclei depending upon its size. At the same time, the protoplasts emerge from their shells and round up (Fig. 4, stage 3). Each protoplast (cell) is then subdivided into as many parts as the number of nuclei. Then, each of them divides to form two gametes, which are drop-shaped (Fig. 4, stage 5). They move around for a while in the space provided by the gamont shells, fusing in pairs to form zygotes. Zygotes are formed only when gametes of different mating types fuse. Mating takes place only between gametes derived from different gamonts.

As mentioned above, fertilization is initiated by aggregation and union of two or more gamonts with haploid nuclei. In the simplest case, two gamonts creep toward each other, touch with their ventral surfaces and extrude a substance to cement them together. Then the inner walls that separate the chambers in the gamont dissolve and provide a common space where the gametes

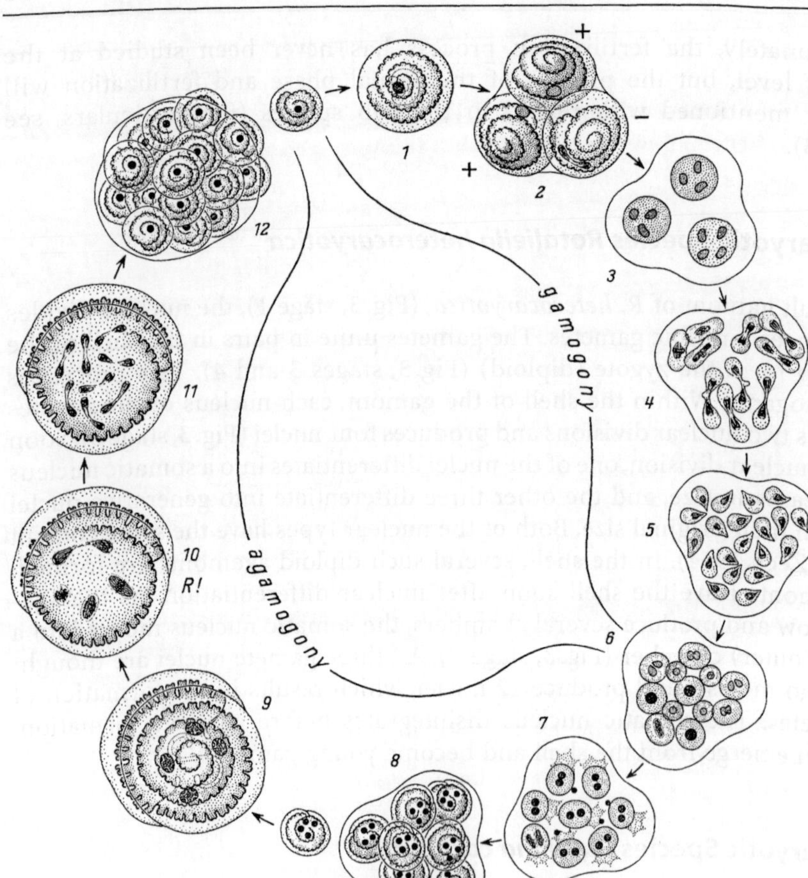

Fig. 4. Life cycle of *Pstellina corrugata*. *1* Gamont; *2* aggregate of three gamonts (two from the + sex, one from the – sex); *3* protoplasts of gamonts at bottom of space which is roofed over by empty shells; *4* last gamogony mitosis and gamete formation; *5* gametes; *6* eight zygotes and four leftover gametes; *7* binucleate agamonts; *8* young agamonts; *9* adult agamont; *10* meiosis I; *11* meiosis II; *12* formation of agametes. (Grell 1973)

fuse. In the aggregate formed, the protoplasts of the gamonts lie at the bottom of the space that is roofed over by the empty shells. When the gamonts have united, nuclear divisions take place (Fig. 4, stage 3). The gamont that has a single haploid nucleus divides mitotically and forms the gametes (Fig. 4, stage 5). In some genera such as *Iridia lucida* and *Discorbis mediterranensis*, flagellated gametes swim around within the shells of the gamonts, and in other genera such as *P. corrugata*, a few amoebic gametes are produced. The gametes fuse in pairs to form the zygotes in the aggregated gamont shells (Fig. 4, stage 6).

5
Conjugation in Ciliates

Conjugation of ciliates has been considered a good model for the study of fertilization (Metz 1954; Hiwatashi 1969; Miyake 1978). Conjugation is easy to induce in the laboratory since Sonneborn (1937) found mating types in *Paramecium*. One of the features of ciliates is the differentiation of nuclei into the somatic macronucleus and germinal micronucleus. Each cell has both types of nuclei, though numbers differ between species. In conjugation, the gamont does not produce gametes but only the micronuclei divide to form gamete nuclei. The macronucleus is polygenomic and transcriptionally very active, while the micronucleus is diploid and almost inert. During the vegetative phase, cells reproduce by binary fission with the nuclear division of both types of nuclei. These cells differentiate to be mating-reactive under certain conditions. Gamonts start the conjugational process when cells of each mating type collide. This developmental process of ciliates was summarized by Nanney (1980) and their sexuality was recently reviewed by Miyake (1996). In this section, the process of conjugation will focus on *Paramecium caudatum*.

5.1
Life Cycle of *Paramecium caudatum*

The life cycle of *P. caudatum* is simple and well defined. A new life cycle starts after conjugation. Cells that have finished conjugation reproduce into clones by binary fissions. The clones pass through periods of sexual immaturity (about 50 fissions after the last conjugation), maturity, and senescence (several hundred fissions after the last conjugation), eventually resulting in clonal death (Takagi 1988). The cells cannot express mating reactivity until they reach the period of sexual maturity after an immaturity period. The length of immaturity is counted in numbers of cell divisions (Takagi 1988), probably the number of rounds of macronuclear DNA replication (K. Mikami, unpubl. observ.). During the immature period, a protein of about 10 kDa called "immaturin" is produced (Haga and Hiwatashi 1981).

Even during the period of sexual maturity, however, cells do not express mating reactivity if they are in well-fed or extremely starved conditions. They can express mating reactivity only when they go into the stationary phase after logarithmic growth.

After the cells have reached the maturity period, reproducing vegetative cells can enter the conjugational process at any time when the environmental conditions are appropriate. The process of conjugation involves (1) an initial adhesion of cells (gamonts) of complementary mating types, (2) meiotic reduction of the chromosome number, (3) fusion (fertilization) of gamete nuclei and (4) differentiation of new somatic and germinal nuclei (reviewed by Hiwatashi 1969, 1981; Hiwatashi and Kitamura 1985; Hiwatashi and Mikami 1989).

The morphospecies *P. caudatum* is subdivided into a number of sibling species called syngens. Clones belonging to each syngen are grouped into complementary mating types. To identify mating types, the occurrence of conjugation between individual clones must be tested. The determination of the mating type is controlled by mating-type genes. Genetic control of mating-type specificity within a sibling species has been reviewed (Hiwatashi 1981; Tsukii 1988).

5.2 Sexual Interaction at the Beginning of Conjugation

A series of events starts with the initial mating type-specific adhesion of gamonts. The cell-to-cell interaction during the mating process has been reviewed by many authors (e.g., Nanney 1977; Miyake 1978; Cronkite 1979; Hiwatashi 1981; Hiwatashi and Kitamura 1985; Fujishima 1988; Watanabe 1990). Initial cell contact of the mating reaction occurs at the ciliary surface of gamonts, but only those cilia on the ventral surface where the oral apparatus is located are reactive. The ciliary location of the mating reaction is supported by the evidence that the cilia artificially detached from complementary mating types can agglutinate with each other (Takahashi et al. 1974). Kitamura and Hiwatashi (1976) succeeded in obtaining membrane vesicles retaining mating substance from the cilia, but the substance has not been isolated. A few minutes after onset of the mating reaction, the velocity of protoplasmic streaming increases. Then, the degeneration of cilia occurs at the anterior tips of the cells about 30 min after the mating reaction. About 1 h after the beginning of the mating reaction, the agglutinates begin to disaggregate into two gamonts paired at their anterior tips (hold-fast union). In pairs of the hold-fast union, cell adhesion occurs not at ciliary surfaces but at the surfaces of cell bodies where cilia degenerate. The degeneration area of cilia extends posteriorly along the ventral surface and covers the region of the cytostome, proceeding finally to the posterior end of the cell (Watanabe 1990). The mating cells finally are contacted with each other at the region of the cytostome and become firmly united (paroral union). The active protoplasmic streaming calms down to the ordinary level of vegetative cells. After the formation of paroral union, cytoplasmic connections occur between the two conjugating cells through many small openings (0.2 to 0.5 μm in diameter). Once this paroral union has been formed, it is very difficult to separate the gamonts by mechanical force. This cytoplasmic continuity may allow the exchange of cytoplasm including gene products from the macronucleus during conjugation. The cytoplasmic exchange could help synchronize nuclear events of both cells of a mating pair.

5.3
Nuclear Events During Conjugation

Nuclear phenomena during the sexual phase of ciliates including *Paramecium* have been reviewed at structural and ultrastructural levels (Raikov 1982; Steinbrük 1986; Mikami 1988; Hiwatashi and Mikami 1989). The basic nuclear events during conjugation are common among species of *Paramecium*, although their details vary.

5.3.1
Nuclear Behavior at Early Stages of Conjugation and Meiosis

The micronucleus relocates from the macronucleus at the beginning of conjugation, while in the vegetative phase the micronucleus usually touches and presses against the surface of the macronucleus. The micronuclear relocation that is called early micronuclear migration takes place soon after the mating reaction (Fujishima 1988). At the same time, activation of protoplasmic streaming begins all over the cell (Kobayashi et al. 1999). The relation between the activation of protoplasmic streaming and the micronuclear migration has not yet been clarified. The increase in free cytosolic calcium concentration or its decrease from an increased level back to a normal level may trigger the micronuclear migration. This is supported by the evidence that the microinjection of Ca^{2+} induced the migration (Xianyu and Haga 1993).

At about the time of forming the paroral union, the micronucleus goes into the premeiotic S phase. After premeiotic S phase, the micronucleus commences meiosis. Treatment with the protease trypsin can separate mating cells in the early stages of conjugation. In separated cells, the micronuclei sometimes go into meiosis when the separation is made about 3 h after mixing the complementary cells. The point of no return on meiosis is estimated to be at this stage. In prophase of the first meiotic division, the spherical micronucleus elongates and forms a large crescent-shaped structure called the crescent, which coincides in time with the parachute stage known in some other ciliates. The crescent is considered to be the zygotene stage. Many intranuclear microtubules lie along with the longitudinal axis of the crescent nucleus. So far, a synaptonemal complex has never been reported in *P. caudatum*, though it was observed in the *P. aurelia* species complex (Raikov 1982). Around this time, the oral apparatus disappears and this area plays an important role for later nuclear differentiation, as described below. The micronucleus changes its structure from crescent to ovoid shape and then goes through metaphase into the first meiotic division. After meiosis, four haploid nuclei are formed in each cell of a mating pair.

In the ciliate *Blepharisma*, at least six proteins are required for the progression of meiosis (Friedl et al. 1983). In *P. caudatum*, micronuclear behavior seems to be under the control of the macronucleus at certain points (Mikami 1992). When the macronucleus is removed during meiosis of *P. caudatum*, the micronuclear behavior stops at several points, but exact stages have not been confirmed yet (K. Mikami, unpubl. observ.).

5.3.2 Random Selection of Meiotic Products

After meiosis, all of the four meiotic products move near to the paroral region. Then, only one of them moves farther into the region and appears attached to the surface of the cell membrane. For the micronuclear migration into the paroral region, microtubules play an important role, because migration was inhibited by the antitubulin reagent Vinblastin (Yanagi and Hiwatashi 1985) and also by microinjection of anti-α-tubulin antibody (Nakajima and Mikami, unpubl. observ.). Moreover, some connection of microtubules is observed between meiotic products and the paroral region. If none of them migrates into the region, all micronuclei become picnotic and then disappear.

In the next stage, one micronucleus in the paroral region is covered with a microtubule meshwork that may prevent the meiotic product from degenerating. The cytoplasm outside the meshwork seems to be ready to make the meiotic products degenerate, because the remaining three nuclei outside the paroral cone degenerate. The phenomenon is considered to be a type of apoptosis. Microinjection experiments revealed that the genetic information required for degeneration of the meiotic products is given by the macronucleus at the stage of micronuclear metaphase of meiosis I (Mikami 1992). To save the micronuclei from the degeneration, injection of the lectin WGA (wheat germ agglutinin) is very effective. Recent work (K. Mikami, unpubl. observ.) suggests that the lectin could bind to lysosome-like cytoplasmic vesicles but not to the nuclear membrane. A lysosomal system may be involved in the degeneration of meiotic products.

5.3.3 Gametogenesis and Fertilization

Generally, the cytoplasm is at a standstill at this stage since the protoplasmic streaming ceases from the stages of degeneration of meiotic products and the exchange of gamete nuclei, until the synkaryon formation. After the disappearance of the microtubule meshwork mentioned above, the surviving nucleus divides and produces gamete nuclei, a 'male' migratory and a 'female' stationary pronucleus. Then, the meshwork structures of the microtubules appear again around the migratory pronucleus (Fig. 5, stage 8). The meshwork structures must be involved in the transfer of gametic pronuclei across the junction of a mating pair. In *Tetrahymena*, Orias et al. (1983) showed that each migratory pronucleus is surrounded by a microtubule meshwork called the fertilization basket, this being required for the transfer of gametic pronuclei across the cell-to-cell junction (reviewed by Orias 1986). Microfilaments as well as microtubules could also be involved in the micronuclear transfer and

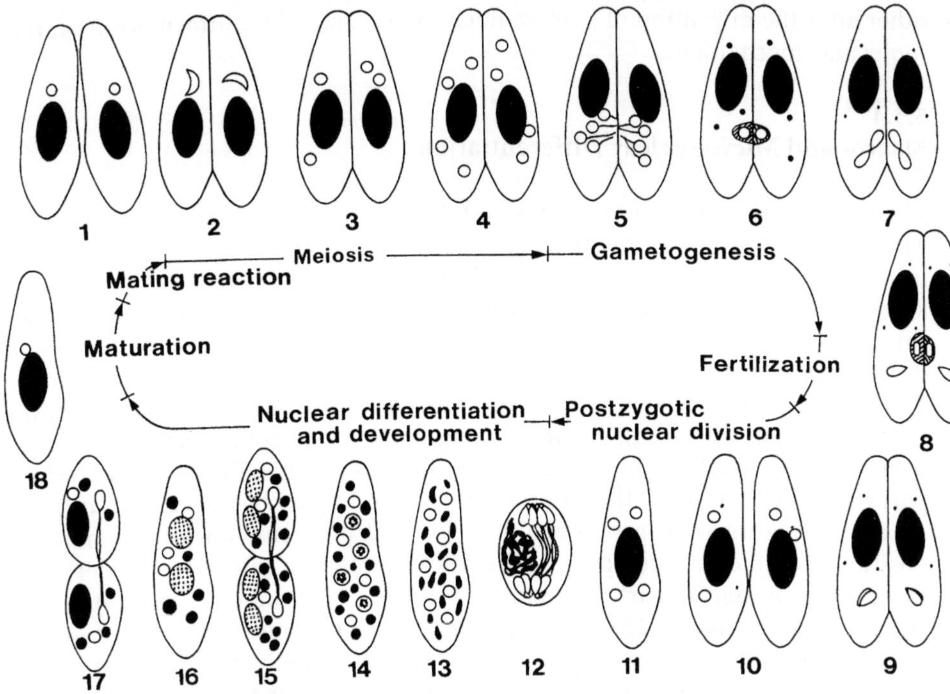

Fig. 5. Conjugational process of *Paramecium caudatum*. Each cell has one polygenomic macronucleus and one diploid micronucleus. When mating reactive cells of complementary mating type are encountered, aggregation of cells occurs and then mating pairs appear (*1, 2*). The germinal micronucleus undergoes meiosis and produces four haploid nuclei (*2–4*). One enters into a bulge of the cell (called paroral cone) and survives there, while other three out of the region degenerate (*5, 6*). Surviving nucleus divides and produces a migratory pronucleus and stationary pronucleus (*5–7*). After reciprocal exchange of migratory pronuclei between mating gamonts, fusion of stationary and migratory nuclei (syngamy) occurs to produce a synkaryon in each of the mating cells (*8, 9*). Mating pairs separate after first division of the synkaryon (*10*). Postzygotic nuclei undergo second division and produce four diploid nuclei (*11*). At third nuclear division, the cell shortens in length and spindles elongate longitudinally so that nuclear division products are localized anteroposteriorly at both ends of cell (*12*). Macronucleus becomes skein-form and then about separates into 50 fragments. Postzygotic nuclei localized in anterior part of cell become presumptive micronuclei and ones localized in posterior part become macronuclear anlagen (new macronuclei) (*13, 14*). At first and second cell division, the four macronuclear anlagen and fragments of old macronuclei are distributed to each daughter cell. Micronuclei are also distributed to daughters, but only one micronucleus divides (*15–17*). Subsequently, the cells follow the ordinary course of binary fission. After an immature period of no mating reactivity, clones can conjugate and next cycle starts (*18*)

fertilization in *Paramecium*, since an intermediate filament protein of 49 kDa appeared around and between the gamete micronuclei in *T. thermophyla* (Numata et al. 1985). In P. *caudatum*, the migratory micronucleus appears pressed against the membrane when cells are observed with anti-α-tubulin antibody. How do the migratory and stationary nuclei meet and fuse with each other after the migration of gamete nucleus? There is little information on the molecular mechanism.

5.3.4 Macro- and Micronuclear Differentiation

The synkaryon divides three times and produces eight nuclei in this species. The differentiation of germ- and somatic nuclei is determined during the third mitosis of the synkaryon. The protoplasmic streaming ceases at this critical stage of macro- and micronuclear differentiation (Kobayashi et al. 1999). At the third division of synkaryon, i.e., the critical division for macro- and micronuclear differentiation, the orientation of nuclear spindles lies parallel to the longitudinal axis of the cell and extends to opposite ends of the cell. At the same time, a marked shortening of the cell length occurs. In consequence, the division products localize anteroposteriorly; four nuclei localize at the anterior part of the cell and the other four at the posterior part of the cell. No difference in structure is seen between anterior and posterior nuclei. The nuclei positioned at the anterior part of the cell differentiate into micronuclei and those at the posterior part differentiate into new macronuclei (Mikami 1980; Hiwatashi and Mikami 1989). This macro- and micronuclear differentiation could be determined by cytoplasmic determinants. The hypothesis is that the cytoplasm of the anterior end determines the nuclei to be the micronuclei and that of the posterior end determines them to differentiate into the macronuclei. However, electron-microscopic studies have never revealed the structural difference in the anteroposterior cytoplasm. The nuclei localized to the posterior part of the cell still have a competency as germ nuclei, because they continue to function as the germinal micronucleus even after transplantation into a vegetative cell.

5.3.5 Development of a New Macronucleus

The nuclei grow through a sequence of developmental stages into postzygotic macronuclei. According to recent morphological work (Mikami et al., unpubl. observ.), six developmental stages can be distinguished. Soon after the critical stage, the eight postzygotic nuclei do not differ from each other (stage I). About 30 min (at 25 °C) after the critical stage, the presumptive macronuclei swell, being distinguished from the other four presumptive micronuclei (stage II). The next morphological change is the appearance of heterochromatic aggregates. The chromosomes are condensed to about ten compact aggregates, called heterochromatic aggregates in the macronuclear anlagen (stage III). An anti-α-tubulin antibody reacts with the micronuclei well but not with the macronucleus during the vegetative phase. The four nuclei that are determined to be new macronuclei react well with the antibody until stage II but lose the feature at this stage. About 12 h after the critical stage, the chromatin aggregates migrate to the central area and make a concentric circle in the anlage (stage IV). According to an electron-microscopic study, macronuclear-specific chromatin bodies appear in the anlage at this stage. Then, the volume of the anlagen increases more and the chromatin aggregates gradually disappear (stage V). The aggregates disappear completely and some nucleolus-like structures appear. It is known from nuclear transplantation experiments that some of the genes in the anlagen have been expressed at this stage (stage VI) (Mikami 1987).

In ciliates, the chromosomal DNAs are different in size between the macronucleus and the micronucleus. For example, in *P. tetraurelia*, the mean size of macronuclear DNA is estimated to be roughly 450 kb, while micronuclear DNA would be much longer than that of the macronucleus (Phan et al. 1989). Reorganization of genes occurs during the development of macronuclear anlagen. The formation of the macronuclear genome is the result of extensive DNA processing of the micronuclear genome. The events include the elimination of specific micronuclear sequences, DNA splicing and fragmentation of germline chromosome into subchromosomal fragments (Blackburn and Karrer 1986; Prescott 1993). In *P. tetraurelia*, DNA rearrangements within the coding regions of genes (Preer et al. 1992) and in the telomere addition sites (Forney and Blackburn 1988; Caron 1992) produce variable DNA rearrangements. Although there is no report so far in *P. caudatum*, it is possible that the same system of gene reorganization occurs in this species.

5.3.6
Determination of the Germinal Micronucleus

New macronuclei and micronuclei are produced always from the micronucleus during the conjugational process. Thus, the germline is continued by the germinal micronucleus. Each cell of *P. caudatum* has only one micronucleus, the number of which is usually characteristic for a species and seems to be regulated somehow. There are four presumptive micronuclei in an exconjugant cell of *P. caudatum*. To arrive at the uni-micronuclear condition, the number of nuclei must be reduced. It was believed that three of the four presumptive micronuclei degenerate and the fourth becomes a germ micronucleus. However, recent work with the anti-α-tubulin antibody as a marker of the micronucleus has revealed that three micronuclei still remain with a dividing micronucleus at two divisions after conjugation. At least in some strains of this species the supernumeral micronuclei remain during some cell cycles after conjugation, but only one of them can divide at postconjugational divisions.

Acknowledgements. I thank Drs. D. Cronkite, K. Heckmann, K. Hiwatashi, H.-W. Kuhlmann, Y. Matsuda and K. Tanabe for suggestions, comments and information. This chapter was written during a stay at Münster University, Germany, as an Alexsander von Humboldt fellow under the auspices of Prof. Heckmann. I also thank Dr. Y. Matsuda, Dr. T. Horii, Igaku-shoin and Springer-Verlag for permission for reproduction of figures.

References

Aikawa M, Carter R, Ito Y, Nijhout M (1984) New observation on gametogenesis, fertilization, and zygote transformation in *Plasmodium gallinaceum.* J Protozool 31:403–413

Armbrust EV (1998) Uniparental inheritance of chloroplast genomes. In: Rochaix J-D, Goldschmidt-Clermont M, Merchant S (eds) The molecular biology of chloroplasts and mitochondria in *Chlamydomonas.* Kluwer Academic, Dordrecht, pp 93–113

Beck CF, Haring MA (1996) Gametic differentiation of *Chlamydomonas.* Int Rev Cytol 168:259–302

Beier JC (1998) Malaria parasite development in mosquitoes. Annu Rev Entomol 43:519–543

Blackburn EH, Karrer KH (1986) Genomic reorganization in ciliated protozoans. Annu Rev Genet 20:501–521

Caron F (1992) A high degree of macronuclear chromosome polymorphism is generated by variable DNA rearrangements in *Paramecium primaurelia* during macronuclear differentiation. J Mol Biol 225:661–678

Cronkite DL (1979) The genetics of swimming and mating behavior in *Paramecium.* In: Lewandowsky M, Hutner SH (eds) Biochemistry and physiology of protozoa, 2nd edn. Academic Press, New York, pp 221–273

Ferris PJ, Woessner JP, Goodenough UW (1996) A sex recognition glycoprotein is encoded by the *plus* mating-type gene tus1 of *Chlamydomonas reinhardtii.* Mol Cell Biol 7:1235–1248

Forney JD, Blackburn EH (1988) Developmentally controlled telomere addition in wild-type and mutant paramecia. Mol Cell Biol 8:251–258

Friedl E, Miyake A, Heckmann K (1983) Requirement of successive protein syntheses for the progress of meiosis in *Blepharisma.* Exp Cell Res 145:105–113

Fujishima M (1988) Conjugation. In: Görtz H-D (ed) *Paramecium.* Springer, Berlin Heidelberg New York, pp 70–84

Garcia GE, Wirtz RA, Rosenberg R (1997) Isolation of a substance from the mosquito that activates *Plasmodium* fertilization. Mol Biochem Parasitol 88:127–135

Gloeckner G, Beck CF (1995) Genes involved in light control of sexual differentiation in *Chlamydomonas reinhardtii.* Genetics 141:937–943

Goodenough UW (1991) *Chlamydomonas* mating interactions. In: Dworkin M (ed) Microbial cell–cell interactions. Am Soc Microbiol, Washington, DC, pp 71–112

Goodenough UW, Ferris PJ (1987) Genetic regulation of development in *Chlamydomonas.* In: Loomis W (ed) Genetic regulation of development. Liss, New York, pp 171–189

Goodenough UW, Detmers PA, Hwang C (1982) Activation for cell fusion in *Chlamydomonas:* analysis of wild-type gametes and nonfusing mutants. J Cell Biol 92:378–386

Goodenough UW, Armburst EV, Campbell AM, Ferris PJ (1995) Molecular genetics of sexuality in *Chlamydomonas.* Annu Rev Plant Physiol Plant Mol Biol 46:21–44

Grell KG (1959) Untersuchungen über die Fortpflanzung und Sexualität der Foraminiferen. IV. *Patellina corrugata.* Arch Protistenk 104:211–235

Grell KG (1967) Sexual reproduction in Protozoa. In: Chen T-T (ed) Research in protozoology, vol 2. Pergamon Press, Oxford, pp 148–213

Grell KG (1973) Protozoology. Springer, Berlin Heidelberg New York, pp 97, 409

Guinet F, Dvorak JA, Fujioka H, Keister DB, Muratova O, Kaslow DC, Aikawa Wand Vaidya AB (1996) A developmental defect in *Plasmodium falciparum* male gametogenesis. J Cell Biol 135:269–278

Haga N, Hiwatashi K (1981) A protein called immaturin controlling sexual immaturity in *Paramecium.* Nature 289:177–179

Hausmann K, Hülsmann N (1996) Protozoology. Thieme, New York
Hiwatashi K (1969) *Paramecium*. In: Metz CB, Monroy A (eds) Fertilization, vol 2. Academic Press, New York, pp 255–293
Hiwatashi K (1981) Sexual interaction of the cell surface in *Paramecium*. In: O'day DH, Horgen PA (eds) Sexual interactions in eukaryotic microbes. Academic Press, New York, pp 351–378
Hiwatashi K, Kitamura A (1985) Fertilization in *Paramecium*. In: Metz CB, Monroy A (eds) Biology of fertilization, vol 1. Academic Press, New York, pp 57–85
Hiwatashi K, Mikami K (1989) Fertilization in *Paramecium*: processes of the nuclear reorganization. Int Rev Cytol 114:1–19
Horii T (1998) Sporozoa. In: Ishii A et al. (eds) Standard textbook of medical zoology, 2nd edn, p 53 (in Japanese). Igaku-Shoin, Tokyo
Kaushal DC, Carter R, Howard R, McAuliffe FM (1983) Characterization of antigens of mosquito midgut stages of *Plasmodium gallinaceum*. I. Zygote surface antigens. J Mol Biochem Parasitol 8:53–69
Kitamura A, Hiwatashi K (1976) Mating-reactive membrane vesicles from cilia of *Paramecium*. J Cell Biol 69:736–740
Kobayashi K, Yoshimura C, Mikami K (1999) Protoplasmic streaming and nuclear differentiation during conjugation of *Paramecium caudatum*. Zool Sci 16:885–892
Matsuda Y (1988) The *Chlamydomonas* cell walls and their degrading enzymes. Jpn J Phycol 36:246–264
Matsuda Y (1998) Gametolysin. In: Barrett AJ, Rawlings ND, Woessner JF (eds) Handbook of proteolytic enzymes. Academic Press, New York
Matsuda Y, Shimada T, Sakamoto Y (1992) Ammonium ions control gametic differentiation in *Chlamydomonas reinhardtii*. Plant Cell Physiol 33:909–914
Metz CV (1954) Mating substances and the physiology of fertilization in ciliates. In: Wenrich DH (ed) Sex in microorganisms. American Association for the Advancement of Science, Washington, DC, pp 284–334
Mikami K (1980) Differentiation of somatic and germinal nuclei correlated with intracellular localization in *Paramecium caudatum* exconjugants. Dev Biol 80:46–55
Mikami K (1987) Micronuclear development and gene expression in the exconjugants of *Paramecium caudatum*. Dev Biol 123:161–168
Mikami K (1988) Nuclear dimorphism and function. In: Görtz H-D (ed) *Paramecium*. Springer, Berlin Heidelberg New York, pp 120–130
Mikami K (1992) Behavior of germinal micronuclei under control of the somatic macronucleus during conjugation in *Paramecium caudatum*. Dev Biol 149:317–326
Miyake A (1978) Cell communication, cell union, and initiation of meiosis in ciliate conjugation. Curr Top Dev Biol 12:37–82
Miyake A (1996) Fertilization and sexuality in ciliates. In: Hausmann K, Bradbury P (eds) Ciliates cells as organisms. Fischer, Stuttgart, pp 243–290
Musgrave A (1993) Mating in *Chlamydomonas*. Prog Phycol Res 9:193–237
Nanney DL (1977) Cell-cell interactions in ciliates: evolutionary and genetic constraints. In: Reissig JI (ed) Microbial interactions. Chapman and Hall, London, pp 353–397
Nanney DL (1980) Experimental ciliatology. Wiley, New York
Nijhout MM (1979) *Plasmodium gallinaceum*: exflagellation stimulated by a mosquito factor. Exp Parasitol 48:75–80
Numata O, Sugai T, Watanabe Y (1985) Control of germ nuclear behavior at fertilization by *Tetrahymena* intermediate filament protein. Nature 314:192–194
Orias E (1986) Ciliate conjugation. In: Gall JG (ed) The molecular biology of ciliated protozoa. Academic Press, New York, pp 45–84
Orias JD, Hamilton EP, Orias E (1983) A microtubule meshwork associated with gametic pronucleus transfer across a cell-cell junction. Science 222:181–184
Phan HL, Forney J, Blackburn EH (1989) Analysis of *Paramecium* macronuclear DNA using pulsed field gel electrophoresis. J Protozool 36:402–408

Preer LB, Hamilton G, Preer JR Jr (1992) Micronuclear DNA from *Paramecium tetraurelia*: serotype 51 A gene has internally eliminated sequences. J Protozool 39:678–682
Prescott DM (1993) Restructuring of DNA sequences in the germline genome of *Oxytricha*. Curr Opin Genet Dev 3:762–769
Quarmby LM (1994) Signal transduction in the sexual life of *Chlamydomonas*. Plant Mol Biol 26:1271–1287
Quesada A, Fernandez E (1994) Expression of nitrate assimilation related genes in *Chlamydomonas reinhardtii*. Plant Mol Biol 24:185–194
Raikov IB (1982) The protozoan nucleus. Springer, Vienna New York
Rener J, Carter R, Rosenberg Y, Miller LH (1980) Antigamete monoclonal antibodies synergistically block transmission of malaria by preventing fertilization in the mosquito. Proc Natl Acad Sci USA 77:6797–6799
Saito T, Inoue M, Yamada M, Matsuda Y (1998) Control of gametic differentiation and activity by light in *Chlamydomonas reinhardtii*. Plant Cell Physiol 39:8–15
Sinden RE, Hartley RH (1985) Identification of the meiotic division of malarial parasites. J Protozool 32:742–744
Sinden RE, Strong K (1978) An ultrastructural study of the sporogonic development of *Plasmodium falciparum* in *Anopheles gambiae*. Trans R Soc Trop Med Hyg 72:477–491
Sinden RE, Canning EU, Bray RS, Smalley ME (1978) Gametocyte and gamete development in *Plasmodium falciparum*. Proc R Soc Lond B Biol Sci 201:375–399
Snell WJ (1993) Signal transduction during fertilization in *Chlamydomonas*. In: Kurjan J (ed) Signal transduction: prokaryotic and simple eukaryotic systems. Academic Press, San Diego, pp 255–277
Sonneborn TM (1937) Sex, sex inheritance, and sex determination in *Paramecium aurelia*. Proc Natl Acad Sci USA 23:378–385
Steinbrück G (1986) Molecular reorganization during nuclear differentiation in ciliates. In: Hennig W (ed) Results and problems in cell differentiation. Springer, Berlin Heidelberg New York, 13, pp 105–174
Takagi Y (1988) Aging. In: Görtz H-D (ed) *Paramecium*. Springer, Berlin Heidelberg New York, pp 131–140
Takahashi M, Takeuchi N, Hiwatashi K (1974) Mating agglutination of cilia detached from complementary mating types of *Paramecium*. Exp Cell Res 87:415–417
Treier U, Fuchs S, Weber M, Wakarchuk WW, Beck CF (1989) Gametic differentiation in *Chlamydomonas reinhardtii*: light dependence and gene expression patterns. Arch Microbiol 152:572–577
Tschermak-Woess E (1959) Extreme Anisogamie und ein bemerkenswerter Fall der Geschlechtsbestimmung bei einer neuen *Chlamydomonas*-Art. Planta 52:606–622
Tsukii Y (1988) Mating-type inheritance. In: Görtz H-D (ed) *Paramecium*. Springer, Berlin Heidelberg New York, pp 59–69
van den Ende H (1994) Vegetative and gametic development in the green alga *Chlamydomonas*. Adv Bot Res 20:125–161
Walker-Jonah A, Dolan SA, Gwadz RW, Panton LJ, Wellems TE (1992) An RFLP map of the *Plasmodium falciparum* genome, recombination rates and favored linkage groups in a genetic cross. Mol Biochem Parasitol 51:313–320
Walliker D, Quakyi IA, Wellems TE, McCutchan TF, Szarfman A, London WT, Corcoran LM, Burkot TR, Carter R (1987) Genetic analysis of the human malaria parasite *Plasmodium falciparum*. Science 236:1661–1666
Watanabe T (1990) The role of ciliary surfaces in mating in *Paramecium*. In: Bloodgood RA (ed) Ciliary and flagellar membranes. Plenum Press, New York, pp 149–171
Wegener D, Beck CF (1991) Identification of novel genes specifically expressed in *Chlamydomonas reinhardtii* zygotes. Plant Mol Biol 16:937–946
Wichterman R (1985) The biology of *Paramecium*, 2nd edn. Plenum Press, New York
Xianyu Y, Haga N (1993) Initiation of the earliest nuclear events in fertilization of *Paramecium* by the microinjection of calcium buffer. Zool Sci 10:857–862

Yanagi A, Hiwatashi K (1985) Intracellular positional control of survival or degeneration of nuclei during conjugation in *Paramecium caudatum.* J Cell Sci 79:237–246
Zech L (1964) Zytochemische Messungen an den Zellkernen der Foraminiferen *Patellina corrugata* und *Rotaliella heterocaryotica.* Arch Protistenk 107:295–330

CHAPTER 2

Fertilization in Invertebrates

HEIDE SCHATTEN[1] AND AMITABHA CHAKRABARTI[2]

1 Introduction

Invertebrates are by far the largest animal group, consisting of a total of ten major and some smaller phyla. The reproductive diversity is enormous. As many as 95% of all known animals are invertebrates, but only a small percent of those has been studied in detail at a cell and molecular level. Most invertebrate oocytes are arrested at specific stages, during which they are fertilized before undergoing complete maturation. The echinoderm sea urchin is one of the few that are spawned and fertilized as mature oocytes after having completed meiotic maturation. The sea urchin is by far the best-studied species of all invertebrates. Much cell and molecular data have been obtained on sea urchins, which has also facilitated the study of higher animal systems and has allowed insights into reproduction of higher vertebrates including humans. Many experiments that have been performed on sea urchins first provided the tools and strategies that could later be applied to less readily available other animal systems.

Invertebrates have been used as experimental research models and the diversity of reproductive systems has resulted in a large body of literature, ranging from classic cell and developmental analyses to histological and cell and molecular biology. Invertebrate models have been used for over a century to study fertilization, cell division, and development. The sea urchin *Arbacia* and the roundworm *Ascaris* are classic examples of the early studies, and more recently the nematode *Caenorhabditis* has been used to study cell polarity, cell and molecular details of cytoskeletal organization, and pattern formation during development. Studies of the fertilization and reproduction of invertebrates have produced increases in reproductive efficiency which are of economic significance. This chapter will focus on selected marine, freshwater, and terrestrial species that have been used as research material to study cell and

[1] Department of Veterinary Pathobiology, University of Missouri-Columbia, 1600 East Rollins Street, Columbia, Missouri 65211, USA

[2] The Cleveland Clinic Foundation, Department of Cell Biology, NC-1, 9500 Euclid Avenue, Cleveland, Ohio 44195, USA

Juan J. Tarín and Antonio Cano (Eds.)
Fertilization in Protozoa and Metazoan Animals

molecular mechanisms of fertilization. The data discussed here will, to some extent, also apply to other species.

Fertilization is a complex process by which both gametes are united into one. The process of sexual reproduction involves gametogenesis and fertilization and most systems also undergo physiological and behavioral adjustments to either copulate or spawn gametes into an environment where both gametes can meet. Once united and fused, the environment must be suitable to allow the development of a single cell into a multicellular organism. The formation of an embryo includes mitosis and cell differentiation, as well as communication with neighboring cells. The cytoskeleton plays crucial roles in many processes, from fertilization throughout embryo development. It is essential for defining the pattern for the cleavage furrows and to serve as transport system to carry out various functions, including transport of organelles such as mitochondria and yolk material. The diversity of reproduction in invertebrate animals is huge. Only a small selection of animal species is treated in greater detail here as representative of a larger group in the specific phylum.

2 Porifera

Sponges are restricted to aqueous environments and gametes are spawned into the water. The Porifera are subdivided into three different classes (Calcarea, Hexactinellida, Demospongiae) on the basis of their skeletal composition.

2.1 Reproductive Strategies and Reproductive Cycle

The reproductive strategies of this phylum of invertebrates span a great variety. Sexual reproduction during part of their life cycle and asexual reproduction are both possible. Individual adults of various species can either produce male or female gametes (gonochorism) or can produce both (hermaphrodism). In hermaphrodites, development of both gamete systems can either be asynchronous with the male system developing prior to the female system (protandry) or occur simultaneously (synchronogamy). Sex reversal is possible in *Spongilla lacustris* (Gilbert and Simpson 1976) which exhibits an alternative type of hermaphrodism, in which a sponge may be either male or female in alternate years. The majority of sponges are viviparous, but little is known about the regulation of reproductive timing, which may be triggered by endogenous chemical stimulation (Simpson 1980).

A number of sponges are oviparous, including *Tetractinomorpha*. Others spawn eggs that are fertilized externally and include *Chondrosia reniformis* and *Tetilla serica*. For many species it is not known whether fertilization takes place before or after release of eggs. A synchronous cycle of oocyte development takes place each year (Fell 1974). In the majority of sponges fertilization is internal and sperm are conveyed from a canal system to the eggs by modified choanocytes (Fell 1974) which function as carrier cells. Choanocytes can be associated with the oocytes and transform into carrier cells for sperm at the appropriate stage. It has been proposed that the oocytes may stimulate cyanocytes to produce specific sperm receptors (Fell 1974; Courréges and Fell 1989).

2.1.1 Design of Reproductive System

Gonads in sponges are not as clearly defined as in all other invertebrate phyla, but there is structural resemblance to those seen in other invertebrate systems. Clusters of sperm cells in various stages of spermatogenesis are encapsulated in a spermatocyst which contains an outer cell wall. In all cases sperm ducts are not apparent, and sperm are released by rupture of the spermatocyst in the aquiferous canals. Similar mechanisms are used for oocytes (in oviparous species) or for larvae (in viviparous species) which are released in the water column. True ovaries are not seen in Porifera. Instead, oocytes develop individually or as clusters in the mesohyl (Gaino et al. 1986; Franzén 1988).

2.1.2 Gametogenesis and Fertilization

The exact origin of sponge germ cells is not clearly determined, although they appear to arise from choanocytes which are uniflagellated collar cells that form the epithelial choanoderm. Spermatogonia form spermatocytes, and spermatids undergo spermiogenesis during which chromatin condensation and flagellar development occur. Cell organelles are formed during this process that vary according to different species. Oocytes which may also be derived from choanocytes undergo a phase of growth that is supported by yolk and nutrients from surrounding cells, called trophocytes or nurse cells which are phagocytosed by the growing oocyte (Gaino et al. 1986, 1987; Paulus and Weisenfels 1986; Saller 1988; Paulus 1989).

The process of fertilization is not well studied in Porifera because of the difficulties of studying internal fertilization, which is typical for most sponges. A sperm carrier cell derived from choanocytes is used that delivers the sperm to the oocyte by either fusion of carrier and sperm with the oocyte or releasing the sperm into the oocyte (Gaino et al. 1984).

2.2
Cellular and Molecular Aspects of Fertilization

2.2.1
Sperm

Spermatozoa have been described in about one-third of the sponges and an acrosome appears to be present as judged by light microscopy and in some cases ultrastructural studies. In *Oscarella lobularis* (Gaino et al. 1986) a flattened acrosomal granule is present. A layer of 30–40 electron-dense vesicles has been shown which originate in close proximity to the Golgi and are considered as homologues to an acrosome. It has been proposed that the acrosome plays a role in a specific interaction between the spermatozoon and the choanocyte/carrier cell. It is not known whether the acrosome homologue contains enzymes and/or binding proteins which play a role during fertilization. A sperm nuclear basic protein (SNBPs) has been identified in *Neofibularia nolitangere* (Ausio et al. 1997) which consists of histones that are very similar to other eukaryotic histones.

2.2.2
Oocytes

Development of oocytes in many sponges is asynchronous which makes it difficult to define the time course of maturation. Meiotic divisions have been observed best in a number of calcareous sponges (Tuzet 1964) but in few others, which may be related to the abundance of reserve substances surrounding the oocytes.

2.2.3
Fertilization

Fertilization has not been observed in most species because of technical problems and there have been only sparse ultrastructural studies of fertilization (Gallissian 1980). Data on cell and molecular biology of sponge fertilization are not available.

In those few cases in which maturational divisions have been observed they occur very soon after spermiocyst incorporation (see Fig. 1 for schematic representation). Incorporation of the spermiocyst into the oocyte has been observed in *Clathrina (Leucosolenia) coriacea* and *Octavella galangaui* in which sperm enter oogonia of about 25–30 μm in diameter. After sperm incorporation the oocytes at prophase I grow to about 100 μm before the meiotic divisions occur, which indicates that protein synthesis and other metabolic activation processes have occurred following sperm incorporation. The sperm nucleus enlarges and forms the male pronucleus. Prominent nucleoli appear in *Grantia compressa* and the mitochondrial body separates from the

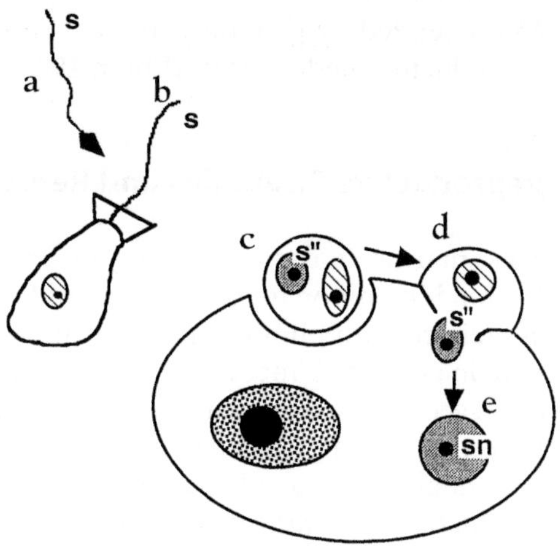

Fig. 1. Composite of stages of sperm incorporation in sponges. **a** Spermatozoon (*s*) before entering a choanocyte; **b** spermatozoon within a choanocyte; **c** spermiocyst (*s*") within a carrier cell; **d** spermiocyst transfer to oocyte; **e** sperm nucleus (*sn*) in oocyte. (Modified from Fell 1989)

nucleus and begins to disintegrate. The female pronucleus forms and increases in size. It also contains nucleoli. The female pronucleus moves and joins the male pronucleus at the center of the egg where they fuse. In *Hippospongia communis* the egg grows from 50 to 300 µm in diameter before cleavage occurs.

Fertilization is associated with a number of egg surface changes. In *Tetilla serica* and *Tetilla japonica* the ovulated eggs are surrounded by a clear vitelline membrane. The vitelline membrane elevates at fertilization into the fertilization membrane. Egg surface changes also include a bundle of fibers of 60 µm with a 17-nm periodicity. They extend through the vitelline layer in unfertilized oocytes and elongate at fertilization. The function of these fibers is unclear, but they may aid in rotation and positioning of the oocyte. The fibers have disappeared at ca. 10 min after fertilization.

The site of sperm entry and the existence of a block to polyspermy in sponges are unclear. Indirect evidence suggests that some mechanisms preventing polyspermy may exist. In *Octavella galangaui* two choanocytes with spermatozoa have been observed, and large numbers of spermatozoa are seen within the canals of the female tract, but there are no indications that this will lead to polyspermy.

3 Cnidaria (Coelenterata)

This phylum of invertebrates includes mostly marine but also a variety of freshwater representatives. Among the marine animals is the class Scyphozoa with the well-known jellyfish *Aurelia*, and the phylum includes sea anemones, corals, and sea fans. Freshwater cnidarians include the well-known genus hydra. Sexual and asexual reproduction alternates in cnidarians and is characterized by generally asexual sessile polyps and the free-living sexually reproducing medusa (Strathman 1987).

3.1 Reproductive Strategies and Reproductive Cycle

Most marine cnidarians employ external fertilization, although internal fertilization is common for freshwater cnidarians such as hydra. Spawning is usually associated with photoperiods during which gametes are released by male and female gonads. A most spectacular spawning of 112 species of anthozoans is observed on the Great Barrier Reef of Australia when release of gametes occurs for only a few nights each year following the spring full moon. Environmental factors also play a role in spawning of corals, related to water temperature. *Hydractinia echinata* spawns after sunrise and is inhibited by keeping animals in the dark. The sensor for light is probably pigment in the epithelial cells of the gonad that are contractile and capable of detecting changes in the light level and causing release of gametes. The variety of gonadochoristic or hermaphrodistic characteristics within this phylum is remarkable and there are even variations within the various species. Asexual reproduction takes place in a variety of different cnidarians but occurs primarily in polyps. It is also possible for some species to undergo simultaneous sexual and asexual reproduction processes.

3.1.1 Design of Reproductive System

Gametogonia are produced by interstitial cells and gonads consist of germ cells in various stages of gametogenesis. Cnidarian gonads have a simple form and consist basically of an assemblage of gametes. Gonads of anthozoans, cubozoans, and scyphozoans mature in the endoderm, and gonads of hydrozoans mature in the ectoderm. As in Porifera, regular gonadal ducts are not formed and gamete release occurs by rupture into the gastrovascular cavity.

3.1.2 Gametogenesis and Fertilization

During gametogenesis, interstitial cells differentiate and testes are produced as spermatogenetic stages develop. Spermatogonia become located nearest the epithelial basal lamina and sperm nearest the apical surfaces. Oogenesis and oocyte maturation are supported by yolk that is produced by interstitial cells. It is influenced by photoperiodicity which depends on regulation by hormones secreted by gonophore somatic cells. Fertilization occurs by sperm binding and fusion at the site of polar body formation after chemoattraction of sperm to the matured oocyte and after completion of polar body formation. As for Porifera, very little is known about the fertilization process itself and about pronuclear apposition and fusion (Honegger 1981; Freeman and Miller 1982; Larkman 1983; Littlefield et al. 1985; Eckelbarger and Larson 1988; Honegger et al. 1989).

3.2 Cellular and Molecular Aspects of Fertilization

Chemical attraction is involved in species-specific recognition of sperm to eggs. Cnidarian oocytes have morphologic diversity and surface characteristics are an important factor in fertilization. The oocytes of *Hydractina* and *Halcampa* are smooth and covered by a jelly coat. Electron microscopy studies of *Metridium* and *Aequorea aequorea* show surfaces covered with microvilli. It has been shown for *Astrangia danae* that cortical granules are located underneath the egg surface. No major discharge of cortical granules indicative of a cortical reaction was observed in the sea anemone *Actina fragacea* (Larkman and Carter 1984). The cortical granules were found to persist well into larval life, and their function is unknown. Cortical reaction was reported in the oocytes of the cnidarian *Bundosoma* (Dewel and Clark 1974). Oocytes are fertilized following polar body emission and it has been shown for *Phialidium* that eggs mature 1 h before sunrise which triggers spawning. Only mature oocytes are spawned in this species. Oocytes of the actinian *Metridium* complete maturation division immediately before spawning and females spawn only after sperm already have been spawned by the male. The spawned oocyte of the hydromedusa *Orthopyxis caliculata* releases a sperm attractant that causes agglutination of sperm over the area close to the female pronucleus (Miller 1978) and fertilization appears to be restricted to the site of the pronucleus (Freeman and Miller 1982). It has been shown (Spaulding 1972) that long spines cover many anthozoan oocytes except over the pronuclear area. In hydrozoan oocytes sperm can only fuse at a small region of the egg surface that corresponds to the site where the polar bodies are extruded (Freeman and Miller 1982; Honeggar 1983). Freeman (1996) has shown that treatment of *Aequorea* eggs with triticum or concanavalin A causes a marked reduction in the ability of these eggs to be fertilized, suggesting that sugar-containing moieties (glycoprotein), to which lectin binds, play a role in fertilization. This moiety may be analogous to the sperm binding receptor in the egg membrane of sea urchins (Foltz et al. 1993).

The cell and molecular biology of fertilization in cnidarians is surprisingly poorly understood. It is not clear whether the sperm exhibits an acrosome (Carré 1984). A 75-μm-thick gelatinous layer forms within 10–15 min after fertilization in the anemone *Halcampa duodecimicircata*. Cortical granules have been reported for *Bunodosoma cavernata* (Dewel and Clark 1974) which are released at fertilization and form a layer around the oocyte. A number of second messengers including Ca^{2+} and pH appear to play a role in mediating changes in egg metabolism following fertilization or activation. Freeman and Ridgway (1993) have shown that Ca^{2+} transient occurs in the hydrozoan oocytes of *Mitrocomella* and *Phialidium* as a consequence of fertilization, whereas no increase in pHi following fertilization has been observed. Freeman and Ridgway (1991) also studied the calcium transient in cnidarians by injecting inositol 1,4,5-trisphosphate (IP_3), which indicates that fertilization in cnidarians may involve calcium release by IP_3 receptors.

4 Ctenophora

This phylum of invertebrates is very small, containing only about 100 species, and very few data exist on the cell and molecular aspects of fertilization. All species are marine and are distinguished by a simple body plan with biradial symmetry, containing eight rows of beating ciliated comb plates (ctenes).

4.1 Reproductive Strategies and Reproductive Cycle

All ctenophores with the exception of two deep-water species are hermaphrodite and release sperm and oocytes during a light period after exposure to dark. It has been reported for *Beroe ovata* (Carré and Sardet 1984) that eggs grow rapidly and this species will produce 50–100 eggs in 2 days which are ca. 1 mm large and very transparent. Most ctenophores shed their gametes, via the mouth, into the surrounding seawater where fertilization takes place. In *Coeloplana* and *Tjalfiella*, fertilization is internal and embryos are brooded.

4.1.1 Design of Reproductive System

The design of the reproductive system is simple and visible gonads are located beneath the comb rows (Strathman 1987). The gonads are entodermal in origin, arising on the walls of the meridional canals.

4.1.2 Gametogenesis and Fertilization

Information on gametogenesis is sparse and only recently has information on fertilization been obtained (Carré and Sardet 1984; Carré et al. 1991), which will be discussed in the following section.

4.2 Cellular and Molecular Aspects of Fertilization

The oocytes of the ctenophore *Beroe ovata* are shed as they are completing the first meiotic division and maturation is completed whether the oocyte is fertilized or not. Fertilization takes place in the mature oocyte. There is a block to self-fertilization which occurs at the levels of the vitelline membrane and the jelly coat. The block to self-fertilization can be overcome when these envelopes are removed (Carré and Sardet 1984). Sperm can enter at any site of the egg surface and the sperm nucleus does not move considerably after incorporation.

Sperm penetration into the egg is preceded by a modification of the sperm tip which may correlate with an acrosome reaction (Carré and Sardet 1984). Upon sperm entry the oocyte surface contracts and a fertilization cone assembles around the entering sperm which is resorbed by 20 min after sperm contact. Polyspermy is the rule in *Beroe ovata* (Carré and Sardet 1984; Carré et al. 1991) and up to 20 sperm can be incorporated into the ctenophore oocyte, but only one male pronucleus finally fuses with the female pronucleus.

There is no obvious membrane block in ctenophore oocytes, and an electrical block to polyspermy as has been reported for many other species (Jaffe and Gould-Somero 1985) does not appear to exist. In the ctenophore *Beroe ovata*, action potentials and sperm-induced oocyte electrical responses were recorded, in unfertilized and in vitro inseminated oocytes, respectively (Goudeau and Goudeau 1993b). The cortex of the oocyte does not appear altered upon fertilization except for the area of sperm incorporation. A local tuft of microvilli, accumulation of mitochondria, and autofluorescent vesicles are recruited to the sperm entry site from the cytoplasm.

The female pronucleus forms under the polar bodies and chooses the sperm nucleus for fusion which is not necessarily the closest one. The movement of the female pronucleus is dependent on microtubules. Microtubules are associated with the egg cortex and with sperm soon after sperm incorporation.

Upon sperm entry, the egg surface of *Bero ovata* contracts and relaxes locally, then a fertilization cone forms and disappears. These events are accompanied by localized exocytosis, growth of a ring of microvilli, thickening of the egg cortex, and gathering of mitochondria and the sperm pronuclei. The female pronucleus then migrates beneath the egg surface toward one or successive stationary male pronuclei in the peripheral cytoplasm that surrounds the yolky interior (Carré et al. 1991; Rouviere et al. 1994, 1998). The sperm become surrounded by a specialized zone (30–50 μm in diameter) beneath the surface, referred to as the sperm pronuclear zone (SPZ), which is devoid of large cortical granules. The oocyte pronucleus can probe several SPZ in polyspermic oocytes, each containing a single sperm nucleus, before it finally enters one SPZ and fuses with the chosen sperm pronucleus. These migrations of the oocyte pronucleus can take hours. The mechanism underlying the motion or how the oocyte pronucleus decides which SPZ to enter is not yet known.

5 Platyhelminthes

This fourth largest animal phylum comprises four classes that live in marine, freshwater, and terrestrial environments. The classes include *Turbellaria*, which are predominantly free-living worms, *Monogenea*, which are obligate parasites in vertebrate hosts, *Trematoda*, which are obligate parasites with asexual reproduction as larvae primarily in molluscs and sexual reproduction primarily in vertebrates, and *Cestoda*, which are obligate parasites in vertebrates that live as juveniles in arthropods (Conn 1985).

5.1 Reproductive Strategies and Reproductive Cycle

Sexual reproduction is most common for Platyhelminthes although asexual reproduction also takes place. Most classes are predominantly hermaphroditic, but gonochoristic species also occur. In schistosomes females are separated from males and are carried by the male in a specialized groove on the ventral surface. Fertilization is internal and may be accomplished by cross-fertilization directly or by transfer of spermatophores. Direct sperm transfer is the rule in Turbellaria and occurs through regular copulation into a well-developed female tract. Indirect sperm transfer occurs by hypodermic injection or dermal impregnation of the female which can occur in a number of variations depending on the species. This form is used in species in which the female reproductive system and a vaginal opening are not well developed. The dermis is either pierced by stylets in the penis (hypodermic injection) or opened by secretions deposited by the sperm on the surface of the partner (dermal impregnation) to transfer sperm. The wound may be sealed again (Hyman 1951). Asexual reproduction can occur in many species and the two reproductive modes can alternate.

5.1.1 Design of Reproductive System

Platyhelminthes contain a structurally complex reproductive system that can remain active throughout a lifetime or be resorbed when environmental conditions are unfavorable. Follicular testes and ovaries are usually surrounded by a cytoplasmic sheath. After full development sperm move from the testes to a short vasa efferentia and a common vas deferens which may enlarge into a seminal vesicle. Oocytes move down from the ovaries to the oviduct into a more or less well-developed uterus. In some species the oocytes may associate with vitellocytes that have come from vitelline glands through vitelloducts, but the role of vitellocytes is not clear. In many species the female ducts can have specialized regions that function as seminal receptacles or bursae. The reproductive system appears complex and detailed studies on the specific cellular organization have not yet been carried out (Nollen 1983; Conn 1985).

5.1.2 Gametogenesis and Fertilization

Two groups of Platyhelminthes can be distinguished that have different features of oogenesis. The *Archoophora* have female germ cells which develop into endolecithal oocytes that can produce or acquire nutrients needed for fertilization and early embryonic development. The *Neoophora* develop ectolecithal oocytes which are uniquely derived from a female germ cell lineage which divided into an ovarium and vitellarium cell line which become spatially and functionally separated into separate organs. The ovarian line forms oocytes with germinal vesicles but without yolk, cortical granules, and other components that are usually found in oocytes. The vitellarium line forms vitellocytes that contain the components needed for later development. Both oocytes and vitellocytes are needed and essential for development of Neoophora embryos. In most Platyhelminthes oogenetic meiosis is arrested in late prophase I or early metaphase I and meiosis is completed upon fertilization.

Spermatogenesis in most Platyhelminthes involves development of numerous sperm from the spermatogonia. Sperm remain as a morula mass until the end of spermiogenesis. Acrosome or Golgi-derived secretory organelles have not been identified in this phylum although it cannot be excluded that acrosome-like structures exist. The number of flagella varies in different groups. While turbellarian sperm are biflagellated, the sperm of nemertodermatid turbellarians and most trematodes, cestodes, and monogeneans are uniflagellated or lack a flagellum. In the trematode genus *Schistosoma*, the doublets in the two flagella lack dynein arms, and the spermatozoa are immotile (Justine et al. 1985), which is a step toward aflagellate sperm. The sperm of platyhelminth gnatostomulids are aflagellate (Graebner and Adam 1970). Chemical factors appear to be necessary for sperm capacitation. The process of fertilization is not well studied and pronuclear events have not been investigated for most species. Pronuclear apposition without fusion is reported for Turbellaria while detailed data are not available for the other three classes (Tyler 1984; Conn 1985; Burt 1986; Hendelberg 1986; Holy and Wittrock 1986; Fischlschweiger 1990).

5.2 Cellular and Molecular Aspects of Fertilization

Because fertilization is internal in Platyhelminthes very few data are available that describe the process of fertilization in detail. A review article has been written by Nollen (1983) describing the patterns of sexual reproduction among parasitic Platyhelminthes with regard to spermatogenic development stages, development and movement of oocytes in the female reproductive tract, and mating behavior. The events of fertilization have been determined in the monogenean skin parasite *Entobdella soleae* by sequential ultrastructural studies of oocytes in the germarium of the parasite which contains many mature oocytes in a coiled tube. Oocytes reach the tube lumen probably by amoeboid movement and penetrate the cytoplasm of a "plug cell" (Tappenden et al. 1993) that is located at the proximal end of the intra-germarial tube.

Cortical granules have been observed in unfertilized oocytes of some Turbellaria and may be excreted upon fertilization to form a fertilization membrane. Monospermic sperm incorporation is the rule, but detailed studies on mechanisms for the prevention of polyspermy have not yet been performed. Multiple sperm entry can occur in Turbellaria and it has been suggested that supernumerary sperm might degenerate in the oocyte's cytoplasm which may represent a form of prevention of polyspermy. The exact moment of fertilization has not yet been established, but it occurs long before the oocytes are deposited (Gremigni et al. 1986; Sopott-Ehlers 1986). The spermatozoon enters the egg before the female germ cell has completed first meiotic division (Benazzi and Benazzi Lentati 1976) and the nucleus enlarges and develops into the pronucleus in the egg cytoplasm. Typically in many turbellarians the male pronucleus does not fuse with the female pronucleus and chromosomes undergo separate cycles. Interestingly, in the species *Dugesia benazzi, Dugesia lugubris-polychroa, Polycelis nigra,* and *Polycelis tenuis*, the spermatozoan not only activates the oocyte (Benazzi Lentati 1970) and develops into the male pronucleus but also either is expelled with the polar body or degenerates in the cytoplasm.

6
Rhynchocoela (Nemertea, Nemertinea)

Rhynchocoela are the least studied of all invertebrates although there are excellent data on the histology of the reproductive systems (Tarpin 1997). The least studied aspect of Rhynchocoela reproduction is fertilization although an interest in the early stages of fertilization is emerging (Kline et al. 1985; Stricker 1986). Most Rhynchocoela are marine animals, with a few exceptions that are found in freshwater and in terrestrial areas. There are over 900 species with size variations ranging from less than 25 cm to ca. 25 m in length in some species (Gibson 1972).

6.1
Reproductive Strategies and Reproductive Cycle

Fertilization in most Rhynchocoela is external and occurs by releasing gametes from gonadal cavities into the seawater. Some species are known to employ internal fertilization or release epithelial secretions to produce strings of eggs embedded in gelatin. Some freshwater species produce fertilization sheaths into which sperm and oocytes are deposited. While most Rhynchocoela are gonochoristic, some are hermaphroditic. Asexual reproduction is rare in this phylum.

6.1.1
Design of Reproductive System

The reproductive system in Rhynchocoela is relatively simple and consists of a series of germ cell aggregations. The ovaries or testes can extend through the length of the body or be more localized to parts of the body and are lined by a thin layer of epithelial somatic cells. They contain a mass of germ cells that are released through transitory gonoducts that form openings through the dorsolateral body wall. This relatively simple system applies to most Rhynchocoela, but others are known to use a more elaborate reproductive design (Stricker 1987).

6.1.2
Gametogenesis and Fertilization

Germ cells that are derived from mesenchyme cells complete development within the gonadal sac. Sperm cells have a distinct morphology with typical head, midpiece, and tail that varies to some extent from species to species. Oocytes are generally fertilized during the arrested metaphase I stage and meiosis is completed after sperm entry. The stages of sperm entry and early fertilization are characterized by a typical sperm acrosomal reaction and fusion of the acrosomal process with the oocyte surface. Little is known about subsequent stages of fertilization (Freeman 1978; Kline et al. 1985; Stricker 1986; Stricker and Cavey 1986; Longo et al. 1988).

6.2
Cellular and Molecular Aspects of Fertilization

Fertilization in Rhynchocoela has only recently gained interest (Kline et al. 1985; Stricker 1986). The eggs of Rhynchocoela have large stores of protein, lipid, and carbohydrate. A single membrane (vitelline) surrounds the oocytes that are embedded in gelatin layers, but more than one membrane is observed to surround oocytes of those species that are free spawners. Oocytes are shed at first meiosis of maturation (Gontcharoff 1951). The first and second polar bodies are extruded within an hour after fertilization and cells begin to divide within 30–40 min. Sperm can enter at any area of the oocyte and a fast block to polyspermy has been reported by Kline et al. (1985) in *Cerebratulus* that lasts for about an hour after fertilization.

The marine nemertean worm *Cerebratulus lacteus* produces oocytes that possess a large nucleus ("germinal vesicle" or GV). No preferred sperm entry location is evident relative to the animal–vegetal axis in this species. The head plus midpiece of the sperm are typically internalized within 5–15 min after attachment. The fertilized oocyte does not undergo a distinct cortical reaction or produce a marked fertilization cone that engulfs the sperm. However, video analysis of this species by Stricker (1996) revealed that a series of contraction waves originate at the sperm attachment site in the form of several contractions of the oocyte cortex that occurred prior to and during sperm incorporation. After 40 min of fertilization the sperm head condensed into a spheroid mass and by 2 h the male pronucleus decondensed. The first and second polar bodies were typically extruded by about 45 and 85 min postfertilization, respectively, and the first cleavage furrow formed 160–180 min after insemination (Stricker 1996). Recently, Ca^{2+} and endoplasmic reticulum dynamics during oocyte maturation and fertilization in *Cerebratulus lacteus* were reported by Stricker et al. (1996). These authors have seen that prophase-arrested immature oocytes of *C. lacteus* fail to develop normally following insemination and typically display a fertilization-induced Ca^{2+} response that lacks the amplitude, kinetics, and spatiotemporal patterns of oocytes that were fertilized after having undergone maturation. It was shown that the endoplasmic reticulum exhibits major structural reorganizations during maturations and fertilization.

7 Nematoda

Nematodes comprise a large phylum of animals that inhabit marine, freshwater, and terrestrial areas. More than 12,000 species are known and many of them are parasitic to a large variety of host cells, including domestic animals and humans. A few species including those of *Parascaris* were studied by cytologists around 1900 in greater detail, which has resulted in information on fertilization and cell division. More recently, *Caenorhabditis elegans* has become a model system to study molecular biology of pattern formation and development (Maggenti 1981; Wharton 1986; Kenyon 1988).

7.1 Reproductive Strategies and Reproductive Cycle

The reproductive system of nematodes is embedded in a pseudocoelom. Reproductive strategies can vary in the different species, depending on different strategies for transmission between hosts. The majority of nematodes reproduce sexually and are gonochoristic, although hermaphrodism is also seen, for instance in the well-studied species *Caenorhabditis elegans* in which self-fertilization takes place. In some cases gonads are used for both spermatogenesis and subsequent oogenesis. Fertilization is internal and is accomplished with the male's hardened copulatory spicules that are inserted into the female's vulva for sperm transfer. Most nematodes are oviparous and the number of oocytes produced varies greatly. Parasitic species generally produce large numbers of oocytes while free-living species produce fewer. In some cases juveniles are developed within the uterus and released directly into the environment (Kemphues 1989).

7.1.1 Design of Reproductive System

Male and female reproductive systems can be single or paired and each gonad consists of an unbranched tube lined by a single epithelium. Male and female cells undergo gametogenesis as they move through the genital aperture which is located at the anterior end for the female and typically at the posterior tip of the organism for males. Ovaries and testes are small tubes. Numerous oocytes are deposited in the ovary which is connected to the uterus by a short oviduct. Depending on the species the uterus may serve additional functions of embryonic development. The distal end of the uterus can also serve as a spermatheca for storage of sperm. The uterine tubes connect to the vagina which is lined with musculature that is involved in oviposition. In the male, the testis is connected to a seminal vesicle that widens into the vas deferens and eventually leads into the genital aperture. Prostatic glands are associated with the male reproductive system and copulatory spicules are embedded in their sacs (Weber 1987).

7.1.2 Gametogenesis and Fertilization

One of the best studied species of this phylum is *Parascaris equorum*, which was explored at the turn of the century by cytologists such as Theodor Boveri. Two forms of gonads exist in nematodes: telogonic gonads, in which all gametes are derived from a single apical germinal cell, and hologonic gonads, in which several progenitor cells give rise to gametes. Primary oocytes undergo extensive growth phases and are arrested in prophase I which is the stage in which fertilization occurs. Nematode spermatozoa are produced during spermiogenesis from secondary spermatocytes that are derived from primary spermatocytes formed in spermatogonia. Nematode spermatozoa generally lack a flagellum and develop pseudopodia for locomotion. Sperm become arrested and stored in the seminal vesicle. After copulation sperm undergo transformation and capacitation in the distal region of the uterus where they fuse with the primary oocyte. Fertilization is accomplished by incorporation of the sperm into the oocyte cytoplasm which also causes release of the oocyte's meiotic arrest. Meiosis is completed after the female nucleus moves to the periphery and after production of two polar bodies and one haploid nucleus that moves back to the center of the oocyte. The haploid sperm and egg pronuclei move in close apposition, but pronuclear membrane fusion and formation of a single zygote nucleus have not been observed (Seidl et al. 1988; Strome 1989).

7.2
Cellular and Molecular Aspects of Fertilization

Focus of nematode research has been on the free-living species *Caenorhabditis elegans* for the ease of cultivation and extensive mapping of both its genome and cell lineage. However, fertilization in nematodes has not been well studied. The little data that exist are limited to *Ascaris lumbricoides* (Foor 1970), *Dipetalonema viteae*, and *Caenorhabditis elegans* (Ward and Carrell 1979) and many of the data are derived from ultrastructural studies.

7.2.1
Sperm

The sperm must move from the seminal receptacle to the vagina in an amoeboid movement. The sperm of *Nematospiroides dubius* (Wright and Sommerville 1977) and *Caenorhabditis elegans* (Ward and Carrell 1979) have to travel long distances in the uterus and exhibit complex movement that use cytoskeletal fibers other than those known for amoeboid microfilament movement. Sperm of *Ascaris lumbricoides* are polarized and contain cell organelles in one end and pseudopodia in the other, which is also true for *Dipetalomena viteae*, *Caenorhabditis elegans* (Ward and Carrell 1979), *Nippostrongylus brasiliensis*, and *Nematospiroides brasiliensis* (Wright and Sommerville 1985b).

The immotile spermatids of *Ascaris suum* extend pseudopods and become rapidly crawling sperm when treated with an extract from the glandular vas deferens of the male under strict anaerobic conditions. Within 9–12 min, a pseudopod develops, elongates rapidly, and exhibits a continuous flow of membrane specializations, the vilipodia, from tip toward base. This amoeboid motility of sperm is the result of independent processes of continuous extension at the leading edge and continuous shortening at the base of the pseudopod which act to propel the cell forward (Sepsenwol and Tuft 1990). Nematode sperm use a major sperm protein (MSP) cytoskeleton in place of an actin cytoskeleton to drive their amoeboid locomotion. Motility is coupled to the assembly of MSP fibers near the leading edge of the pseudopod plasma membrane. The MSP of *Ascaris suum* forms filaments constructed from two helical subfilaments (Stewart et al. 1994) with no overall polarity and it is unlikely that molecular motors play a direct role in the generation of protrusive forces in these amoeboidal cells (Bullock et al. 1998).

In the nematode worm *Caenorhabditis elegans*, individuals with mutations in the spe-9 gene produce spermatozoa with wild-type morphology and motility that cannot fertilize oocytes even after contact between gametes (Singson et al. 1998). The spe-9 gene encodes a sperm transmembrane protein with an extracellular domain that contains ten epidermal growth factor-like repeats. These epidermal growth factor-like motifs are involved in extracellular functions such as adhesive and ligand-receptor interactions required for fertilization. In *C. elegans* Browning and Strome (1996) reported that the protein spe-11 produced only in the sperm plays a crucial role during the first few minutes after the egg is fertilized. When fertilized by sperm lacking the protein spe-11, *C. elegans* oocytes yield embryos that fail to complete fundamental tasks and never grow into multicelled embryos – a fate that can be reversed by engineering the egg to express spe-11 on its own (Roush 1996).

7.2.2 Oocytes

The number of oocytes produced per day in nematodes varies considerably and ranges from 60 in *Ditylenchus myceliophagus* to 200,000 in *Ascaris lumbricoides* (Rogers 1962). *C. elegans* lays an average of about 300 eggs/day. Egg laying is associated with the production of a gelatinous matrix in which DNA is the most active component. This has been shown for *Meloidogyne*.

Two types of cells can be distinguished in the nematode's uterus: uterine cells located just behind the spermatotheca, which contain large intracytoplasmic spaces bounded by ER and refringent bodies, and other types of cells located further down the uterus, distinguished by compact cytoplasm with dense ER and numerous mitochondria. The mature oocyte is bounded by the oolemma which becomes the vitelline layer of the oocyte shell. A chitinous layer is formed in the space between the oocyte membrane and the vitelline layer, and another lipid layer is formed beneath the chitinous layer. The lipid layer is made up of lipoprotein.

7.2.3 Fertilization

The pseudopodia of sperm make first contact with the oocyte (Foor 1970; Harada et al. 1970; Wright and Sommerville 1985a,b) and the cell coat separates from the oocyte by either mechanical forces or enzyme activity. It is not clear whether an acrosome reaction takes place in nematode sperm before fertilization. Sperm and oocyte plasma membranes fuse, but little is known about how sperm becomes translocated into the egg. Images of *Parascaris equorum* indicate that pronuclear apposition but not pronuclear fusion occurs before mitosis takes place.

Hermaphrodite self-fertilization is the primary mode of reproduction in the nematode *C. elegans*. However, when a hermaphrodite is crossed with a male, nearly all of the oocytes are fertilized by male-derived sperm and this results in a functional suppression of hermaphrodite self-fertility (Singson et al. 1999). LaMunyon and Ward (1999) found that the sperm size is a determinant of sperm competitiveness in free-living *C. elegans*. Larger sperm crawl faster and physically displace smaller sperm to take fertilization priority. It was found that the increased competitiveness of *C. briggasae* sperm is due to its male X chromosome over nullo-X sperm (LaMunyon and Ward 1997).

8
Mollusca

This phylum is the second largest phylum of invertebrates, with eight classes including Polyplacophora, Gastropoda, Bivalvia, Scaphopoda, Cephalopoda, and more than 100,000 different species described. All classes have marine, freshwater, and terrestrial representatives and span a great diversity of lifestyles. The diversity ranges from the minute pond snail through nudibranchs, slugs, and the gigantic clam *Tridacna* (Tompa et al. 1983).

8.1
Reproductive Strategies and Reproductive Cycle

The reproductive system reflects the different lifestyles and includes a great variety of reproductive strategies. A large number of groups including opisthobranchs and pulmonates are hermaphrodites, while others including chitons, prosobranchs, and cephalopods are gonadochoristic, and a third group including bivalves are both. The diversity of reproductive strategies includes free spawning of gametes into the water, where fertilization and development of the embryos occur. These strategies are common for chitons and bivalves. Most gastropods and cephalopods undergo internal fertilization after transfer of sperm by copulation, or pseudocopulation with spermatophores. Many species brood their young or deposit the fertilized eggs in hardened masses for protection. Brooding can occur either externally or within the mantle cavity. Some species undergo ovoviviparous reproduction. In a few cases, natural parthenogenesis has been reported.

Asexual reproduction is rare in this phylum and is limited to a few freshwater species (Strathmann 1987).

8.1.1
Design of Reproductive System

Gonads of molluscs are either follicular or acinar and have germinal and accessory cells accumulated in clusters. The diversity of reproductive systems spans hermaphroditic species in which separate testes and ovaries are found to those that are combined into ovotestes. In most cases oocytes are surrounded by supportive follicle cells while sperm can be surrounded by nutritional or regulatory cells. Gametes are spawned through a system of ducts that are combined into one or two gonoducts that lead to a gonopore. For copulation sperm are either released through the gonopore or deposited in special transfer organs (Eyster 1983).

8.1.2 Gametogenesis and Fertilization

The production of sperm follows a typical scheme by which each mother spermatogonium divides mitotically and produces 64 secondary spermatogonia that differentiate into primary spermatocytes. Through further meiotic divisions 256 spermatids are produced which undergo spermiogenesis and are spawned into the water or spermatophore. The production of oocytes is supported by accessory cells. Typically, ovarian follicle cells are involved in yolk production or processing. Because most higher molluscs fertilize after copulation, data on in-vitro fertilization are rare. Fertilization occurs in the germinal stage prophase I or metaphase I arrested oocyte. A typical acrosome reaction in sperm precedes fertilization which is followed by sperm penetration of the external oocyte layers. Oogenetic meiosis and polar body production continues, followed by apposition of the two haploid pronuclei. Monospermic fertilization is the rule although the mechanisms for prevention of polyspermy are not clearly understood and a distinct fertilization membrane to prevent polyspermy is not apparent (Labordus et al. 1986; Claveria and Etges 1988; Buckland-Nicks et al. 1990).

8.2 Cellular and Molecular Aspects of Fertilization

8.2.1 Sperm

Sperm morphology varies considerably between different species. In particular, sperm in species that employ external fertilization differ from those with internal fertilization. A variety of primitive sperm have been reported, including those of *Oxystele* (Hodgson et al. 1990). Sperm of Pelecypoda, Amphineura, and Scaphopoda have a rounded shape and the primitive sperm of Spisula, Acmaea, and Tegula have rounded or elongated heads, and elongated tails. Mitochondria are located at the base of the nucleus. Beta-glycogen has been localized to the matrix between mitochondria and surrounding basal body. The sperm in those species that employ internal fertilization have mitochondria in a sheath surrounding the middle piece and tail.

8.2.2 Oocytes

In most molluscs, maturation of the oocytes is independent of fertilization. Cortical granules are seen underneath the egg surface and, depending on the species, insemination of oocytes can occur at completion of meiosis, second meiotic metaphase, first meiotic metaphase, and germinal vesicle stage.

8.2.3 Fertilization

Fertilization in molluscs can occur by external fertilization, internal insemination, and hypodermic insemination. It can be accomplished by either transfer of spermatophores, through spermatozeugmata, or direct sperm transfer. Self-fertilization has been reported in the freshwater snails *Helisoma duryi* and *Helisoma trivolvis* (Paraense and Correa 1988). External fertilization is used in certain marine bivalves and primitive gastropods that shed sperm freely in the water. Internal insemination is used by cephalopods, mesogastropods, neogastropods, and Neritidae. In this case, eggs passing through the oviduct are fertilized by sperm stored in the reproductive tract. Hypodermic insemination is used frequently and refers to transfer of spermatozoa to the partner indirectly through the skin. This form of insemination, which is also used by Turbellaria among Platyhelminthes and Hirudinea among annelids, is often seen in opisthobranch molluscs (Reid 1964; Purchon 1968). This mechanism is described for at least three species – *Limpontia*, *Actaeonia*, and *Alderia* – in which a hollow style of the penis functions as a hypodermic syringe to inject sperm into body segments of the female. Piercing with stylets has also been described in a variety of other species.

Fertilization is best studied in the surf clam *Spisula solidissima* (Longo and Anderson 1970) and occurs at the germinal vesicle stage which is the stage in which *Spisula* oocytes are arrested. Longo and Anderson (1970) showed that within 6 min of insemination, the structure of the germinal vesicle surface became altered. Protein phosphorylation by protein kinase C is involved in the pathway that regulates germinal vesicle breakdown (GVBD) in *Spisula* oocytes (Dube et al. 1987; Eckberg et al. 1987). Artificial activation causes an increase in intracellular pH in *Spisula solidissima* that is driven by Na^+/H^+ exchange (Dube and Eckberg 1997) which might also be the case during normal fertilization. The nuclear envelope becomes vesiculated and after breakdown cisternae and vesicles are seen in ultrastructural studies to surround the condensed chromatin. It appears that centrosomes are present but not described in this system while two asters are described that migrate to either side of the chromatin and form the first meiotic apparatus. First and second polar bodies are formed. The events of germinal vesicle breakdown, disruption of the nucleus and completion of meiosis can also be induced by the divalent calcium ionophore A23187, which argues for a role of calcium in this process. These events do not occur when the medium is free of calcium. Deguchi et al. (1996) have shown that the extracellular Ca^{2+} entry and Ca^{2+} release from inositol 1,4,5-triphosphate-sensitive stores function at fertilization in the marine bivalve *Mytilus edulis*.

At the site of sperm entry membrane fusion occurs (Longo and Anderson 1970) and a fertilization cone is formed. The sperm's nuclear envelope becomes vesiculated and male and female pronuclei migrate into close apposition, but no zygote nucleus appears to be formed. Sperm–egg interactions typically involve four stages that are described as sperm agglutination, sperm activation, acrosomal reaction, and chemotaxis. In *Unio elongatus* the 220-kDa vitelline coat glycoprotein is concentrated in a restricted region of the vegetal pole, the crater region, where the sperm–egg interaction occurs. Binding assays indicate that this glycoprotein interacts with the sperm, and the protein bound in the apical region triggers acrosomal changes in sperm (Focarelli and Rosati 1995). The chemical interactions with regard to sperm and egg receptors are not known in Mollusca.

A distinct acrosome reaction is seen in spermatozoa of *Mytilus edulis* which forms a long filamentous process that interacts with the vitelline coat lysin (Niijima and Dan 1965a,b). These studies have been performed with electron microscopy. An ATPase activity is associated with the formation of the acrosome process (Mabuchi and Mabuchi 1973) which appears to be activated by calcium from seawater. The contents of the acrosome have been studied in *Mytilus edulis* with silver methanamine (SM) and phosphotungstic acid (PTA) and it has been shown that the PTA-stained area is sensitive to pronase digestion. The axial strand containing lysin is heavily stained with SM. Studies in the abalone *Haliotis* show that the acrosomal process can also be extruded by using the calcium-ionophore A23187, arguing for a role of calcium in this process which is discussed in Longo (1983).

A block to polyspermy is not common in molluscs and more than one sperm can enter the micropyle, although in most species supernumerary sperm are not observed in the oocyte cytoplasm. In *Mytilus edulis* an aminopeptidase-like protease participates in the suppression of the acrosome reaction, thereby preventing polyspermy (Togo and Morisawa 1997). Cortical granules underneath the egg surface are not extruded upon sperm–egg contact in *Spisula* (Longo 1973) and in this case supernumerary sperm were found in the egg. A fertilization cone is formed at the site of sperm entry and the entire sperm is incorporated into the egg which includes the nucleus, centrioles, mitochondria, and flagellum. Pronuclear apposition but not fusion is known for *Mytilus edulis* and *Crepidula* and may be representative for other species.

9 Annelida

Three classes (Polychaeta, Oligochaeta, and Hirudinea) are grouped under this phylum which live in marine, freshwater, and terrestrial environments (Strathmann 1987).

9.1 Reproductive Strategies and Reproductive Cycle

The phylum Annelida is characterized by either gonochoristic or hermaphrodistic reproductive systems. The majority of annelids are oviparous although there are many exceptions. Some annelid species brood their developing embryos close to their body, while most species of this phylum undergo basic reproductive strategies and spawn their gametes freely into the water. There are distinct features that are unique for some species. A reproductive phenomenon described as epitokous transformation occurs in many marine polychaetes and involves morphological and behavioral changes when the animals become sexually mature and spawn gametes that develop into epitokous adults. More common for polychaete reproduction is copulation or transfer of spermatophores. Most oligochaetes transfer semen in a fluid medium to a special receptacle, the spermatotheca in the female. Transdermal impregnation is also possible and has been observed in leeches in which the semen is directly transferred to the female antrum. In these cases a muscular tube developed from the male antrum is used to transfer semen into the female.

Asexual reproduction is not common in polychaetes and is restricted to a few sessile groups. Leeches also do not employ asexual reproduction strategies. In contrast, asexual reproduction is frequently used in oligochaetes where fission is the primary mode of reproduction (Qian and Chia 1989).

9.1.1 Design of Reproductive System

The design of the reproductive system in annelids is simple compared to most other invertebrate phyla. The gonads are located between the somatic peritoneum and the body wall musculature. In most cases the germ cells are released from the gonad into the coelom and release gametes through rupture of the body wall or through special expulsion organs. The male system contains testes, seminal vesicles, and prostatic organs while the female system contains one or more ovaries and specialized glands that are associated with nutrient production. Spermatozoa are frequently stored in seminal receptacles placed in the female (Jamieson and Rouse 1989).

9.1.2 Gametogenesis and Fertilization

Spermatogenesis in annelids involves the production of spermatogenic morulae (baskets) in which sperm are stored and released after detaching from the morula. Clitellates have seminal vesicles in which spermatogenesis takes place. Oogenesis occurs in the ovary of clitellates and oocyte growth is followed by vitellogenesis. In most species oogenetic meiosis is temporarily arrested in prophase I and oocytes are spawned at the germinal vesicle stage. Fertilization takes place at metaphase I in most species but can also occur at prophase I in a few species. There is still little information on the early processes of fertilization in most species, including insemination reactions, and gamete surface components. Fertilization processes include an acrosome reaction, sperm binding, and translocation through the oocyte layers. In most cases sperm binding is localized to specific regions on the oocyte and microvilli surround the entering sperm, resulting in a distinct fertilization cone (Block and Goodnight 1980; Eckberg and Anderson 1985; Fischer and Dhainaut 1985; Swalla et al. 1985; Eckelberger and Grassle 1987; Eckelberger and Rice 1988; Basa and Brafield 1989; Smith 1989; Hsieh and Simon 1990).

9.2 Cellular and Molecular Aspects of Fertilization

The events of calcium signaling during fertilization and egg activation have been recently reviewed by Stricker (1999). In annelids as in most other invertebrate animals fertilization-induced calcium waves are induced in which IP_3-mediated calcium signaling is triggered by external calcium influx and internal calcium release. The polychaete annelid, *Chaetopterus pergamentaceus*, has been used for a variety of polychaete annelid fertilization studies and has revealed multiple brief Ca^{2+} transients in eggs (Eckberg and Miller 1995) that occur through IP_3-gated channels (Thomas et al. 1998). In *Urechis caupo*, oocyte activation is reported to occur without waves (Stephano and Gould 1997) while nuclear and cytoplasmic pH increase occur at fertilization in *Urechis caupo* (Gould and Stephano 1993).

The vitelline envelope does not appear to be required for gamete binding in most species and does not appear to have sperm receptors that are necessary for sperm–egg interactions. Fertilization occurs normally in vitelline-envelope free eggs (Polk et al. 1987). The vitelline envelope's function is primarily to prevent polyspermy. Sperm attach and fuse with egg microvilli and penetrate the vitelline envelope. A cortical reaction accompanied by a massive structural reorganization has been observed in the polychaete *Platynereis dumerlii* (Kluge et al. 1995). In this system the process occurs during a period of about 25 min during which the membranes of the cortical granules are set free into the perivitelline space. Thirty thousand microvilli are attached to the vitelline envelope.

A review on polychaete sperm has been written by Jamieson and Rouse (1989). Polychaete sperm can be divided into ect-aquasperm, ent-aquasperm, and introsperm. Most polychaete sperm are ect-aquasperm and display acrosome, nucleus, mitochondria, and centrioles with the associated anchoring apparatus.

The fertilization process of the polychaete *Chaetopterus* has been described in detail by Eckberg and Anderson (1997). *Chaetopterus* sperm possess an acrosome that consists of two components, a cup-like vesicle at the apex of the nucleus and a granular region between the inner acrosomal membrane and the apex of the nucleus. During fertilization the plasma and the outer acrosomal membrane fuse, which exposes the inner acrosome membrane for interactions with the oocyte surface. The acrosome reaction takes place after contact with the egg microvilli. Microvilli develop during oogenesis at the oocyte surface and increase in number as the oocyte develops. Interestingly, sperm must interact with complexes attached to microvilli to fuse with and

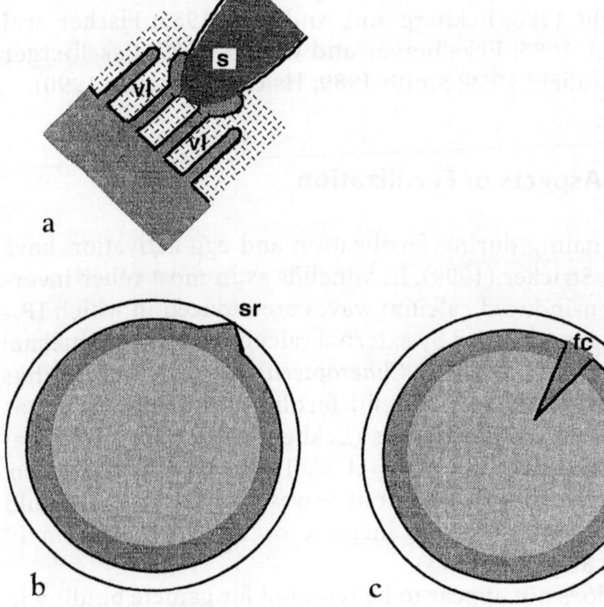

Fig. 2. Fertilization in annelids displaying **a** sperm (*s*) attachment and fusion with microvillar tips within vitelline layer (*vl*) of oocyte; **b** development of sperm receptacle (*sr*); **c** inverted fertilization cone (*fc*) that incorporates sperm into oocyte

fertilize the oocyte. The receptor for sperm has not yet been identified, but it is likely to be concentrated on microvilli. The vitelline envelope is complex in this species and consists of four different layers that can be distinguished morphologically, but the vitelline layer plays no obligatory role in the fertilization process. The fertilizing sperm attach to the receptor complexes at the tip of the microvilli and this interaction induces the acrosome reaction. The microvillar tip fuses with the inner acrosomal membrane and the fertilizing sperm enters the egg while microvilli elongate into a fertilization cone around the fertilizing sperm (see Fig. 2 for schematic representation). The microvilli that have not fused with the sperm retract from the egg surface and are no longer available for supernumerary sperm; this may present a block to polyspermy, which is different from other invertebrate systems in which cortical granules are involved in the physical block to polyspermy. Fertilization involving tips of microvilli is also seen in *Tylorrhynchus* (Sato and Osani 1986) and in *Neanthes* oocytes (Sato and Osani 1986).

Cellular and molecular data on the process of fertilization in oligochaete annelids are only sparsely available. Many oligochaete annelids use parthenogenesis as the preferred mode of proliferation and many others use special fertilization and activation mechanisms. A combination of oocyte self-fertilization and activation by sperm also occurs. In the case of oocyte activation by sperm a pseudomeiosis is observed that is arrested at anaphase. Sperm and egg chromatin are brought into close apposition by a U-shaped elongated spindle and each group of chromatin forms a zygote nucleus which is followed by cell division and development. It is common for sperm to only activate the egg without a true process of fertilization. In *Cognettia glandulosa* sperm triggers activation and arrests the oocyte at the second maturation division at anaphase. A presumptive second polar body is formed that returns and fuses with the female pronucleus and restores the polyploid somatic number of chromosomes.

During the fertilization process in the lumbricolid *Rhynchelmis*, a receptacle cone is formed that engulfs the sperm and enlarges into the fertilization cone which moves sperm deep into the oocyte. Through mechanisms not understood the male pronucleus is moved to the germinal vesicle and first maturation begins after loss of sperm membrane. In *Eisenia* the male pronucleus contributes the centrosome. In *Tubifex hattai* oocyte maturation is initiated after sperm incorporation. Microfilaments probably play a role in orienting the meiotic spindle and are observed at the cortex during the first meiotic maturation but are mainly lost during the second meiotic maturation process. The events following this process are described by Shimizu (1984) and involve restructuring of the cortical layer with components contributed by cell organelles such as mitochondria and the ER.

An acrosome reaction occurs (Olive and Clark 1978), though cellular and molecular details concerning this process are not yet available. It is also not well known if sperm triggers the elevation of a fertilization membrane. Oligochaetes appear to lack a block to polyspermy and polyspermy appears to occur commonly (Avel 1959). Multiple sperm incorporation has been observed in *Eisenia*.

10 Arthropoda

Approximately 80% of all animal species are grouped in this invertebrate phylum, which is divided into the three subphyla Uniramia, Crustacea, and Chelicerata. As can be expected, the reproductive system varies greatly in this phylum and is based on different functions that are required for adaptations to marine, freshwater, and terrestrial lifestyles. While the Chelicerata are primarily aquatic and lack copulatory organs, Arachnids are terrestrial and all have internal fertilization although the mechanisms vary greatly.

10.1 Reproductive Strategies and Reproductive Cycle

Most arthropods are gonochoristic, but hermaphrodism applies to some species. Internal fertilization is the rule and direct spermatophore transfer is most common (reviewed in Weygoldt 1990), though there are many exceptions to these basic reproductive strategies. As reproduction itself has many variations, arthropods also engage in complex and diverse mating behavior which ranges from pheromonal attraction to courtship behavior. Multiple mating is known for some Chelicerata even when spermatophores already have been successfully transferred to females. Terrestrial arthropods generally deposit and leave their eggs in protected areas while most marine crustaceans spawn fertilized eggs into the seawater. Other arthropod species carry their embryos close to their body and some care well for their brood. Environmental factors often play a role in stimulating reproductive cycles which can be triggered by factors such as temperature, tides and photoperiodicity. Asexual reproduction is rare in arthropods. Polyembryonic reproduction is known for parasitic wasps and parthenogenesis occurs in some insects (King 1970; Strathman 1987).

10.1.1
Design of Reproductive System

Most arthropods have one single gonad or bilaterally paired gonads that hold spermatogonia or oogonia. A vas deferens or oviduct is connected to the gonads that open to the outside by one opening or bilaterally paired openings. Gonads are either follicular or lobular, with the main characteristic that follicular gonads contain usually more germ cells while lobular gonads contain less but are composed of more ducts. Also, arthropods have a variety of accessory reproductive organs, which carry out functions of secretion for the production of egg coatings or which contribute to sperm transfer and involve prostatic organs (Zeh et al. 1989; Couche and Gillott 1990; Jawale and Ranade 1990).

10.1.2
Gametogenesis and Fertilization

Arthropods produce discrete packets of spermatocysts which are derived from primary spermatocytes produced in spermatogonia. Spermatogenesis is completed as spermatocysts pass through the vasa efferentia or vas deferens. The completion of spermiogenesis is followed by release into the gonadal ducts where they can be stored or further capacitated. A great variety of sperm exists and most sperm are motile, but those of crustaceans are immotile. Oocyte development is aided by follicular epithelial cells or can also proceed when two oocytes are joined and only one develops while the other serves as nurse cell, producing yolk and maternal RNA. In many species of insects and crustaceans somatic tissue also contributes to the development of oocytes which includes production of fat bodies and other supplemental nutrients. Oocytes are usually arrested during prophase I and meiosis resumes usually in the ovary prior to fertilization. Oocyte activation occurs in most cases by sperm and fertilization usually occurs before the completion of oogenic meiosis. Little is known about the fertilization process itself for most arthropods. It is known that sperm must enter through a micropyle, the only opening in the chorion-surrounded oocyte. Fertilization in insects usually takes place in the vagina. In spiders it takes place in the uterus externus. At fertilization a cortical reaction occurs and gamete fusion is followed by polar body extrusion and pronuclear fusion (Papathanassiou and King 1984; Witalínski 1985; Foe 1989; Weygold 1990).

10.2
Cellular and Molecular Aspects of Fertilization

The literature on arthropod fertilization is sparse despite this group being among the largest of all invertebrate species. Most studies have been devoted to reproductive behavior and embryogenesis, but the fertilization process in most species remains largely unknown. This is surprising because arthropods contain dangerous species such as scorpions as well as those used for food sources such as crustaceans. The lack of our knowledge on arthropod fertilization may be related to the fact that eggs and sperm are not readily available because fertilization is internal and artificial fertilization has not yet been attempted. The crustaceans are highlighted in this section because of the few detailed studies that are available.

10.2.1
Sperm

Ultrastructural studies are available on sperm and include sperm of *Mystacocarida, Cirripedia,* and *Branchiura* for which motile sperm with a flagellum are described. Other sperm are non-motile and do not have a flagellum. The variety of sperm morphology reflects the diversity of reproduction. For instance, sperm heads of *Natantia* have acrosomal and nuclear material and also a spine, for which function has not been described. *Anomuran* sperm have an acrosome, mitochondria, rough ER, and arms filled with microtubules, as well as an interesting nucleus that often forms small arms or projections (Hinsch 1980). *Brachyuran* sperm have a large portionated acrosome and a nucleus with cytoplasm and mitochondria embedded between them (Hinsch 1973).

In *Drosophila subobscura* the male produces two classes of motile spermatozoa that differ in total length and nuclear length. The phenomenon is known as polymegaly. Both types of sperm are potentially functional in egg penetration and karyogamy (Pasini et al. 1996). Sperm polymorphism is also reported in *Drosophila pseudoobscura* by Snook et al. (1994).

The spermatozoon of the shrimp *Rhynchocinetes* is atypical because it is non-motile and lacks a head and tail. The body has a rigid spike. Neither an acrosome-like structure nor changes during gamete interactions which could be interpreted as an acrosome reaction have been observed in this species

(Rios and Barros 1997). The spermatozoon exerts a lytic effect on the extracellular envelope of the egg by the secretion of trypsin-like enzyme and makes its way through the egg coats.

The important roles played by glycoconjugates in the recognition of gametes and in fertilization are well documented in *Drosophila*. Two glycosidases, beta-*N*-acetylglucosaminidase (GlcNAcase) and alpha-mannosidase (Manase), have been detected in *Drosophila melanogaster* spermatozoa which might be involved in the recognition of the carbohydrate residues of the egg shell at the site of sperm entry (Cattaneo et al. 1997). The molecular weight of the GlcNAcase and mannosidase was estimated as 158 and 317 kDa, respectively (Pasini et al. 1999). Both glycosidases are mannosylated integral membrane proteins of *Drosophila* spermatozoa which would function as exoenzymes with their active sites accessible in the extracellular space.

10.2.2 Oocytes

The unfertilized oocyte of most crustaceans is enveloped by the chorion and the vitelline membrane although this is not the rule (reviewed in Hinsch 1990). Little is known about the morphology, ultrastructure, and molecular biology of the egg envelopes.

Few studies report on the cytoskeleton in activated arthropod oocytes. Riparbelli et al. (1998) studied a hymenopteran, *Muscidifurax uniraptor*. Their observations suggest that the cytoplasm of *Muscidifurax* eggs contains a pool of inactive centrosomal precursor proteins becoming able to nucleate microtubules into well-defined asters containing centrioles after activation.

10.2.3 Fertilization

Several reports describe the early stages of fertilization, but most studies were conducted with light microscopy and did not allow detailed analysis of sperm–egg interactions. However, an explosion and eversion of the acrosome have been reported in ultrastructural studies for *Eupagurus* and *Carcinus*, *Callinectes*, and *Homarus* (Talbot and Chanmanon 1978). The early stages of fertilization are reported for the naturally spawned *Libinia* in which the acrosome, the acrosomal filament, and part of the nucleus were viewed within the chorion (Hinsch 1971). No studies of crustacean fertilization showing membrane fusion are available. Typically maturation division is initiated or resumed upon sperm–oocyte contact. Cortical granules are exocytosed where present and a fertilization membrane is formed which serves as a block to polyspermy. It is also possible to find more than one fertilization membrane. For instance, the isopod *Hemioniscus balani* produces two fertilization membranes (Goudeau 1976) that result in four membranes surrounding the fertilized embryo.

The fertilized lobster oocyte of *Homarus americanus* and *Homarus gammarus* consists of two discrete coats, i.e., the outer and inner coat. The outer coat forms inside the ovary while the inner one originates from a complex cortical reaction. The outer coat is comprised of small electron-dense granules and structures resembling "bottlebrushes". The cortex of ovarian oocytes contains four different types of vesicles, i.e., high-density vesicles (HDV), low-density vesicles (LDV), moderate-density vesicles (MDV), and ring vesicles (RV). The contents of the HDV and LDV appeared first in the perivitelline space, but their fate could not be determined at later times. The contents of MDV and RV are released in the perivitelline space to form the outer coat (Talbot and Goudeau 1988). The multilayer coat enveloping eggs of *Pontella mediterranea* (Crustacea, Copepoda) also results from the exocytosis of four types of cortical vesicles that sequentially release their contents in the perivitelline space upon egg activation (Santella and Ianora 1992).

The fertilization process of the shrimp *Rhynchocinetes typus* is among those studied in greater detail (Barros et al. 1986). In this species the sperm displays a main body with 11 radial arms and a single spike that are well shown by scanning electron microscopy. The tip of the spike contacts the egg first and penetrates by lytic activity. A fertilization cone arises around the sperm at the sperm–egg fusion side. From these scanning electron micrographs it is not clear whether an acrosomal reaction occurs.

In the rock shrimp *Syciona ingestis* the acrosomal process is biphasic (Clark et al. 1984) as described later in more detail for echinoderms. Two glycoprotein components are exposed after the spike if the acrosomal tip is decised. In *Syciona ingestis* sperm bind to the vitelline envelope and in a second step to the surface of the ovum (Clark et al. 1981, 1984). The acrosomal filament formation is associated with a decrease in pH and an efflux of K^+ ions (Griffin et al. 1987). The plasma membrane of the oocytes becomes more permeable to K^+ ions and the membrane potential becomes altered, resulting in meiotic division. It is known for *Palaemon serratus* (Goudeau and Goudeau 1986) that Mg^{2+} is required for hyperpolarization of the oocyte membrane which also allows meiotic maturation.

The acrosome reaction (AR) was induced in sperm from the brachyuran crustacean *Uca tangeri* either by mixing male and female gametes in filtered seawater or by treating the spermatozoa with the divalent cation ionophore A23187 (Medina and Rodriguez 1992). This latter method provided a sufficient number of reacted spermatozoa to allow detailed ultrastructural study of the AR. The process consists of two separate phases: (1) initial release of the acrosomal vesicle contents, and (2) further elongation of the acrosomal filament. In *Sicyonia ingentis*, Baldwin et al. (1998) have shown that after acrosomal vesicle exocytosis, the sperm generates an acrosomal filament over a period of 4–6 min which is about 10 μm in length. Actin or normal microtubules have not been detected in acrosomal filaments at the ultrastructural level of this unusual filament. Antibodies to tau, MAP2, and neurofilament medium and heavy subunits were all localized in this region of sperm. Baldwin et al.

(1998) have also shown that proteins localized to the acrosomal filament are of a unique sequence and may represent a new type of protein of centrosomal origin.

In contrast to most systems in which oocyte activation is triggered by the fertilizing sperm, *Sicyonia ingentis* oocytes are activated by seawater Mg^{2+} during spawning. Lindsay et al. (1992) suggested that Mg^{2+} induces an intracellular Ca^{2+} wave in regulating sperm entry into the oocyte.

The events of fertilization have also been studied in greater detail in the crab *Carcinus maenas* in which a fertilization cone is formed and cortical granules are exocytosed to form a fertilization envelope (Goudeau 1984). In these studies fine granular material and ring-shaped structures are described that are involved in the formation of the fertilization envelope.

11 Echinodermata

Five classes, Asteroidea (starfish), Ophiuroidea (brittle stars), Echinoidea (sea urchins), Holothuroidea (sea cucumbers), and Crinoidea (sea lilies), comprise this phylum which is exclusively confined to marine habitats. The sea urchin and starfish species have been explored for more than a century and have provided most of the information that we now have on echinoderm fertilization and development. They have provided a wealth of information on cell, molecular, and developmental mechanisms that has also been useful as a model for higher vertebrate systems. Echinoderms have a pentaradial symmetry and gonads are the predominant organs (Strathmann 1987; Buznikov and Podmarev 1990; Daydov et al. 1990). This section will highlight sea urchins as the most widely explored of all echinoderms.

11.1 Reproductive Strategies and Reproductive Cycle

Echinoderms are largely gonochoristic and only a few hermaphroditic cases are known. Most echinoderms are free spawners and release their gametes through pores directly into the seawater depending on environmentally favorable conditions. The triggers for gamete release are not well known and differ for different species. Most starfish and sea urchins have an annual breeding season that seems to be induced by changes in water temperature and can be different for different species within the same class. The seasonal fertilizability of sea urchin gametes is of consideration for experimental purposes. For instance, the United States West Coast sea urchin *Strongylocentrotus purpuratus* is fertile from ca. October to April, while *Lytechinus pictus* which lives in the same waters is fertile from ca. April through October. Once released into the ocean, male and female gametes are united and fuse to form embryos. Asexual reproduction is rare in echinoderms and is limited to some ophiuroids. Regeneration occurs in a number of species with best-known examples in the *Asterias* species (Pearse and Walker 1985; Mladenov 1987; McClary and Mladenov 1988; Smiley 1988).

11.1.1 Design of Reproductive System

The pentaradial symmetry of the gonads defines the design of the reproductive system. Each gonad is located within the coelom and is bounded by a thin layer of visceral peritoneum. The fully formed gametes are spawned through five gonopores and this is induced by contraction of muscular tissue (Schoenmakers et al. 1981).

11.1.2 Gametogenesis and Fertilization

Spermatogenesis in Echinodermata is preceded by germ cells that move from the basal region of the germinal layer toward the gonad lumen. The mature sperm contains an acrosomal process, a large nucleus, a midpiece with a ring-shaped mitochondrium, and a single flagellar axoneme. Oogenesis begins with differentiation of primary oocytes from oogonia in the basal region of the germinal layer. The primary oocyte accumulates RNA transcripts and nutrients for heterosynthetic vitellogenesis. Fully developed oocytes are ovulated from the surrounding follicle cells and are stored in the ovary lumen. While most species are temporarily arrested in prophase I, echinoid oocytes typically spawn oocytes at the germinal vesicle stage and many complete meiosis and are spawned as mature eggs. Fertilization occurs at the polar body I or polar body II stage. For the purpose of distinguishing unfertilized from fertilized eggs we will use the term oocytes for mature eggs, although the literature usually does not make this distinction. A jelly coat typically surrounds the oocytes and may serve to attract sperm in the ocean. A typical acrosome reaction occurs when sperm are near an egg. Sperm activation triggers cortical granule release and elevation of a fertilization coat which represents a physical block to polyspermy. A pronounced fertilization cone consisting of large microvilli surrounds the fertilizing sperm and anchors the sperm to the egg (Schatten and Schatten 1980). Pronuclear fusion occurs after sperm and egg nuclei are moved in close apposition by a microtubule-based sperm aster (Schatten 1982; Buckland-Nicks et al. 1984; Kyozuka and Osanai 1988).

11.2 Cellular and Molecular Aspects of Fertilization

Of all the five classes of Echinodermata the starfishes and sea urchins are by far the best studied. This section will focus on the sea urchin and will also include data on starfish oocyte activation.

11.2.1 The Unfertilized Egg

The sea urchin egg is different from most invertebrates in that the female germ cells are released into seawater as haploid, metabolically inert, mature egg cells after having completed their meiotic divisions inside the ovary. The egg size can range from 80 to over 100 μm and a mature egg has stored all the material that is needed for development until the pluteus stage and is not dependent on nutrients. The unfertilized egg contains a large amount of yolk platelets that contain lipoglycoproteins and glycoproteins (reviewed by Guidice 1986). Ribosomes and mitochondria are produced in the oocyte and stored in the mature egg as well as cytoskeletal proteins such as tubulin, actin, and cytoskeletal regulatory proteins. The mature egg's internal pH is rather acidic and measures 6.84 in *Lytechinus pictus* (Whitaker and Steinhardt 1985). The nucleus is typically eccentric but can be found at any place in the unfertilized egg.

The unfertilized egg surface is stubbed with short microvilli of ca. 0.25 μm in length which are lined with microfilaments internally (Chandler and Heuser 1981; Sardet 1984). A layer of cortical granules is underneath the egg surface, which varies in size, but average cortical granules are typically 1 μm in diameter. Cortical granules and microvilli undergo dramatic changes after fertilization which affect the architecture of the egg surface (see Fig. 3 for schematic representation). In the unfertilized egg, a thin layer of glycoproteins surrounds the egg surface and is called the vitelline layer. Sperm receptors are localized in the vitelline layer which is surrounded by a gelatinous jelly coat.

The receptor on the surface of the egg vitelline envelope (VE) of the sea urchin *Strongylocentrotus purpuratus* which mediates species-specific binding of sperm is a 350-kDa cell-surface glycoprotein (Hirohashi and Lennarz 1998). There are two oligosaccharide chains in glycoprotein. The sulfated O-linked oligosaccharide chains act as an egg receptor for sea urchin sperm during fertilization (Dhume and Lennarz 1995; Kitazume-Kawaguchi et al. 1997). Correa and Carroll (1997) have shown that out of three major VE glycoproteins (i.e., VE-A, VE-B, and VE-C) and one minor component (i.e., VE-D), VE-B plays a role in gamete binding.

11.2.2 Sperm

The sea urchin sperm is metabolically inert like the mature egg but contains a variety of proteins that are important for sperm–egg interactions. The sperm plasma membrane has been isolated (Trimmer and Vacquier 1986) and found to contain an egg jelly binding a 21,000- and a related 80,000-Mr glycoprotein (Podell and Vacquier 1985), a 77-kDa speract receptor for *Strongylocentrotus purpuratus* (Dangott and Garbers 1984), and a guanylate cyclase for *Arbacia*

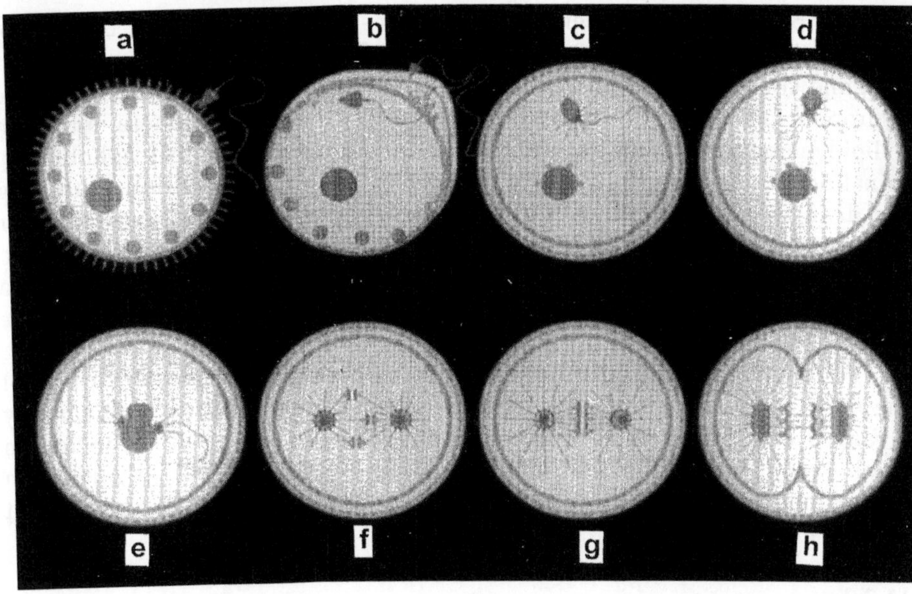

Fig. 3. Fertilization and cell division in sea urchin eggs. Unfertilized egg (**a**) is stubbed with short microvilli at plasma membrane while cortical granules line interior of plasma membrane. Upon sperm attachment and insemination (**b**) cortical granules discharge around entering sperm, lifting vitelline layer off plasma membrane into a fertilization coat membrane while microvilli enlarge around entering sperm and aid in incorporation of sperm into cytoplasm. A sperm aster forms (**c**) from centrosomal material associated with developing sperm pronucleus, while microtubules grow toward female pronucleus and move both pronuclei toward egg center (**d**). Centrosomal material expands around both pronuclei and becomes bipolarized (**e**). Metaphase spindle is formed from compacted centrosomal materal that has moved to the two mitotic poles where it organizes a metaphase spindle (**f**) and anaphase spindle (**g**) which results in cell division (**h**). [Composite drawings after original work by Schatten et al. (1982, 1986, 1987). Printed with permission from G. Schatten]

punctulata (Ward et al. 1985). The sea urchin sperm contains a large nucleus and a membrane-bound acrosomal vacuole that contains several enzymes and proteins including the egg-binding protein bindin (Vacquier and Moy 1977). Sea urchin sperm bindin, a 30,000-Mr polypeptide, is a key protein in fertilization, constituting the major content of the acrosome granule (Glabe and Vacquier 1977; Vacquier and Moy 1977). After its exposure on sperm surface, the multifunctional protein mediates the species-selective gamete adhesion and presumably fusion with the egg membrane (Vacquier et al. 1995). Bindin has been shown to engage in a number of interactions: with itself, with lipid bilayers (Glabe 1985), with carbohydrate components of the egg vitelline envelope (DeAngelis and Glabe 1988, 1990a), and with Zn^{2+} ions (DeAngelis and Glabe 1990b). Sequence alignment among different sea urchin species shows that the protein shares a central, highly conserved hydrophobic domain of 70–80 residues. The N and C terminal parts of the protein are strongly divergent and contain glycine- and proline-rich elements (Glaser et al. 1999). The hydrophobic central part of bindin is responsible for its peripheral adhesion to lipid membranes (Miraglia and Glabe 1993). The divergent hydrophobic terminal domains of bindin mediate the species-specific recognition of the egg via distinct carbohydrate patterns or via a proteinaceous sperm receptor (Minor et al. 1993; Just and Lennarz 1997; Mauk et al. 1997). A profilactin cap is located between the nucleus and the acrosomal vesicle which contains a filamentous actomere that has been described in more detail for *Thyone* sperm (Tilney 1978).

The ring-shaped sperm mitochondrion is located at the posterior end of the nucleus and surrounds the centriolar fossa, containing two orthogonally arranged centrioles with centrosomal material associated with it (Schatten et al. 1986). A flagellum extends from it in a typical "9+2" organization whose motility has been studied in great detail by a number of different investigators.

11.2.3 Sperm Activation

The sperm is kept immotile at a pH of 7.2. Alkalinization of pH is required for activation of sperm motility which is related to stimulation of the axonemal dynein ATPase, a pH-sensitive enzyme (Trimmer and Vacquier 1986). A decrease in the intracellular levels of ATP and a corresponding increase in the level of ADP concentration results in mitochondrial respiration that drives sperm motility. In a related reaction, high-energy phosphate is produced by respiration in the mitochondria and becomes uniformly distributed throughout the axoneme which explains why respiration and sperm motility are always related.

A number of other complex molecular changes are associated with the activation of sperm before fertilization preceding the acrosome reaction. The egg jelly coat can induce changes in sperm that include ion movements with an Na^+ and Ca^{2+} influx and H^+ and K^+ efflux, a transient pH alkalinization, a transient increase in intracellular levels of cAMP and cGMP, membrane depolarization, increase in activity of adenylate cyclase and of cAMP-dependent protein kinase, inactivation of guanylate cyclase, histone phosphorylation, increased fatty acid oxidation, and transient activation of respiration and motility. Two highly specific classes of jelly components are well characterized as inducers of sperm activation. Speract and resact have been identified in different species and have been shown to interact with homologous sperm receptors (Ramarao and Garbers 1985).

11.2.4
Fertilization

Acrosome and Acrosome Reaction. The acrosome reaction of sperm is the most apparent change that is induced by the previously described molecular changes. The exocytosis of the acrosomal vacuole is followed by the assembly of a very thin acrosomal process that will allow sperm attachment and penetration into the egg.

The acrosome reaction (AR) was discovered in the early 1950s in sea urchins, starfish, and several other marine invertebrates. To induce AR, three egg jelly components act in concert on the spermatozoa: a highly sulfated glycoprotein named acrosome reaction-inducing substance (ARIS), a group of sulfated steroidal saponins named CO-ARIS, and a group of glutamine-rich tetratriacontapeptides named sperm activating peptide (SAP). Nishigaki et al. (1996) isolated 12 SAPs from the egg jelly of starfish *Asterias amurensis*. In *Hemicentrotus pulcherrimus*, fucose-rich glycoprotein isolated from the jelly coat was found to induce AR (Mikami-Takei et al. 1991). In the egg jelly layer of sea urchin *Echinometra lucenter* a homopolymer of 2-sulfated 3-linked alpha-l-galactan and in *Arabacia lixula* and *Lytechinus variegatus* linear sulfated alpha-l-fucans with regular tetrasaccharide repeating units were detected (Alves et al. 1997). All of the sulfated polysaccharides induce acrosome reactions in conspecific but not in heterospecific spermatozoa.

Sperm causes egg activation which results in the exocytosis of approximately 15,000 cortical granules which are detected at the plasma membrane. Cortical granules contain several important proteins such as synaptotagmin, syntaxin, and vesicle-associated membrane protein (VAMP). Schulz et al. (1997) recently reported syntaxin and VAMP homologues in sea urchin sperm. During acrosomal exocytosis, the plasma membrane over the acrosomal vesicle fuses at multiple points with the acrosomal membrane (vesiculation) and syntaxin and VAMP are shed from the sea urchin sperm to form membrane vesicles. Thus, syntaxin and VAMP play a role in acrosome reaction. Conner and Wessel (1998) reported that rab3, a small GTPase of the ras superfamily, is also associated with cortical granules. This rab3 functions in the regulation of cortical granule exocytosis.

The acrosomal reaction requires an external pH and the presence of Ca^{2+} and Na^{+} in seawater. Na^{+} and Ca^{2+} influx, H^{+} and K^{+} efflux, and a transient alkalinization are required for the acrosome reaction which is always associated with membrane depolarization. The acrosome reaction can be separated into two distinct events that have been described by Tilney (1978). The assembly of actin monomers is the first event which requires precise control of pH and influx of water. The extrusion of the acrosomal vacuole is the second event and requires Ca^{2+} and fusion of the acrosomal vacuole with the sperm plasma membrane. The vacuole contains mostly bindin and several lytic enzymes. As mentioned above, bindin is a 30,000-Mr polypeptide which covers the tip of the acrosomal process and is crucial for species-specific attachment of sperm to the egg plasma membrane (Vacquier and Moy 1977).

Sperm–Egg Fusion, Calcium Events, and Blocks to Polyspermy. In sea urchins, sperm can enter at any site and the number of sperm attachment sites has been estimated to be large, with an average of 1 per 28.2 μm (Giudice 1986). Although sperm receptors have been shown to be located in the vitelline layer, sperm do attach and fertilize the egg when the vitelline layer is removed with dithiothreitol (Schatten and Schatten 1980).

The fusion of sperm with the egg requires external calcium, and sperm incorporation does not occur in calcium-free seawater. Directly after sperm–egg binding the sperm membrane fuses with the egg plasma membrane and inserts ionic channels from the sperm membrane into the egg plasma membrane. The opening of Na^{+} channels and activation of voltage-dependent Ca^{2+} channels in the egg plasma membrane results in a depolarization of the membrane and prevents further spermatozoa from fusing with the egg plasma membrane; this is referred to as fast block to polyspermy (Jaffe and Gould 1985). This event causes a depolarization from −70 to +10 mV and occurs within 0.1–1 s.

Gamete interaction triggers a variety of responses within the egg, collectively referred to as egg activation. In addition to the hallmarks of Ca^{2+} release from the endoplasmic reticulum and cortical granules and fertilization envelope elevation, there are cytoskeletal rearrangements, protein tyrosine phosphorylation and an increase in pH, among other events. Tyrosine phosphorylation of several egg proteins within minutes of sperm–egg binding and inhibition studies have shown that tyrosine kinase activity is required for many aspects of egg activation. Tyrosine of high molecular weight cortical protein (HMWCP) localized in the egg cortex becomes phosphorylated by an egg tyrosine kinase during a small alkaline shift in pH at fertilization, indicating a novel mechanism by which a tyrosine protein kinase may act as a pH-sensitive transducer (Jiang et al. 1991). Abassi and Foltz (1994) have shown that the sperm receptor is phosphorylated by an egg cortical tyrosine kinase in response to sperm or purified ligand (bindin) binding. Maximal phosphorylation was reached within 20 s. These tyrosine phosphorylations are signals for egg activation. Wright and Schatten (1995) using monoclonal antiphosphotyrosine antibody have detected several phosphotyrosyl proteins during sea urchin fertilization. Kinsey (1996) demonstrated the presence of p59c-fyn kinase in the sea urchin egg. An eightfold activity of this kinase was found between 5 and 15 min post-insemination, thereby proving that it may play a role in the egg activation process. An egg src family tyrosine kinase was found to act as a potential upstream regulator of phospholipase C gamma (PLC-gamma) in the activation of starfish eggs (Giusti et al. 1999). The initiation of Ca^{2+} release of fertilization in the eggs of most animals relies on the production of IP_3, implicating the activation of phospholipase C. Recent work has demonstrated that injection of PLC-gamma SH2 domain fusion proteins into starfish oocytes specifically inhibits the initiation of calcium release in response to sperm, indicating that PLC-gamma is necessary for Ca^{2+} release at fertilization (Carroll et al. 1997). PLC-gamma activity has also been detected by De Nadai et al. (1998a) in sea urchin eggs.

These experiments show that the cascade of calcium events during early fertilization is well documented and involves phosphatidyl inositol 4,5-biphosphate (PIP_2) hydrolysis triggered by Ca^{2+} release (Whitaker and Steinhardt 1985). These events are exceedingly fast and by about 15 s post-insemination a large amount of PIP_2 is turned over, resulting in a 40% increase in the amount of inositol 1,4,5-triphosphate (IP_3). The transient rise in Ca^{2+} in sea urchins during fertilization requires the production of IP_3 (Lee and Shen 1998). PIP_2 is present in the egg plasma membrane and is hydrolyzed by phospholipase C to produce IP_3 and 1,2-diacylglycerol (1,2-DG). Further processing of compounds eventually restores PIP_2. The release of calcium from intracellular stores is triggered by IP_3 which induces exocytosis of cortical granules. A rise of $[Ca^{2+i}]$ is first detected at the site of sperm entry, which has been demonstrated by measuring microinjected aquorin into the eggs. This wave propagates around the egg and indicates the release of calcium from internal sources, possibly the endoplasmic reticulum which is closely associated with egg cortices (McPherson et al. 1992) and often encircles cortical granules (Terasaki et al. 1991). Calcium induces the calmodulin-mediated activation of a kinase which generates $NADP^+$ which is reduced to NADPH in subsequent reactions. This cascade of events eventually is involved in the activation of the egg's oxidative pathways, lipid metabolism, nicotinamide nucleotide reduction, and protein and DNA syntheses (Whitaker and Steinhardt 1985).

The elevation of the fertilization coat is referred to as a slow and permanent block to polyspermy. Ovoperoxidase is one of several oocyte-specific proteins which are stored within sea urchin cortical granules, released during the cortical reaction, and incorporated into the newly formed fertilization envelope. Ovoperoxidase plays a particularly important role in this process, crosslinking the envelope into a hardened matrix that is insensitive to biochemical and mechanical challenges, thus providing a permanent block to polyspermy (Lafleur et al. 1998).

Trypsin-like activity is secreted from eggs of many species at fertilization and this activity is critical for the block to polyspermy. Cortical granule serine protease 1 (CGSP1) has been detected by zymography assay in *Strongylocentrotus purpuratus* (Haley and Wessel 1999). This CGSP1 is activated at fertilization and is responsible for the protease-mediated reactions that follow cortical granule exocytosis and contribute to the permanent block to polyspermy.

Cortical Reaction and Changes of the Egg Surface. The cortical reaction is one of the best studied events in sea urchins and has been used as a model for studying higher vertebrate systems including that of humans. Exocytosis of cortical granules results in the block to polyspermy that is generally referred to as slow block or permanent block to polyspermy (Jaffe and Gould 1985). Cortical granules are derived from the Golgi apparatus and move to the egg periphery during maturation. Several enzymes and proteins are continued within cortical granules, including a protein called hyalin, a serine protease, a peroxidase, and several other sulfated mucopolysaccharides. The attachment of sperm to the vitelline envelope initiates the exocytosis of cortical granules which rapidly spreads around the egg surface and is completed within 1 min. The discharge of cortical granules results in several changes of the egg surface. The hyaline secreted from the cortical granules forms a layer around the egg which is called the hyaline layer. Other contents of the cortical granules are discharged between the plasma membrane and the vitelline layer which elevates the vitelline layer into the fertilization coat. The peroxidase secreted from cortical granules causes the hardening of the fertilization coat by cross-linking tyrosine residues of the vitelline layer, aided by H_2O_2 generated in a reaction during sperm–egg fusion.

Another event readily seen with transmission electron microscopy is the fusion of cortical granule membranes with the plasma membrane. The increase in plasma membrane serves to elevate the short microvilli which now become longer and enlarge in particular around the entering sperm, forming a structure called a fertilization cone (Schatten and Schatten 1980). The fertilization cone engulfs the entering sperm and is important for sperm incorporation. It is internally lined with microfilaments which maintain the shape of the elongated microvilli. When treated with the microfilament inhibitor cytochalasin B, sperm incorporation does not take place because microfilaments in the fertilization cone become disassembled and cannot physically enlarge and strengthen the microvilli of the fertilization cone (Schatten and Schatten 1980). The transient fertilization cone structure is disassembled after sperm incorporation at which time restructuring of the egg surface occurs and membrane material from the plasma membrane is internalized by endocytosis. A second phase of membrane internalization takes place at ca. 20 min post-insemination before another wave of microvilli elongation occurs.

Centrosomes and Cytoskeletal Organization. Centrosomes play crucial roles during fertilization and are contributed by sperm in the sea urchin system. The use of immunological probes to centrosome proteins has made it possible to analyse centrosomes during sea urchin fertilization (Schatten et al. 1985, 1986, 1987, 1988, 1992). Unincorporated sperm contains densely compacted centrosome material which surrounds a pair of centrioles located in the sperm's centriolar fossa. This centrosomal material becomes decondensed shortly after sperm is incorporated into the egg cytoplasm. It then organizes the sperm aster which contacts the egg nucleus and is responsible for moving sperm and egg nuclei into close apposition. Centrosomes therefore are crucial for fertilization and have recently regained the appreciation that was devoted to them about a century ago. Centrosome research is now in an explosive phase and more centrosome proteins are being discovered rapidly. The sea urchin system has provided new information on centrosomes which can be utilized for other cell systems. Experiments performed on sea urchin eggs and embryos have revealed that centrosomes are composed of fibrous material that can compact and decompact to form the rapidly changing microtubule formations. When sperm enters the sea urchin egg the compacted sperm centrosome forms a very focused sperm aster. The centrosome structure then expands in an arc around the decondensing sperm nucleus to form an aster that grows toward the female pronucleus and eventually attaches to it. These experiments show that centrosomes are important in order to form the dynamic microtubule structures that are necessary to move sperm and egg nuclei together for karyogamy. Once sperm and egg nuclei are fused centrosome material surrounds the zygote nucleus and duplicates during the S-phase. At the beginning of prophase, centrosome structure becomes separated from the mitotic poles and organizes the bipolar mitotic apparatus (Schatten et al. 1986, 1987). The shape of centrosome structure defines the shape of microtubule formations. Compacted centrosomes result in well-focused spindles, while expanded centrosomes organize wider microtubule formations. Gamma tubulin has been identified as the centrosome-specific protein that nucleates microtubules (Oakley and Oakley 1989) and is therefore crucial for proper centrosome functions. This includes the union of parental genomes during fertilization (Schatten et al. 1986, 1987) and the equal separation of the genome to the daughter cells after cell division. It has been shown in mammalian fertilization that defects in centrosomes can be causes of cases of infertility (Simerly et al. 1995).

The unfertilized sea urchin egg contains latent centrosome material that can be aggregated by increasing pH (with ammonium), increasing calcium (with ionophore A23187), and by inducing microtubule polymerization with D_2O or taxol (Schatten et al. 1992; Schatten et al. 2000). It is most likely that but not entirely clear whether the egg contributes centrosomal material during fertilization which is united with the sperm's centrosome during fertilization. However, it is clear that the sperm contributes the dominant microtubule organizing material.

The finding that maternal centrosomes can be activated in unfertilized eggs has also been confirmed in starfish (Saiki and Hamaguchi 1998). These studies found that the maternal centrosome possesses the ability to form asters but generally does not have the capacity for reproduction.

Pronuclear Movements. The process of fertilization is completed when paternal and maternal nuclei have fused and the genomic material is mixed. The dynamic pronuclear movements that allow pronuclear apposition during sea urchin fertilization have been well documented by video microscopy (Schatten 1982; Schatten et al. 1982). After sperm incorporation and translocation into the cytoplasm, microtubules begin to assemble from the pool of tubulin that is already present in the unfertilized egg. The amount of unpolymerized tubulin is high in the unfertilized egg and amounts to 5% of the total egg proteins.

The sperm nucleus is transformed into the male pronucleus during its movement to the egg center and the male pronucleus becomes enveloped with new material from the egg cytoplasm. A breakdown of the sperm's nuclear envelope allows decondensation of the highly compacted sperm chromatin which becomes exposed to the egg cytoplasm. Phosphorylation of Sp H1 and the two H2Bs is followed by chromatin decondensation which occurs at 15 min post-insemination. The phosphorylated Sp H1 is replaced by new histones which replace the sperm-specific histones. A new nuclear envelope, probably derived from endoplasmic reticulum (ER) of the egg, is assembled by fusion with pre-existing sperm membrane fragments while sperm chromatin is decondensing. Further assembly of enzymes and other components that constitute a functional male pronucleus are necessary before the paternal genome can start to replicate and transcribe. The male nucleus becomes also equipped with lamins that are not present in the sperm's nuclear envelope (Schatten et al. 1985). The process of male pronuclear formation is dependent on a rise in internal pH and is completed within a very short time, typically within 8–12 min, depending on the specific sea urchin species.

Protein kinase C (PKC) is also activated during the early steps of sea urchin egg fertilization. Collas et al. (1997) have reported that nuclear lamina disassembly in interphase sea urchin egg cytosol is a result of lamin B phosphorylation mediated by PKC. Shen and Buck (1990) demonstrated that PKC activation was necessary to induce the pH rise characteristic of the fertilization process. It has been also shown that a massive Ca^{2+}-dependent tyrosine phosphorylation occurs in sea urchin eggs during the first steps of fertilization and is triggered by PKC activity (Ciapa and Epel 1991). A calcium-dependent PKC isoform has also been cloned in sea urchin eggs (Rakow and Shen 1994). Olds et al. (1995) and DeBarry et al. (1997) studied PKC activation during the early stages of sea urchin egg fertilization by using three different fluorescent probes specific for PKC activation, and their results appear in complete agreement with the putative functions previously suggested for fertilization.

While the male pronucleus is formed and centrosomes spread around the decondensing chromatin the sperm aster forms and unites sperm and egg nuclei. After pronuclear fusion and mixing of the genome, DNA and the centrosome–centriole complex become multiplied during the S phase. The zygote nuclear envelope breaks down and centrioles migrate to the mitotic poles during prophase through mechanisms only partly explored (Paweletz et al. 1984). Centrosome material surrounds the zygote nucleus and becomes relocalized to the mitotic poles where it surrounds the pair of centrioles. Successful mitosis and cell division are the beginning of development into sea urchin embryos.

11.3 MPF and MAP Kinases During Meiotic Maturation and Fertilization

Immature starfish oocytes are arrested at the G2-M-phase border of meiosis I and fertilization occurs naturally at late meiosis I. Exposure to the hormone 1-methyladenine induces meiotic maturation and leads to the activation of the mitosis-promoting factor (MPF, which is composed of the $p34^{cdc2}$ catalytic-cyclin B regulatory complex), resumption of the meiotic cell cycle, and fertilization competency. A Cdc2 kinase is important for the early stages of activation and MAPK is activated after GVBD. A MAP kinase is involved in the arrest at the female pronucleus stage and inactivation of MAP kinase is required for the G2-M-phase transition (Fisher et al. 1998).

It has recently been shown in early sea urchin embryos (Philipova and Whitaker 1998) that MAP kinase activity not only is confined to meiosis but also occurs during mitotic cell cycles. MAP kinase activity also increases immediately after fertilization in sea urchin eggs and experiments using the calcium ionophore A23187 have demonstrated that a calcium signal can activate MAP kinase and is likely to be due to the fertilization-induced increase in cytoplasmic-free calcium concentration (Philipova and Whitaker 1998). Further evidence for MAP kinase activity during mitotic division in sea urchins comes from experiments using inhibitors of the MAP kinase signaling cascade. Wortmannin, an inhibitor of PI_3 kinase, arrests the fertilized sea urchin eggs at the prometaphase stage and inhibits the inactivation of MAP kinase which is normally observed during division in sea urchin eggs (Chiri et al. 1998; De Nadai et al. 1998b).

12 Protochordata

Three different subphyla are grouped under the phylum Chordata: Urochordata, Cephalochordata, and Vertebrata. The Urochordata and Cephalochordata are grouped together as Protochordata and will be discussed here. Both groups inhabit the world's oceans and most species are found in coastal waters (Cloney 1987; Galt 1987).

12.1 Reproductive Strategies and Reproductive Cycle

Most urochordates are hermaphroditic which includes self-fertile and self-sterile species, while most cephalochordates are gonochoristic. Most urochordates including *Ciona intestinalis* and all cephalochordates are free spawners and fertilization and embryo development occur in the ocean. Many of the colonial urochordates are ovoviviparous with internal fertilization being the rule. In some of these species, such as *Clavelina,* brooding occurs. Asexual reproduction is rare in protochordates but can occur in sessile colonies.

12.1.1 Design of Reproductive System

The gonads of urochordates are branched and ovaries contain numerous follicles which surround the developing oocytes. Gonoducts are connected to the gonads through which gametes reach an atrium from where they are released through an atrial siphon or similarly specialized chambers. The gonads of the two cephalochordate families are organized in rows of sacs which are located in the coelom. A myoepithelial visceral peritoneum lines the outer gonadal wall. The germinal epithelium contains somatic and sperm cells and spermatogonia give rise to spermatocytes, spermatids, and spermatozoa. The female reproductive organs contain primary oocytes that are released into the ovarian lumen. Gametes are directly released into the atrial lumen after bursting of the gonadal wall (Mukai 1977; Irons and Walker 1984; Sugino et al. 1987; Holland and Holland 1989b).

12.1.2 Gametogenesis and Fertilization

Spermatogenesis in Protochordata occurs as in other systems through development of spermatozoa from spermatogonia after progressive maturation from spermatocytes and spermatids. Oogenesis is achieved by maturation of oocytes in the ovary supported by vitellogenesis. Oocytes are arrested in metaphase I in most urochordates and in metaphase II in most cephalochordates. For fertilization to occur sperm must penetrate a thick layer of follicle cells, a chorion, and, in some cases, test cells. Upon sperm–egg contact cortical granules are exocytosed into the perivitelline space. The oocyte is released from the meiotic block at fertilization and development of the embryo takes place (DeSantis et al. 1980; Jeffery 1984; Honegger 1986; Villa and Patricolo 1987; Holland 1988; Holland et al. 1988; Sardet et al. 1989).

12.2 Cellular and Molecular Aspects of Fertilization

Cellular and molecular data on fertilization are best available for the subphylum Urochordata. This section will discuss two of the three classes of Urochordata, the Urochordata-Ascidiacea and the Urochordata-Larvacea. The interactions of sperm and egg in the Urochordata-Ascidiacea have been reviewed by Rosati (1985) and the genetics of fertility in ascidians have been investigated by Oka (1970), Mukai and Watanabe (1978), and Schofield et al. (1982) using the genus *Bortryllus* as an experimental system. This genus uses cross-fertilization as a rule, but self-fertilization can occur under experimental conditions.

12.2.1
Sperm

A typical solitary sperm of ascidians consists of a nucleus containing a nuclear envelope with pores, a single basal body, and a typical flagellum, a single mitochondrion and a prominent glycocalyx (Cloney and Abbott 1980; Fukumoto 1983, 1984, 1985, 1986; Rosati et al. 1985). It may or may not contain an acrosome (Baccetti 1979). Ascidian sperm have apical vesicles which may or may not be equivalent to acrosomes in other invertebrates (Cloney and Abbott 1980). In the larvacean *Oikopleura dioica* the sperm contains a tail and a spherical head, 1 μm in diameter, which contains the acrosome, nucleus, and a single centriole (Flood and Afzelius 1978). Different from ascidians, the larvacean contain a distinct midpiece (Franzén 1970).

12.2.2
Oocytes

The size of Ascidiacea oocytes varies in different species and can range from 110 μm as in the species *Molgula* to 280 μm, with 730 μm as an exception in *Polycarpa tinctor*. The oocytes of most species are unpigmented and clear for microscopy studies. While cortical granules are not observed in most species such as *Ascidia, Ascidiella, Corella,* and *Ciona* (LaSpina D'Anna 1974), they contain typical cell organelles such as yolk granules, lipid droplets, glycogen granules, free ribosomes, rough ER, Golgi, annulate lamellae, and mitochondria. Mitochondria are typically concentrated in the cortical cytoplasm but can also be organized around lipid droplets (Berg and Humphreys 1960).

Shed larvacean eggs are smaller and have little or no yolk material and no jelly or follicle cells. In larvaceans a sparse layer of cortical granules underlining the plasma membrane has been observed in unfertilized oocytes of *Oikopleura dioica* (Holland et al. 1988).

The Ascidiacea oocyte is typically invested in a chorion which is also called the 'vitelline coat' (Cotelli et al. 1981) and can be composed of one or more layers. It measures 60 nm in *Ciona intestinalis* and consists of a fine meshwork of fibers composed of carbohydrates and fucose-binding proteins (DeSantis et al. 1980; Cotelli et al. 1981; Rosati 1985) as well as other components. The chorions of *Ascidia nigra, Ascidia malaca,* and *Phallusia mammillata* contain three distinguishable layers (Ursprung and Schabtach 1964; LaSpina D'Anna 1974). The chorion of all species is important for sperm recognition, but details are not well known on an ultrastructural and biochemical level. For experimentation, the chorion is typically removed manually although enzymatic methods are also possible (Zalokar and Sardet 1984). In some cases the chorion is covered by follicle cells that contain vacuoles filled with ammonium ions (Lambert and Lambert 1978) which allows floating of the invested oocyte. Test cells can compose another layer surrounding the unfertilized oocyte (Cloney and Cavey 1982; Satoh et al. 1982; Honegger 1986).

Passing through the oviduct induces germinal vesicle breakdown. The meiotic spindle forms near the animal pole and remains arrested until fertilization takes place (Mukai and Watanabe 1978).

In ascidians little work has been done on sperm recognition systems present on the egg plasma membrane. Flannery and Epel (1998) have shown that wheatgerm agglutinin (WGA) interferes with the sperm-interacting system on the plasma membrane of eggs of the ascidian *Ascidia ceratodes*. In the ascidian *Halocynthia roretzi*, sperm–egg binding is probably mediated through the interaction between alpha-L-fucosidase present on the sperm surface and anionic saccharide chains of the egg vitelline coat (Baginski et al. 1999). Matsuura et al. (1995) isolated N-acetylglucosaminidase from the vitelline coat of ascidian eggs which is a candidate for sperm receptor. Downey and Lambert (1994) found that ascidian sperm bind to vitelline coat N-acetylglucosamine groups of the egg via sperm surface N-acetylglucosaminidase. This sperm surface egg receptor remains anchored throughout penetration. Hoshi (1986) concluded that in the *Ciona intestinalis* sperm alpha-L-fucosidase is a recognition protein for the sperm receptor in the vitelline envelope.

12.2.3 Spawning of Gametes and Fertilization

Spawning of ascidian gametes is triggered by light stimulation followed by a period of darkness (Woollacott and Porter 1977; Young and Braithwaite 1980). The events of fertilization in ascidians include sperm recognition by the chorion and sperm head penetration into the chorion which is facilitated by proteolytic enzymes including proteases (Sawada et al. 1982, 1984; Fukumoto 1986; Honegger 1986). It has been suggested that the egg induces changes in the acrosome (Fukumoto 1984) although the acrosome process itself is not well studied. Interestingly, the mitochondrion moves down the tail as the sperm enters the perivitelline layer and is translocated to the end of the tail and is not incorporated in the oocyte (Ursprung and Schabtach 1965). Mitochondrial shedding has been observed in many species (Lambert and Epel 1979; Lambert 1982) and can also be induced by egg water, high pH, or Na seawater. It is inhibited by cytochalasin B (Lambert 1982; Lambert and Lambert 1983). Finally, the membrane of the sperm head fuses with the oocyte membrane near the vegetal pole of the oocyte (Jeffery and Meier 1983). The interactions between sperm and the follicle and chorion are important for sperm penetration. Sperm penetration is impaired when chorion and follicle cells are removed, although there are variations in different species. For instance, in *Ascidia malaca*, *P. mammilata*, and *Ciona intestinalis* fertilization can occur after removal of the chorion (Rosati and DeSantis 1978; Honegger 1986).

Fertilization in Larvacea includes the release of cortical granules. Cortical granules migrate to the inner surface of the vitelline layer and remain intact through several cleavages (Holland et al. 1988). The final meiotic divisions of the oocyte take place after fertilization. During fertilization, species-specific gamete binding must be followed by sperm penetration of egg vestments before gamete fusion can occur. Sperm proteases, called lysins, aid this process. Chymotrypsin-like protease is exposed on the external surface of the plasma membrane of the head and was detected in *Ascidia ceratodes*, *A. callosa*, and *A. paratropa* (Pinto et al. 1990; Saitoh et al. 1993; Koch et al. 1994). This chymotrypsin plays a role in sperm penetration.

The unfertilized oocyte of *Molgula occidentalis* possess a small meiotic spindle. In *Ciona savignyi* and *Halocynthia roretzi* a typical meiotic spindle is found when stained with antitubulin antibody (Sawada and Schatten 1988). At fertilization, a sperm aster first appears near the cortex and it enlarges into a large aster. The sperm aster grows after the formation of the second polar body. The female pronucleus migrates to the male pronucleus at the center of the sperm aster. The microtubule latticework and the sperm aster disappear towards the end of the first interphase, with only a small bipolar structure remaining until first mitosis. At mitosis the asters enlarge tremendously, while the mitotic spindle remains remarkably small (Sawada and Schatten 1988).

Polar bodies are formed after sperm incorporation and movements in the cytoplasm are observed that relate to ooplasmic segregation. In an interesting process soon after fertilization the cortical cytoplasm begins to move to the vegetal pole. Movements of the sperm nucleus are accompanied by pigmented cytoplasm shifts that form a crescent in the future posterodorsal part of the embryo. The ooplasmic segregation is important for setting up patterns for blastomere formation. These movements within the cytoplasm and those of the cortical cytoplasm are calcium dependent and can also be induced by the calcium ionophore A23187 (Jeffery 1982). The cortical cytoplasm changes are normally induced by sperm that release calcium and trigger ooplasmic segregation. The segregation is also microfilament driven and can be inhibited by cytochalasin B (Zalokar 1974; Reverberi 1975). It has been proposed (Jeffery and Meier 1983) that actin filaments at the egg plasma membrane surround pigmented droplets that are attracted to the vegetal pole at the side of sperm entry and form the polar cap. These experiments demonstrate the importance of calcium and microfilaments for successful fertilization and ooplasmic segregation.

12.2.4 Calcium Oscillations

More recent work has focused on the role of calcium oscillations in activation of ascidian eggs (Speksnijder et al. 1989; Wilding and Dale 1998; Yoshida et al. 1998). A calcium wave and cortical contraction associated with intensive microfilament reorganization at the cell surface are triggered by the entering sperm (Roegiers et al. 1995) which induces a cascade of kinase activities (McDougall and Lavasseur 1998). It has been proposed that two intracellular calcium-release channels, the ryanodine receptor and inositol 1,4,5-

triphosphate receptor, play different roles in ascidian fertilization (Albrieux et al. 1997).

Changes in Ca^{2+} are an essential factor regulating oocyte activation. Matured ascidian eggs are arrested at metaphase I of the first meiotic division. Fertilization triggers the completion of meiosis without any further arrest. Two series of Ca^{2+} transients have been observed after fertilization: the first series of Ca^{2+} waves occurs just after fertilization and the second series of Ca^{2+} oscillations occurs between the first and second polar body extrusion. The fertilizing spermatozoon induces intracellular calcium release by activating the IP_3 and calcium-induced calcium release (CICR) pathways. Shortly after insemination, the free Ca^{2+} concentration rises within a minute from a resting level of about 90 nM in the unfertilized egg to a peak level of about 7 μM in *Phallusia* and about 10 μM in *Ciona*. The total duration time of this Ca^{2+} wave is 2–3 min. It is immediately followed by a series of 12–25 briefer Ca^{2+} waves with peak levels of about 1–4 μM. These postfertilization pulses occur at regular intervals of 1–3 min during the completion of meiosis, and they stop as soon as the second polar body is formed at about 25 min (Speksnijder et al. 1989). In *Phallusia* and other species, the first fertilization-induced calcium wave originates from the point of sperm entry and directly precedes a microfilament-mediated cortical contraction (Speksnijder et al. 1990; McDougall and Sardet 1995; Roegiers et al. 1995).

Ca^{2+} signaling in cells generally involves both its influx from the extracellular medium and its release from intracellular stores. The two families of intracellular Ca^{2+}-release channels have been characterized, i.e., the inositol 1,4,5-triphosphate receptor and ryanodine receptor. Ryanodine receptors were found to localize in the cortex of unfertilized *Ciona* eggs (Arnoult et al. 1997). Both of the receptors play distinct roles during fertilization in ascidians (Albrieux et al. 1997). Albrieux et al. (1998) have shown that three independent Ca^{2+} messengers, cyclic ADP-ribose (CADPR), nicotinic acid-adenine dinucleotide phosphate (NAADP), and inositol 1,4,5-triphosphate (IP) are involved in Ca^{2+}-signaling functions in ascidian oocytes. Kyozuka et al. (1998) proposed that a cytosolic sperm protein factor is introduced to the oocyte cortex and induces Ca^{2+} waves and thereby meiotic resumption, leading to cell-cycle-related Ca^{2+} oscillations. Wilding and Dale (1998) suggest that ascidian sperm trigger the hydrolysis of nicotinamide nucleotides in the oocyte to ADP-ribose and this mechanism is responsible for the production of the fertilization current.

12.2.5 Block to Polyspermy

To ensure normal development, most animals have evolved a number of mechanisms to block polyspermy. A block to polyspermy exists in ascidians and is caused by a factor released by the fertilized egg that modifies the chorion by removing *N*-acetylglucosamine (GlcNAc). The block to polyspermy occurs at ca. 21 s after fertilization (Lambert and Lambert 1981) while the process of fertilization takes 2.2 min on average. Ascidian sperm bind to vitelline coat (VC) glycosides. In the genus *Ascidia*, GlcNAc is the ligand to which sperm bind. Eggs of *Ascidia ceratodes* and *Phallusia mammillata* release GlcNAc into seawater. This has been shown to inactivate VC GlcNAc groups, thus preventing polyspermy (Lambert 1989). In *Halocynthia roretzi* the GlcNAcase was purified by Matsuura et al. (1993). They observed that GlcNAcase is released from the egg and following fertilization is bound to the VC in order to prevent polyspermy. Lambert et al. (1997) proposed that the eggs of ascidians have two blocks to polyspermy that operate in sequence: an early first block resulting from enzymatic modification of the VC by GlcNAc released primarily from follicle cells and a second electrical block operating at the egg plasma membrane level. The extracellular calcium requirement for the glycosidase release was observed by McDougall et al. (1995) in the eggs of *Phallusia mammillata*. In the eggs of the ascidian *Phallusia mammillata*, removal of external Ca^{2+} modifies the fertilization potential and induces polyspermy (Goudeau and Goudeau 1993a). Release of this glycosidase is thought to be under the control of a membrane-bound phospholipase C (PLC). Goode et al. (1997) demonstrated that ascidian egg PLC activity is located on the surface of either the egg or follicle cell, and half is located within the egg (Robert et al. 1999).

Electrical responses to fertilization have been described by Dale et al. (1983) and Pecorella et al. (1993) which have related activation currents and surface contractions to sperm entry.

12.3 MPF and MAP Kinases During Meiotic Maturation and Fertilization

The calcium and phosphorylation requirements for ascidian maturation and fertilization have recently been reviewed by Russo et al. (1998). Maturation of oocytes and fertilization are associated with two cell cycle kinases: maturation promoting factor (MPF) activity and microtubule-associated protein kinase (MAPK) activity. MPF activity is elevated in metaphase-I (M-I)-arrested unfertilized oocytes and inactivation of MPF is observed at M-I release. The activity falls 5 min after fertilization to remain at low levels for 5 min. The first polar body is extruded when the activity is low and becomes elevated again 15 min after fertilization when the second polar body is extruded. Progression to M-II is dependent on the intracellular calcium oscillations. MAPK activity increases to a maximal activity at 10 min after fertilization and falls to a low level at 25 min after fertilization. MAPK activity begins to decrease at M-1 exit and remains low during the completion of meiosis.

Acknowledgements. It is a great pleasure to acknowledge stimulating discussions with Drs. Elke Bloh and Geoffrey Gobert who contributed ideas to the section on Platyhelminthes. Special thanks are extended to Dr. Bert Menco for the computer drawings of Figs. 1 and 2.

References

Abassi YA, Foltz KR (1994) Tyrosine phosphorylation of the egg receptor for sperm at fertilization. Dev Biol 164(2):430–443

Albrieux M, Sardet C, Villaz M (1997) The two intracellular Ca^{2+} release channels, ryanodine receptor and inositol 1,4,5-trisphosphate receptor, play different roles during fertilization in ascidians. Dev Biol 189(2):174–185

Albrieux M, Lee HC, Villaz M (1998) Calcium signaling by cyclic ADP-ribose, NAADP, and inositol trisphosphate are involved in distinct functions in ascidian oocytes. J Biol Chem 273(23):14566–14574

Alves AP, Mulloy B, Diniz JA, Mourao PA (1997) Sulfated polysaccharides from the egg jelly layer are species-specific inducers of acrosomal reaction in sperm of sea urchins. J Biol Chem 272(11):6965–6971

Arnoult C, Albrieux M, Antoine AF, Grunwald D, Marty I, Villaz M (1997) A ryanodine-sensitive calcium store in ascidian eggs monitored by whole-cell patch-clamp recordings. Cell Calcium 21(2):93–101

Ausio J, Van Veghel MLJ, Gomez R, Barreda D (1997) The sperm nuclear basic proteins (SNBPs) of the sponge *Neofibularia nolitangere*: implications for the molecular evolution of SNBPs. Mol Evol 45:91–96

Avel M (1959) Oligochaeta. In: Grassé P-P (ed) Traité de zoologie, vol V, part 1. Masson, Paris, pp 224–470

Baccetti B (1979) The evolution of the acrosomal complex. In: Fawcett DW, Bedford JM (eds) The spermatozoan, maturation, motility, surface properties, and comparative aspects. Urban and Schwarzenberg, Baltimore, pp 305–329

Baginski T, Hirohashi N, Hoshi M (1999) Sulfated O-linked glycans of the vitelline coat as ligands in gamete interaction in the ascidian, *Halocynthia roretzi*. Dev Growth Differ 41(3):357–364

Baldwin JD, Griffin FJ, Clark WH Jr (1998) Immunological characterisation of the acrosomal filament in the marine shrimp *Sicyonia ingentis*. Zygote 6(4):329–339

Barros C, Dupre E, Viveros L (1986) Sperm–egg interactions in the shrimp *Rhynchocinetus typus*. Gamete Res 14:171–180

Basa NR, Brafield AE (1989) A scanning electron microscopical study of sperm development and activation in *Arenicola marina* L (Annelida: Polychaeta). Invert Reprod Dev 15:211–219

Benazzi M, Lentati G (1970) Gametogenesis and egg fertilization in planetarians. Int Rev Cytol 27:101–179

Benazzi M, Benazzi Lentati G (1976) Platyhelminthes. In: John HB (ed) Animal cytogenetics. Borntraeger, Berlin, pp 1–82

Berg WE, Humphreys WJ (1960) Electron microscopy of four-cell stages of the ascidians *Ciona* and *Styela*. Dev Biol 2:42–60

Block EM, Goodnight CJ (1980) Spermatogenesis in *Limnodrilus hoffmeisteri* (Annelida: Tubificidae): a morphological study of the development of two sperm types. Trans Am Microsc Soc 99:368–384

Boveri T (1900) Zellen-Studien: über die Natur der Centrosomen. Fisher, Jena

Brinkley BR (1985) Microtubule organizing centers. Annu Rev Cell Biol 1:145–172

Browning H, Strome S (1996) A sperm-supplied factor required for embryogenesis in *C. elegans*. Development 122:391–404

Buckland-Nicks J, Walker CW, Chia FS (1984) Ultrastructure of the male reproductive system and of spermatogenesis in the viviparous brittle star, *Amphipholis squamata*. J Morphol 179:243–262

Buckland-Nicks H, Chia F-S, Koss R (1990) Spermiogenesis in Polyplacophora, with special reference to acrosome formation. Zoomorphology 109:179–188

Bullock TL, McCoy AJ, Kent HM, Roberts TM, Stewart M (1998) Structural basis for amoeboid motility in nematode sperm. Nat Struct Biol 5(3):184–189

Burt MDB (1986) Early morphogenesis in the Platyhelminthes with special reference to egg development and development of cestode larvae. In: Howell MJ (ed) Parasitology – quo vadit? Australian Academy of Science, Canberra, pp 241–253

Buznikov GA, Podmarev VI (1990) The sea urchin *Strongylocentrotus drobachiensis*, *S. nudus*, and *S. internedius*. In: Dettlaff TA, Vassetzky SG (eds) Animal species for developmental studies: invertebrates, vol 1. Consultants Bureau, New York, pp 253–285

Calarco-Gillam PC, Siebert MC, Hubble R, Mitchison T, Kirschner M (1983) Centrosome development in early mouse embryos as defined by an autoantibody against pericentriolar material. Cell 35:621–629

Carré D (1984) Existence d'un complexe acrosomal chez les spermatozoides du cnidaire *Muggiaceae kochi* (Siphonophore Calycophore): differenciation et réaction acrosomale. Int J Inv Reprod Dev 7:95–103

Carré D, Sardet C (1984) Fertilization and early development in *Beroe ovata*. Dev Biol 105:188–195

Carré D, Sardet C, Rouviere C (1990) Fertilization in ctenophores. In: Dale B (ed) Mechanisms of fertilization: plants to humans. Springer, Berlin Heidelberg New York

Carré D, Rouviere C, Sardet C (1991) In vitro fertilization in ctenophores: sperm entry, mitosis, and the establishment of bilateral symmetry in *Beroe ovata*. Dev Biol 147:381–391

Carroll DJ, Ramarao CS, Mehlmann LM, Roche S, Terasaki M, Jaffe LA (1997) Calcium release at fertilization in starfish eggs is mediated by phospholipase C gamma. J Cell Biol 138(6):1303–1311

Cattaneo F, Pasini ME, Perotti ME (1997) Glycosidases are present on the surface of *Drosophila melanogaster* spermatozoa. Mol Reprod Dev 48(2):276–281

Chandler DE, Heuser J (1981) Postfertilization growth of microvilli in the sea urchin egg: new views from eggs that have been quick-frozen, freeze-fractured, and deeply etched. Dev Biol 82:393–400

Chiri S, De Nadai C, Ciapa B (1998) Evidence for MAP kinase activation during mitotic division. J Cell Sci 111:2519–2527

Ciapa B, Epel D (1991) A rapid change in phosphorylation on tyrosine accompanies fertilization of sea urchin eggs. FEBS Lett 295 (1–3):167–170

Clark WH Jr, Kleve MG, Yudin AI (1981) An acrosome reaction in *Natantian* sperm. J Exp Zool 218:279–291

Clark WH Jr, Yudin AI, Griffin FJ, Shikegawa K (1984) The control of gamete activation and fertilization in the marine Penaeidae *Siconia ingentis*. Adv Inv Reprod 3:459–472

Claveria FG, Etges FJ (1988) Spermatogenesis in *Oncomelania hupensis quadrasi*, a molluscan host of *Schistosoma japonicum*. Malacologia 28:81–94

Cloney RA (1987) Phylum Urochordata, class Ascidiacea. In: Strathmann MF (ed) Reproduction and development of marine invertebrates of the northern Pacific Coast. Seattle University of Washington Press, Seattle, pp 607–639

Cloney RA, Abbott LC (1980) The spermatozoa of ascidians, acrosome and nuclear envelope. Cell Tissue Res 206:261–270

Cloney RA, Cavey MJ (1982) Ascidian larval tunic: extraembryonic structures influence morphogenesis. Cell Tissue Res 222:547–562

Collas P, Thompson L, Fields AP, Poccia DL, Courvallin JC (1997) Protein kinase C-mediated interphase lamin B phosphorylation and solubilization. J Biol Chem 272(34):21274–21280

Conn DB (1985) Life-cycle and postembryonic development of *Oochoristica anolis* (Cestoda: Linstowiidae). J Parasitol 71:10–16

Conner S, Wessel GM (1998) ab3 mediates cortical granule exocytosis in the sea urchin egg. Dev Biol 203(2):334–344

Correa LM, Carroll EJ Jr (1997) Identification of a new sea urchin vitelline envelope sperm binding glycoprotein. Dev Growth Differ 39(6):773–786

Cotelli F, Andronico F, DeSantis R, Monroy A, Rosati F (1981) Differentiation of the vitelline coat in the ascidian *Ciona intestinalis*. An ultrastructural study. Wilhelm Rouxs Arch Dev Biol 190:252–258

Couche GA, Gillott C (1990) Structure of the accessory reproductive glands of the male migratory grasshopper, *Melanoplus sanguinipes*. J Morphol 203:219–245

Courregès VC, Fell PE (1989) Sexual and asexual reproduction by the freshwater sponge *Anheteromeyenia ryderi*, with emphasis on spermatogenic activity. Trans Am Microsc Soc 108:127–138

Dale B, de Santis A, Ortolani G (1983) Electrical response to fertilization in ascidian oocytes. Dev Biol 99(1):188–193

Dangott LJ, Garbers DL (1984) Identification and partial characterization of the receptor for speract. J Biol Chem 259:13712–13716

Daydov PV, Shrubravyi OI, Vassetzky SG (1990) The starfish *Asterina pectinifera* (Muller and Troschel 1842). In: Dettlaff TA, Vassetzky SG (eds) Animal species for developmental studies: invertebrates, vol 1. Consultants Bureau, New York, pp 287–311

DeAngelis PL, Glabe CG (1988) Role of basic amino acids in the interaction of bindin with sulfated fucans. Biochemistry 27(21):8194

DeAngelis PL, Glabe CG (1990a) Specific recognition of sulfate esters by bindin, a sperm adhesion protein from sea urchins. Biochim Biophys Acta 1037(1):100–105

DeAngelis PL, Glabe CG (1990b) Zinc specifically stimulates the selective binding of a peptide analog of bindin to sulfated fucans. Pept Res 3(2):62–68

DeBarry J, Kawahara S, Takamura K, Janoshazi A, Kirino Y, Olds JL, Lester DS, Alkon DL, Yoshioka T (1997) Time-resolved imaging of protein kinase C activation during sea urchin egg fertilization. Exp Cell Res 234(1):115–124

Deguchi R, Osanai K, Morisawa M (1996) Extracellular Ca^{2+} entry and Ca^{2+} release from inositol 1,4,5-triphosphate-sensitive stores function at fertilization in oocytes of the marine bivalve *Mytilus edulis*. Development 122(110):3651–3660

De Nadai C, Cailliau K, Epel D, Ciapa B (1998a) Detection of phospholipase C gamma in sea urchin eggs. Dev Growth Differ 40(6):669–676

De Nadai C, Huitorel P, Chiri S, Ciapa B (1998b) Effect of wortmannin, an inhibitor of phosphatidylinositol 3-kinase, on the first mitotic divisions of the fertilized sea urchin egg. J Cell Sci 111:2507–2518

DeSantis R, Jamunno G, Rosati (1980) A study of the chorion and the follicle cells in relation to sperm–egg interaction in the ascidian, *Ciona intestinalis*. Dev Biol 74:490–499

Dewel WC, Clark WH Jr (1974) A fine structural investigation of surface specializations and the cortical reaction in eggs of the cnidarian *Bunodosoma cavernata*. J Cell Biol 60:78–91

Dhume ST, Lennarz WJ (1995) The involvement of O-linked oligosaccharide chains of the sea urchin egg receptor for sperm in fertilization. Glycobiology 5(1):11–17

Downey JC, Lambert CC (1994) Attachment of the ascidian sperm surface egg receptor N-acetylglucosaminidase to the cell membrane. Mol Reprod Dev 38(4):453–458

Dropkin VH, Bird AF (1978) Physiological and morphological studies on secretion of a protein-carbohydrate complex by a nematode. Int J Parasit 8:225–232

Dube F, Eckberg WR (1997) Intracellular pH increase driven by an Na^+/H^+ exchanger upon activation of surf clam oocytes. Dev Biol 190(1):41–54

Dube F, Golsteyn R, Dufresne L (1987) Protein kinase C and meiotic maturation of surf clam oocytes. Biochem Biophys Res Commun 142(3):1072–1076

Eckberg WR, Anderson WA (1985) Blocks to polyspermy in *Chaetopterus*. J Exp Zool 233:253–260

Eckberg WR, Anderson WA (1997) The initiation of development in *Chaetopterus*. In: Adiyodi KG, Adiyodi RG, Collier JR (eds) Reproductive biology of invertebrates, vol VII: progress in developmental biology Wiley, New York, pp 110–135

Eckberg WR, Miller AL (1995) Propagated and nonpropagated calcium transients during egg activation in the annelid, *Chaetopterus*. Dev Biol 172:654–656

Eckberg WR, Szuts EZ, Carroll AG (1987) Protein kinase C activity, protein phosphorylation and germinal vesicle breakdown in *Spisula* oocytes. Dev Biol 124(1):57–64

Eckelberger KJ, Grassle JP (1987) Spermatogenesis, sperm storage and comparative sperm morphology in nine species of *Capitella, Capitomastus* and *Capitellides* (Polychaeta: Capitellidae). Mar Biol 95:415–429

Eckelbarger KJ, Larson RL (1988) Ovarian morphology and oogenesis in *Aurelia aurita* (Scyphozoa: Semaeostomae): ultrastructural evidence of heterosynthetic yolk formation in a primitive metazoan. Mar Biol 100:103–115
Eckelberger KJ, Rice SA (1988) Ultrastructure of oogenesis in the holopelagic polychaetes *Rhynchonerella angelini* and *Alciopa reynaudii* (Polychaeta: Alciopidae). Mar Biol 98:427–439
Evans T, Rosenthal ET, Youngblom J, Distel D, Hunt T (1983) Cyclin: a protein specified by maternal mRNA in sea urchin eggs that is destroyed at each cleavage division. Cell 33(2):389–396
Eyster LS (1983) The embryonic capsules of nudibranch molluscs: literature review and new studies on albumen and capsule wall ultrastructure. Am Malacol Bull 4:205–216
Fell PE (1974) Porifera. In: Giese AC, Pearse JS (eds) Reproduction of marine invertebrates: acoelomate and pseudocoelomate metazoans, vol 1. Academic Press, New York, pp 51–132
Fell PE (1989) Porifera. In: Adiyodi KG, Adiyodi RG (eds) Reproductive biology of invertebrates. Wiley, New York
Fischer A, Dhainaut A (1985) The origin of yolk in the oocytes of *Nereis virens* (Annelida, Polychaeta). Cell Tissue Res 240:67–76
Fischlschweiger W (1990) Ultrastructure of the sperm duct and penis bulb of *Dugesia tigrina* (Platyhelminthes: Tricladida). Trans Am Microsc Soc 109:141–151
Fisher D, Abrieu A, Simon MN, Keyse S, Verge V, Doree M, Picard A (1998) MAP kinase inactivation is required only for G2-M phase transition in early embryogenesis cell cycles of the starfishes *Marthasterias glacialis* and *Astropecten aranciacus*. Dev Biol 202:1–13
Flannery B, Epel D (1998) Effects of wheat germ agglutinin on tunicate egg activation and fertilization: is there a plasma membrane sperm receptor system on *Ascidia ceratodes* eggs. Dev Growth Differ 40(3):297–306
Flood PR, Afzelius BA (1978) The spermatozoon of *Oikoplura dioica* Fol (Larvacea, Tunicata). Cell Tissue Res 191:27–37
Focarelli R, Rosati F (1995) The 220-kDa vitelline coat glycoprotein mediates sperm binding in the polarized egg of *Unio elongatus* through O-linked oligosaccharides. Dev Biol 171(2):606–614
Foe VE (1989) Mitotic domains reveal early commitment of cells in *Drosophila* embryos. Development 107:1–22
Foltz KR, Parotin JS, Lennarz WJ (1993) Sea urchin egg receptor for sperm: sequence similarity of binding domain to hsp 70. Science 259:1421–1425
Foor WE (1970) Spermatozoan morphology and zygote formation in nematodes. Biol Reprod (Suppl) 2:177–202
Franzén A (1970) Phylogenetic aspects of the morphology of spermatozoa and spermiogenesis. In: Baccetti B (ed)Comparative spermatology. Academic Press, New York, pp 29–46
Franzén W (1988) Oogenesis and larval development of *Scypha ciliata* (Porifera, Calcarea). Zoomorphology 107:349–357
Freeman G (1978) The role of asters in the localization of the factors that specify the apical tuft and the gut of the nemertine *Cerebratulus lacteus*. J Exp Zool 206:81–108
Freeman G (1996) The role of localized cell surface-associated glycoproteins during fertilization in the hydrozoan *Aequorea*. Dev Biol 179(1):17–26
Freeman G, Miller RL (1982) Hydrozoan eggs can only be fertilized at the site of polar body formation. Dev Biol 94:142–152
Freeman G, Ridgway EB (1991) Endogenous photoproteins as calcium indicators in hydrozoan eggs and larvae. Zool Sci 8:225–233
Freeman G, Ridgway EB (1993) The role of intracellular calcium and pH during fertilization and egg activation in the hydrozoan *Phialidium*. Dev Biol 156:176–190
Fukumoto M (1983) Fine structure and differentiation of the acrosome-like structure in the solitary ascidians, *Pyura haustor* and *Styela plicata*. Dev Growth Differ 25:503–515
Fukumoto M (1984) Fertilization in ascidians: acrosome fragmentation in *Ciona intestinalis* spermatozoa. J Ultrastr Res 87:252–262

Fukumoto M (1985) Acrosome differentiation in *Molgula manhattensis* (Ascidiacea, Tunicata). J Ultrastruct Res 92:158–166
Fukumoto M (1986) The acrosome in ascidians, I. Pleurogona. Int J Inv Reprod Dev 10:335–346
Gaino E, Burlando B, Zunino L, Pansini M, Buffa P (1984) Origin of male gametes from choanocytes in *Spongia officinalis* (Porifera, Demospongiae). Int J Invert Reprod Dev 7:83–93
Gaino E, Burlando B, Buffa P, Sarà M (1986) Ultrastructural study of spermatogenesis in *Oscarella lobularis* (Porifera, Demospongiae). Int J Invert Reprod Dev 10:297–305
Gaino E, Burlando B, Buffa P (1987) Ultrastructural study of oogenesis and fertilization in *Sycon ciliatum* (Porifera, Calcispongiae). Int J Invert Dev 11:73–82
Gallissian M-F (1980) Etude ultrastructurale de la fécondation chez *Grantia compressa* F. Int J Invert Reprod 2:321–329
Galt CP (1987) Phylum Chordata, subphylum Urochordata, class Larvacea (Appendicularia). In: Strathmann MF (ed) Reproduction and development of marine invertebrates of the northern Pacific Coast. University of Washington Press, Seattle, pp 640–646
Gibson R (1972) Nemerteans. Hutchinson University Library, London, 224 pp
Gilbert JJ, Simpson TL (1976) Sex reversal in a fresh water sponge. J Exp Zool 195:145–151
Giusti AF, Carroll DJ, Abassi YA, Foltz KR (1999) Evidence that a starfish egg src family tyrosine kinase associates with PLC-gamma 1 SH2 domains at fertilization. Dev Biol 208(1):189–199
Glabe CG (1985) Interaction of the sperm adhesive protein, bindin, with phospholipid vesicles. I. Specific association of bindin with gel-phase phospholipid vesicles. J Cell Biol 100(3):794–799
Glabe CG, Vacquier VD (1977) Species specific agglutination of eggs by bindin isolated from sea urchin sperm. Nature 267(5614):836–838
Glaser RW, Grune M, Wandelt C, Ulrich AS (1999) Structure analysis of a fusogenic peptide sequence from the sea urchin fertilization protein bindin. Biochemistry 38(8):2560–2569
Gontcharoff M (1951) Biologie de la regeneration et de la reproduction chez quelques Lineidae de France. Ann Sci Nat (Zool) Paris Ser II 13:149–235
Goode CA, Gamboa-Pinto AJ, Cruz R, Gough LL, Lund CV, Lambert CC (1997) Evidence for cell surface and internal phospholipase activity in ascidian eggs. Dev Growth Differ 39(5):655–660
Goudeau M (1976) Secretion of embryonic envelopes and embryonic molting cycles in *Hemioniscus balani* Buchholz, Isopoda, Epicaridae. J Morphol 148:427–452
Goudeau M (1984) Fertilization in a crab. III. Cytodifferentiation of vesicles enclosing ring-shaped elements involved in the cortical reaction. Gamet Res 9:409–424
Goudeau M, Goudeau H (1986) The resumption of meiotic maturation of the oocyte of the prawn *Palaemon serratus* is regulated by an increase in extracellular Mg^{2+} during spawning. Dev Biol 118:361–370
Goudeau M, Goudeau H (1993a) In the egg of the ascidian *Phallusia mammillata*, removal of external Ca^{2+} modifies the fertilization potential, induces polyspermy, and blocks the resumption of meiosis. Dev Biol 160(1):165–177
Goudeau M, Goudeau H (1993b) Successive electrical responses to insemination and concurrent sperm entries in the polyspermic egg of the Ctenophore *Beroe ovata*. Dev Biol 156:537–551
Gould MC, Stephano JL (1993) Nuclear and cytoplasmic pH increase at fertilization in *Urechis caupo*. Dev Biol 159(2):608–617
Graebner I, Adam H (1970) Electron microscopical study of spermatogenesis and sperm morphology in gnathostomulids. In: Baccetti B (ed) Comparative spermatology. Academic Press, New York, pp 375–382
Gremigni V, Nigro M, Settembrini MS (1986) Ultrastructural features of oogenesis in some marine neoophoran turbellarians. Hydrobiologia 132:145–150
Griffin FJ, Clark WH Jr, Crowe WH, Crowe LM (1987) Intracellular pH decreases during the in vitro induction of the acrosome reaction in the sperm of *Sicyonia ingentis*. Biol Bull 173:311–323
Guidice G (1986) The sea urchin embryo: a developmental biology system. Springer, Berlin Heidelberg New York, 246 pp

Haley SA, Wessel GM (1999) The cortical granule serine protease CGSP1 of the sea urchin, *Strongylocentrotus purpuratus*, is autocatalytic and contains a low-density lipoprotein receptor-like domain. Dev Biol 211(1):77–87

Harada R, Maeda T, Nakashima A, Sadakata Y, Ando M, Yonamine K, Otsuji Y, and Sato H (1970) Electron microscopical studies on the mechanism of oogenesis and fertilization in *Dirofilaria immitis*. In: Sasa M (ed) Recent advances in research on filariasis and schistosomasis in Japan. University Park Press, Baltimore, pp 99–121

Hendelberg J (1986) The phylogenetic significance of sperm morphology in the Platyhelminthes. Hydrobiologia 132:53–58

Hinsch G (1971) Penetration of the oocyte envelope by spermatozoa in the spider crab. J Ultrastruct Res 35:86–97

Hinsch G (1973) Sperm structure of Oxyrhyncha. Can J Zool 51:421–426

Hinch GW (1980) Spermiogenesis in *Coenobita clypeatus*. I. Sperm structure. Int J Invert Reprod 2:189–198

Hinsch G (1990) Arthropoda – Crustacea. In: Adiyodi KG, Adiyodi RG (eds) Reproductive biology of invertebrates. Fertilization, development, and parental care. Wiley, New York

Hirohashi N, Lennarz WJ (1998) The 350-kDa sea urchin egg receptor for sperm is localized in the vitelline layer. Dev Biol 204(1):305–315

Hodgson AN,Heller J, Bernard RT (1990) Ultrastructure of the sperm and spermatogenesis in five South African species of the trochid genus *Oxystele* (Mollusca, Prosobranchia). Mol Reprod Dev 25(3):263–271

Holland LZ (1988) Spermatogenesis in the salps *Thalia democratica* and *Cyclosalpa affinis* (Tunicata: Thaliacea): an electron microscopic study. J Morphol 198:189–204

Holland ND, Holland LZ (1989a) Fine structural study of the cortical reaction and formation of the egg coat in a lancelet (=amphioxus), *Branchiostoma floridae* (phylum Chordata: subphylum Cephalochordata=Acrania). Biol Bull 176:111–122

Holland ND, Holland LZ (1989b) The fine structure of the testis of a lancelet (=amphioxus), *Branchiostoma floridae* (phylum Chordata: subphylum Cephalochordata=Acrania). Acta Zool (Stockh) 70:211–219

Holland LZ, Gorsky G, Fenaux R (1988) Fertilization in *Oikoploira dioica* (Tunicata, Appendicularia): acrosome reaction, cortical reaction, and sperm–egg fusion. Zoomorphology 108:229–243

Holy JM, Wittrock DD (1986) Ultrastructure of the female reproductive organs (ovary, vitellaria, and Mehlis' gland) of *Halipegus eccentricus* (Trematoda: Derogenidae). Can J Zool 64:2201–2212

Honegger TG (1981) Light and scanning electron microscopic investigations of sexual reproduction in *Hydra carnea*. Int J Invert Reprod 3:245–255

Honegger TG (1983) Ultrastructural and experimental investigations of sperm–egg interactions in fertilization in *Hydra carnea*. Rouxs Arch Dev Biol 192:13–20

Honegger TG (1986) Fertilization in ascidians: studies on the egg envelope, sperm and gamete interactions in *Phallusia mammillata*. Dev Biol 118:118–128

Honegger TG, Zürrer D, Tardent P (1989) Oogenesis in *Hydra carnea*: a new model based on light and electron microscopic analyses of oocyte and nurse cell differentiation. Tissue Cell 21:381–393

Hoshi M (1986) Sperm glycosidase as a plausible mediator of sperm binding to the vitelline envelope in Ascidians. Adv Exp Med Biol 207:251–60

Hsieh H-L, Simon JL (1990) The sperm transfer system in *Kinbergonium simoni* (Polychaeta: Onuphidae). Biol Bull 187:85–93

Hyman LH (1951) The invertebrates, vol II: Platyhelminthes and Rhynchocoela. The acoelomate bilateria. McGraw-Hill, New York

Irons KP, Walker CW (1984) The histology, ultrastructure, and functional morphology of the male reproductive system of *Molgula manhattensis*. Int J Invert Reprod Dev 7:13–21

Jaffe LA, Gould-Somero M (1985) Polyspermy-preventing mechanisms. In: Metz CB, Monroy A (eds) Biology of fertilization, vol 3. Academic Press, New York, pp 223–250

Jamieson BGM, Rouse GW (1989) The spermatozoa of the Polychaeta (Annelida): an ultrastructural review. Biol Rev 64:93–157

Jawale SM, Ranade DR (1990) Morphology of the ovaries of *Sphaerodema* (*Diplonychus*) *rusticum* (Heteroptera, Belostomatidae). J Morphol 205:183–192

Jeffery W (1982) Calcium ionophore polarizes ooplasmic segregation in ascidian eggs. Science 216:545–547

Jeffery W, Meier (1983) A yellow crescent cytoskeletal domain in ascidian eggs and its role in early development. Dev Biol 96:125–143

Jeffery WR (1984) Pattern formation by ooplasmic segregation in ascidian eggs. Biol Bull 166:277–298

Jiang WP, Veno PA, Wood RW, Peaucellier G, Kinsey WH (1991) pH regulation of an egg cortex tyrosine kinase. Dev Biol 146(1):81–88

Just ML, Lennarz WJ (1997) Re-examination of the sequence of the sea urchin egg receptor for sperm: implications with respect to its properties. Dev Biol 184(1):25–30

Justine JL, Lambert A, Mattei X (1985) Spermatozoon ultrastructure and phylogenetic relationships in the monogeneans (Platyhelminthes). Int J Parasitol 15:601–608

Kallenbach RJ (1983) The induction of de novo centrioles in sea urchin eggs: a possible common mechanism for early cytaster formation in parthenogenetically stimulated sea urchin eggs. Eur J Cell Biol 30:159–166

Kemphues KJ (1989) *Caenorhabditis*. In: Glover DM, Hames ED (eds) Frontiers in molecular biology: genes and embryos. IRL Press, London, pp 95–126

Kenyon C (1988) The nematode *Caenorhabditis elegans*. Science 240:1448–1453

King RC (1970) Ovarian development in *Drosophila melanogaster*. Academic Press, New York

Kinsey WH (1996) Biphasic activation of fyn kinase upon fertilization of the sea urchin egg. Dev Biol 174(2):281–287

Kitazume-Kawaguchi S, Inoue S, Inoue Y, Lennarz WJ (1997) Identification of sulfated oligosialic acid units in the O-linked glycan of the sea urchin egg receptor for sperm. Proc Natl Acad Sci USA 94(8):3650–3655

Kline D, Jaffe LA, Tucker RP (1985) Fertilization potential and polyspermy prevention in the egg of the nemertean, *Cerebratulus lacteus*. J Exp Zool 236:45–52

Kluge B, Lehmann-Greif M, Fischer A (1995) Long-lasting exocytosis and massive structural reorganisation in the egg periphery during cortical reaction in *Platynereis dumerlii*. Zygote 3:141–156

Koch RA, Norton ML, Vazquez H, Lambert CC (1994) Sperm-surface chymotrypsin-like protease activity required for fertilization in ascidians. Dev Biol 162(2):438–450

Kyozuka K, Osanai K (1988) Fertilization cone formation in starfish oocytes: the role of egg cortex actin microfilaments in sperm incorporation. Gamete Res 20:275–285

Kyozuka K, Deguchi R, Mohri T, Miyazaki S (1998) Injection of sperm extract mimics spatiotemporal dynamics of Ca^{2+} responses and progression of meiosis at fertilization of ascidian oocytes. Development 125(20):4099–4105

Labordus V, Hooftman RN, van der Wal UP (1986) The formation of pronuclei and their associated perinuclear cytoplasm; the determination of the phases of the first cleavage cycles in *Crepidula fornicata* (mollusca: Gastropoda). Int J Invert Reprod Dev 10:89–104

LaFleur GJ Jr, Horiuchi Y, Wessel GM (1998) Sea urchin ovoperoxidase: oocyte-specific member of a heme-dependent peroxidase superfamily that functions in the block to polyspermy. Mech Dev 70(1–2):77–89

Lambert CC (1982) The ascidian sperm reaction. Am Zool 22:841–849

Lambert CC (1989) Ascidian eggs release glycosidase activity which aids in the block against polyspermy. Development 105(2):415–420

Lambert CC, Epel D (1979) Calcium-mediated mitochondrial movement in ascidian sperm during fertilization. Dev Biol 69:296–304

Lambert CC, Lambert G (1978) Tunicate eggs utilize ammonium ions for flotation. Science 200:64–65

Lambert CC, Lambert G (1981) Formation of the block to polyspermy in ascidian eggs: time course, ion requirements, and role of accessory cells. J Exp Zool 217:291–295

Lambert CC, Lambert G (1983) Mitochondrial movement during ascidian sperm reaction. Gamete Res 8:295–307

Lambert C, Goudeau H, Franchet C, Lambert G, Goudeau M (1997) Ascidian eggs block polyspermy by two independent mechanisms: one at the egg plasma membrane, the other involving the follicle cells. Mol Reprod Dev 48(1):137–143

LaMunyon CW, Ward S (1997) Increased competitiveness of nematode sperm bearing the male X chromosome. Proc Natl Acad Sci USA 94(1):185–189

LaMunyon CW, Ward S (1999) Evolution of sperm size in nematodes: sperm competition favours larger sperm. Proc R Soc Lond B Biol Sci 266(1416):263–267

Larkman AU (1983) An ultrastructural study of oocyte growth within the endoderm and entry into the mesoglea in *Actinaria fragacea* (Cnidaria: Anthozoa). J Morphol 178:155–177

Larkman AU, Carter MA (1984) The apparent absence of a cortical reaction after fertilization in a sea anemone. Tissue Cell 16(1):125–130

LaSpina D'Anna R (1974) Light and electron microscopic study of unfertilized egg of *Ascidia malaca* (Tunicata). Acta Embryol Exp 1:3–17

Lee SJ, Shen SS (1998) The calcium transient in sea urchin eggs during fertilization requires the production of inositol 1,4,5-trisphosphate. Dev Biol 193(2):195–208

Lindsay LL, Hertzler PL, Clark WH Jr (1992) Extracellular Mg^{2+} induces an intracellular Ca^{2+} wave during oocyte activation in the marine shrimp *Sicyonia ingentis*. Dev Biol 152(1):94–102

Littlefield CL, Dunne JF, Bode HR (1985) Spermatogenesis in *Hydra oligactis*. I. Morphological description and characterization using a monoclonal antibody specific for cells of the spermatogenic pathway. Dev Biol 110:308–320

Longo FJ (1973) An ultrastructural analysis of polyspermy in the surf clam *Spisula*. J Exp Zool 183:153–180

Longo FJ (1983) Meiotic maturation and fertilization. In: Verdonk NH, van den Biggelaar JAM, Tompa AS (eds) The mollusca, vol III. Academic Press, New York, pp 49–89

Longo FJ, Anderson E (1970) An ultrastructural analysis of artificial parthenogenesis in the surf clam *Spisula solidissima*. J Ultrastr Res 36:806–823

Longo F, Clark WH, Hinsch GW (1988) Gamete interactions and sperm incorporation in the nemertean, *Cerebratulus lacteus*. Zool Sci 5:573–584

Mabuchi Y, Mabuchi I (1973) Acrosomal ATPase in star fish and bivalve mollusc spermatozoa. Exp Cell Res 82:271–279

MacKinnon BM (1987) A light and electron microscopical investigation of fertilization, chromosome behaviour, and early cleavage patterns in *Heligmosomoides polygyrus* (Nematode: *Trichostrongyloidea*). Int J Invert Reprod Dev 11:89–108

Maggenti A (1981) General nematology. Springer, Berlin Heidelberg New York, pp 131–157

Matsuura K, Sawada H, Yokosawa H (1993) Purification and properties of N-acetylglucosaminidase from eggs of the ascidian, *Halocynthia roretzi*. Eur J Biochem 218(2): 535–541

Matsuura K, Sawada H, Yokosawa H (1995) N-acetylglucosaminidase inhibitor isolated from the vitelline coat of ascidian eggs is a candidate sperm receptor. Biochem Biophys Res Commun 213 (1):311–316

Mauk R, Jaworski D, Kamei N, Glabe CG (1997) Identification of a 97-kDa heat shock protein from *S. franciscanus* ovaries with 94% amino acid identity to the *S. purpuratus* egg surface receptor for sperm. Dev Biol 184(1):31–37

Mazia D (1984) Centrosomes and mitotic poles. Exp Cell Res 153:1–15

McClary DJ, Mladenov PV (1988) Brood and broadcast: a novel mode of reproduction in the sea star *Pteraster militaris*. In: Burke et al. (eds) Echinoderm biology. Balkema, Rotterdam, pp 163–168

McDougall A, Levasseur M (1998) Sperm-triggered calcium oscillations during meiosis in ascidian oocytes first pause, restart, then stop: correlation with cell cycle kinase activity. Development 125:4451–4459

McDougall A, Sardet C (1995) Function and characteristics of repetitive calcium waves associated with meiosis.Curr Biol 5(3):318–328
McDougall A, Sardet C, Lambert CC (1995) Different calcium-dependent pathways control fertilization-triggered glycoside release and the cortical contraction in ascidian eggs. Zygote (3):251–258
McPherson SM, McPherson PS, Mathews L, Campbell KP, Longo FJ (1992) Cortical localization of a calcium release channel in sea urchin eggs. J Cell Biol 116(5):1111–1121
Medina A, Rodriguez A (1992) Structural changes in sperm from the fiddler crab, *Uca tangeri* (Crustacea, Brachyura), during the acrosome reaction. Mol Reprod Dev 33(2):195–201
Mikami-Takei K, Kosakai M, Isemura M, Suyemitsu T, Ishihara K, Schmid K (1991) Fractionation of jelly substances of the sea urchin egg and biological activities to induce acrosome reaction and agglutination of spermatozoa. Exp Cell Res 192(1):82–86
Miller RI (1978) Site-specific sperm agglutination and the timed release of a sperm chemo-attractant by the egg of the leptomedusan, *Othopyxis calliculata.* J Exp Zool 205:385–392
Minor JE, Britten RJ, Davidson EH (1993) Species-specific inhibition of fertilization by a peptide derived from the sperm protein bindin. Mol Biol Cell 4(4):375–387
Miraglia SJ, Glabe CG (1993) Characterization of the membrane-associating domain of the sperm adhesive protein bindin. Biochim Biophys Acta 1145(2):191–198
Mladenov PV (1987) Phylum Echinodermata, class Crinoidea. In: Strathmann MF (ed) Reproduction and development of marine invertebrates of the Northern Pacific Coast. University of Washington Press, Seattle, pp 597–606
Mukai H (1977) Comparative studies on the structure of reproductive organs of four botryllid ascidians. J Morphol 152:363–379
Mukai H, Watanabe H (1978) Studies on some aspects of ovulation and embryogenesis in the compound ascidian *Botryllus primigenus.* Ann Zool Japon 51:211–221
Niijima L, Dan J (1965a) The acrosome reaction in *Mytilus edulis* I. Cell Biol 25:243–248
Niijima L, Dan J (1965b) The acrosome reaction in *Mytilus edulis* II. Cell Biol 25:249–259
Nishigaki T, Chiba K, Miki W, Hoshi M (1996) Structure and function of asterocaps, sperm-activating peptides from the jelly coat of starfish eggs. Zygote 4(3):237–245
Nollen PM (1983) Patterns of sexual reproduction among parasitic platyhelminths. Parasitology 86:99–120
Oakley CD, Oakley BR (1989) Identification of g-tubulin, a new member of the tubulin superfamily encoded by mipA gene of *Aspergillus nidulans.* Nature 338:662–664
Oka H (1970) Colony specificity in compound ascidians: the genetic control of fusibility. In: Yikawa H (ed) Profiles of Japan and Japanese scientists. Kodansha, Tokyo, pp 196–206
Okano-Uchida T, Sekiai T, Lee K, Okumura E, Tachibana K, Kishimoto T (1998) In vivo regulation of cyclin A/Cdc2 and cyclin B/Cdc2 through meiotic and early cleavage cycles in starfish. Dev Biol 197:39–53
Olds JI, Favit A, Nelson T, Ascoli G, Gerstein A, Cameron M, Cameron L, Lester DS, Rakow T, De Barry J (1995) Imaging protein kinase C activation in living sea urchin eggs after fertilization. Dev Biol 172(2):675–682
Olive PJW, Clark RB (1978) Physiology of reproduction. In: Mill PJ (ed) Physiology of Annelides. Academic Press, London, pp 271–368
Papathanassiou E, King PE (1984) Ultrastructural studies on gametogenesis of the prawn *Palaemon serratus* (Pennant). I. Oogenesis. Acta Zool (Stockh) 65:17–31
Paraense WL, Correa LR (1988) Self fertilization in the freshwater snails *Helisoma duryi* and *Helisoma trivolvis.* Mem Inst Oswaldo Cruz 83(4):405–409
Pasini ME, Redi CA, Caviglia O, Perotti ME (1996) Ultrastructural and cytochemical analysis of sperm dimorphism in *Drosophila subobscura.* Tissue Cell 28(2):165–175
Pasini ME, Cattaneo F, Pinto MR, De Santis R, Perotti ME (1999) Plasma membrane association and preliminary characterization of *Drosophila* sperm surface glycosidases. Mol Reprod Dev 52(2):166–173
Paulus W (1989) Ultrastructural investigation of spermatogenesis in *Spongilla lacustris* and *Ephydatia fluviatilis* (Porifera, Spongilidae). Zoomorphology 109:123–130

Paulus W, Weissenfels N (1986) The spermatogenesis of *Ephydatia fluviatilis* (Porifera). Zoomorphology 106:155–162

Paweletz N, Mazia D, Finze E-M (1984) The centrosome cycle in the mitotic cycle of sea urchin eggs. Exp Cell Res 152: 47–65

Pearse JS, Walker CW (1985) Photoperiodic regulation of gametogenesis in a North Atlantic sea star, *Asterias vulgaris*. Int J Invert Reprod Dev 9:71–78

Pecorella C, Tosti E, Kyozuka K, Dale B (1993) Activation currents, sperm entry and surface contractions in ascidian eggs. Zygote 1(2):113–119

Philipova R, Whitaker M (1998) MAP kinase activity increases during mitosis in early sea urchin embryos. J Cell Sci 111(17):2497–2505

Pinto MR, Hoshi M, Marino R, Amoroso A, De Santis R (1990) Chymotrypsin-like enzymes are involved in sperm penetration through the vitelline coat of *Ciona intestinalis* egg. Mol Reprod Dev 26(4):319–323

Podell SB, Vacquier VD (1985) Purification of the Mr 80,000 and Mr 210,000 proteins of the sea urchin sperm plasma membrane. Evidence that the Mr 210,000 protein interacts with egg jelly. J Biol Chem 260:2715–2718

Polk PM, Anderson WA, Eckberg WR (1987) The *Chaetopterus* vitelline envelope is not necessary for the gamete interactions that lead to fertilization. J Exp Zool 243:323

Purchon RD (1968) The biology of Mollusca. Pergamon Press, Oxford

Qian PY, Chia FS (1989) Sexual reproduction and larval development *of Rhaphidrilus nemasoma* Monticelli 1910 (Polychaeta: Ctenodrilidae). Can J Zool 67:2345–2351

Rakow TL, Shen SS (1994) Molecular cloning and characterization of PKC from sea urchin *L. pictus*. Dev Growth Differ 36:489–497

Ramarao CS, Garbers DL (1985) Receptor-mediated regulation of guanylate cyclase activity in spermatozoa. J Biol Chem 260:8390–8396

Reid K (1964) Reproduction in sacoglossan opisthobranch *Elysia maoria*. Proc Zool Soc (Lond) 143:365–393

Reverberi G (1975) On some effects of cytochalasin B on the eggs and tadpoles of the ascidians. Acta Embryol Exp 2:137–158

Rios M, Barros C (1997) Trypsin-like enzymes during fertilization in the shrimp *Rhynchocinetes typus*. Mol Reprod Dev 46(4):581–586

Riparbelli MG, Stouthamer R, Dallai R, Callaini G (1998) Microtubule organization during the early development of the parthenogenetic egg of the hymenopteran *Muscidifurax uniraptor*. Dev Biol 195(2):89–99

Robert LK, Lucio-Gough LM, Goode CA, Mckinney K, Lambert CC (1999) Activation of follicle cell surface phospholipase by tyrosine kinase dependent pathway is an essential event in ascidian fertilization.Mol Reprod Dev 54(1):69–75

Roegiers F, McDougall A, Sardet C (1995) The sperm entry point defines the orientation of the calcium-induced contraction wave that directs the first phase of cytoplasmic reorganization in the ascidian egg. Development 121:3457–3466

Rogers WP (1962) The nature of parasitism. Academic Press, New York

Rosati F (1985) Sperm–egg interaction in ascidians. In: Metz CB, Monroy A (eds) Biology of fertilization. Academic Press, New York, pp 361–388

Rosati F, DeSantis R (1978) Studies on fertilization in the ascidians. I. Self sterility and specific recognition between gametes of *Ciona intestinalis*. Exp Cell Res 112:111–119

Rosati F, Pinto MR, Casazza G (1985) The acrosomal region of the spermatozoon of *Ciona intestinalis*: its relationship with the binding to the vitelline coat of the egg. Gamete Res 11:379–389

Roush W (1996) Sperm protein makes its mark upon the worm embryo. Science 271(5245):33

Rouviere C, Houliston E, Carré D, Chang P, Sardet C (1994) Characteristics of pronuclear migration in *Beroe ovata*. Cell Motil Cytoskeleton 29(4):301–311

Rouviere C, Houliston E, Carré D, Chang P, Sardet C (1998) Fertilization and pronuclear migration in the ctenophore, *Beroe ovata*. Video supplement. Cell Motil Cytoskel 39(4):340–342

Russo GL, Wilding M, Marino M, Dale B (1998) Ins and outs of meiosis in ascidians. Semin Cell Dev Biol 9(5):559–567

Saiki T, Hamaguchi Y (1998) Aster-forming abilities of the egg, polar body, and sperm centrosomes in early starfish development. Dev Biol 203:62–74

Saitoh Y, Sawada H, Yokosawa H (1993) High-molecular-weight protease complexes(proteasomes) of sperm of the ascidian, *Halocynthia roetzi*: isolation, characterization, and physiological roles in fertilization. Dev Biol 158(1):238–244

Saller U (1988) Oogenesis and larval development of *Ephydatia fluviatilis* (Porifera, Spongillidae). Azoomorphology 108:23–28

Santella L, Ianora A (1992) Fertilization envelope in diapause eggs of *Pontella mediterranea* (Crustacea, Copepoda). Mol Reprod Dev 33(4):463–469

Sardet C (1984) The ultrastructure of the sea urchin egg cortex isolated before and after fertilization. Dev Biol 105:196–210

Sardet C, Speksnijder J, Inoue S, Jaffe L (1989) Fertilization and ooplasmic movements in the ascidian egg. Development 105:237–249

Sato M, Osani K (1983) Sperm reception by an egg microvillus in the polychaete *Tylorrhynchus heterochaetus*. J Exp Zool 227:459–469

Sato M, Osani K (1986) Morphological identification of sperm receptors above egg microvilli in the polychaete *Neanthes japonica*. Dev Biol 113:263–270

Satoh N, Numakunai T, Hori R (1982) Behavior and cellular morphology of the test cells during embryogenesis of the ascidian *Halocynthia roretzi*. J Morphol 171:219–223

Sawada T, Schatten G (1988) Microtubules in ascidian eggs during meiosis, fertilization, and mitosis. Cell Motil Cytoskel 9(3):219–230

Sawada H, Yokosawa H, Hoshi M, Ishii S (1982) Evidence for acrosin-like enzyme in sperm extract and its involvement in fertilization of the ascidian *Halocynthia roretzki*. Gamete Res 5:291–309

Sawada H, Yokosawa H, Someno T, Saino T, Ishii S (1984) Evidence for the participation of two sperm proteases, spermosin and acrosin, in fertilization of the ascidian *Halocynthia roretzki*: inhibitory effects of leupeptin analogs in enzyme activities and fertilization. Dev Biol 105:246–249

Schatten G (1982) Motility during fertilization. Int Rev Cytol 79:59–163

Schatten H (1994) Dithiothreitol prevents membrane fusion but not centrosome or microtubule organization during the first cell cycles in sea urchins. Cell Motil Cytoskel 27:59–68

Schatten H, Schatten G (1980) Surface activity at the egg plasma membrane during sperm incorporation and its cytochalasin B sensitivity. Dev Biol 78:435–449

Schatten H, Schatten G, Petzelt C, Mazia D (1982) Effects of griseofulvin on fertilization and early development in sea urchins. Independence of DNA synthesis, chromosome condensation, and cytokinesis cycles from microtubule-mediated events. Eur J Cell Biol 27:74–87

Schatten G, Maul G, Schatten H, Chaly N, Simerly C, Balczon R, Brown D (1985) Nuclear lamins and peripheral nuclear antigens during fertilization and embryogenesis in mice and sea urchins. Proc Natl Acad Sci USA 82:4727–4731

Schatten H, Schatten G, Mazia D, Balczon R, Simerly C (1986) Behavior of centrosomes during fertilization and cell division in mouse oocytes and in sea urchin eggs. Proc Natl Acad Sci USA 83:105–109

Schatten H, Walter M, Mazia D, Biessmann H, Paweletz N, Coffe G, Schatten G (1987) Centrosome detection in sea urchin eggs with a monoclonal antibody against *Drosophila* intermediate filament proteins: characterization of stages of the division cycle of centrosomes. Proc Natl Acad Sci USA 84:8488–8492

Schatten H, Walter M, Biessmann H, Schatten G (1988) Microtubules are required for centrosome expansion and positioning while microfilaments are required for centrosome separation in sea urchin eggs during fertilization and mitosis. Cell Motil Cytoskel 11(4):248–259

Schatten H, Walter M, Biessmann H, Schatten G (1992) Activation of maternal centrosomes in unfertilized sea urchin eggs. Cell Motil Cytoskel 23:61–70

Schatten H, Hueser CN, Chakrabarti A (2000) From fertilization to cancer. The role of centrosomes in the union and separation of genomic material. Microsc Res Tech 49:420–427

Schoenmakers H, Colenbrander P, Peute J, van Oordt P (1981) Anatomy of the ovaries of the starfish *Asterias rubens*. A histological and ultrastructural study. Cell Tissue Res 217:577–598

Schofield VL, Schlumpberger JM, Weissman IL (1982) Colony specificity in the colonial tunicate *Botryllus* and the origins of vertebrate immunity. Am Zool 22:783–794
Schulz JR, Wessel GM, Vacquier VD (1997) The exocytosis regulatory proteins syntaxin and VAMP are shed from sea urchin sperm during the acrosome reaction. Dev Biol 191(1):80–87
Seidl C, Bauer M, Moritz KB (1988) Chromatin diminution and early cleavage in *Parascaris univalens* (Nematoda). Rouxs Arch Dev Biol 197:307–320
Sepsenwol S, Taft S (1990) In vitro induction of crawling in the amoeboid sperm of the nematode parasite, *Ascaris suum*. Cell Motil Cytoskel 15(2):99–110
Shen SS, Buck WR (1990) A synthetic peptide of the pseudosubstrate domain of protein kinase C blocks cytoplasmic alkalinization during activation of the sea urchin egg. Dev Biol 140(2):272–280
Shimizu T (1984) Dynamics in the actin microfilament system in the *Tubifex hattai* egg during ooplasmic segregation. Dev Biol 106:414–426
Simerly C, Wu GJ, Zoran S, Ord T, Rawlins R, Jones J, Navara C, Gerrity M, Rinehart J., Binor Z, Asch R, Schatten G (1995) The paternal inheritance of the centrosome, the cell's microtubule-organizing center, in humans, and the implications for infertility. Nature Med 1:47–52
Simpson TL (1980) Reproductive processes in sponges: a critical evaluation of current data and views. Int J Invert Reprod Dev 2:251–269
Singson A, Mercer KB, L'Hernault SW (1998) The *C. elegans* spe-9 gene encodes a sperm transmembrane protein that contains EGF-like repeats and is required for fertilization. Cell 93(1):71–79
Singson A, Hill KL, L'Hernault SW (1999) Sperm competition in the absence of fertilization in *Caenorhabditis elegans*. Genetics 152(1):201–208
Smiley S (1988) The dynamics of oogenesis and the annual ovarian cycle of *Stichopus californicus* (Echinodermata: Holothuroidea). Biol Bull 175:79–93
Smith RI (1989) Observations on spawning behavior of *Eupolymnia nebulosa*, and comparisons with *Lanice conchilega* (Annelida, Polychaeta, Terebellidae). Bull Mar Sci 45:406–414
Snook RR, Markow TA, Karr TL (1994) Functional nonequivalence of sperm in *Drosophila pseudoobscura*. Proc Natl Acad Sci USA 91(23):11222–11226
Sopott-Ehlers B (1986) Fine-structural characteristics of female and male germ cells in Proseriata Otoplanidae (Platyhelminthes). Hydrobiologia 132:137–144
Spaulding JG (1972) The life cycle of *Peachia quinquecapicitata*, an anemone parasitic on medusae during its larval development. Biol Bull 143:440–453
Speksnijder JE, Corson DW, Sardet C, Jaffe LF (1989) Free calcium pulses following fertilization in the ascidian egg. Dev Biol 135:182–190
Speksnijder JE, Sardet C, Jaffe LF (1990) The activation wave of calcium in the ascidian egg and its role in ooplasmic segregation. J Cell Biol 110(5):1589–1598
Stephano JL, Gould MC (1997) The intracellular calcium increase at fertilization in *Urechis caupo* oocytes: activation without waves. Dev Biol 191:53–68
Stewart M, King KL, Roberts TM (1994) The motile major sperm protein (MSP) of *Ascaris suum* forms filaments constructed from two helical subfilaments. J Mol Biol 243(1):60–71
Strathmann MF (1987) Reproduction and development of marine invertebrates of the Northern Pacific Coast. Seattle University of Washington Press, Seattle, 670 pp
Stricker SA (1986) An ultrastructural study of oogenesis, fertilization, and egg laying in a nemertean ectosymbiont of crabs, *Carcinomertes epialti* (Nemertea, Hoplonemertea). Can J Zool 64:1256–1269
Stricker SA (1987) Phylum Nemertea. In: Strathmann MF (ed) Reproduction and development of marine invertebrates of the Northern Pacific Coast. University of Washington Press, Seattle, pp 129–137
Stricker SA (1996) Repetitive calcium waves induced by fertilization in the nemertean worm *Cerebratulus lacteus*. Dev Biol 176:243–263
Stricker SA (1999) Comparative biology of calcium signaling during fertilization and egg activation in animals. Dev Biol 211:157–176

Stricker SA, Cavey MJ (1986) An ultrastructural study of spermatogenesis and the morphology of the testis in the nemertean worm *Tetrastemma phyllospadicola* (Nemertea: Hoplonemertea). Can J Zool 64:2187–2202

Strome S (1989) Generation of cell diversity during early embryogenesis in the nematode *Caenorhabditis elegans*. Int Rev Cytol 114:81–123

Sugino YM, Tominaga A, Takashima Y (1987) Differentiation of the accessory cells and structural regionalization of the oocyte in the ascidian *Ciona savignyi* during early oogenesis. J Exp Zool 242:205–214

Swalla BJ, Moon RT, Jeffrey WR (1985) Developmental significance of a cortical cytoskeletal domain in *Chaetopterus* eggs. Dev Biol 111:434–450

Talbot P, Chanmanon P (1978) The acrosome reaction of sperm from the lobster *Homarus americana*. J Cell Biol 79:158a

Talbot P, Goudeau M (1988) A complex cortical reaction leads to formation of the fertilization envelope in the lobster, *Homarus*. Gamete Res 19(1):1–18

Tappenden T, Kearn GC, Evans-Gowing R (1993) Fertilization and the functional anatomy of the germarium in the monogenean *Entobdella soleae*. Int J Parasitol 23:901–911

Terasaki M, Henson J, Begg D, Kaminer B, Sardet C (1991) Characterization of sea urchin egg endoplasmic reticulum in cortical preparations. Dev Biol 148(1):398–401

Thomas TW, Eckberg WR, Dube F, Galione A (1998) Mechanisms of calcium release and sequestration in eggs of *Chaetopterus pergamentaceus*. Cell Calcium 24:285–292

Tilney LG (1978) Polymerization of actin. V. A new organelle, the actomere, that inhibits the assembly of actin filaments in Thyone sperm. J Cell Biol 78:551–564

Togo T, Morisawa M (1997) Aminopeptidase-like protease released from oocytes affects oocyte surface and suppresses the acrosome reaction in establishment of polyspermy block in oocytes of the mussel *Mytilus edulis*. Dev Biol 182(2):219–227

Trimmer JS, Vacquier VD (1986) Activation of sea urchin gametes. Ann Rev Cell Biol 2:1–26

Tuzet O (1964) L'origine de la lignée germinale et la gametogenese chez le Spongiaires. In: Wolff E (ed) L'origine de la lignee Germinale. Hermann, Paris, pp 79–111

Tyler S (1984) Ciliogenesis in embryos of the acoel turbellarian *Archaphanostoma*. Trans Am Microsc Soc 103:1–15

Tyler S, Rieger RM (1975) Uniflagellate spermatozooa in Nemertoderma (Turbellaria) and their phylogenetic significance. Science 188:730–732

Ursprung H, Schabtach E (1964) The fine structure of the egg of a tunicate *Ascidia nigra*. J Exp Zool 156:253–268

Ursprung H, Schabtach E (1965) Fertilization in tunicates: loss of the paternal mitochondrion prior to sperm entry. J Exp Zool 159:379–384

Vacquier VD, Moy GW (1977) Isolation of bindin: the protein responsible for adhesion of sperm to sea urchin. Proc Natl Acad Sci USA 74(6):2456–2460

Vacquier VD, Swanson WJ, Helberg ME (1995) What have we learned about sea urchin sperm bindin? Dev Growth Differ 37:1–10

Van Beneden E (1876) Contribution à l'histoire de la vesiculaire germinative et du premier embryonnaire. Bull Acad R Belg 42:35–97

Villa L, Patricolo E (1987) A scanning electron microscope study of *Ascidia malaca* egg (Tunicate). Changes in the cell surface morphology at fertilization. Biol Bull 173:355–366

Ward GE, Garbers DL, Vacquier VD (1985) Effects of extracellular egg factors on sperm guanylate cyclase. Science 227:768–770

Ward S, Carrell JS (1979) Fertilization and sperm competition in the nematode *Caenorhabditis elegans*. Develop Biol 73:304–321

Weber P (1987) The fine structure of the female reproductive tract of adult *Loa loa*. Int J Parasitol 17:927–934

Weygoldt P (1990) Arthropoda-Chelicerata: sperm transfer. In: Adiyodi KG, Adiyodi RG (eds) Reproductive biology of invertebrates. Fertilization, develoment, and parental care. Wiley, New York

Wharton D (1986) A functional biology of nematodes. Johns Hopkins University Press, Baltimore, pp 60–131
Wheatley D (1982) The centriole: a central enigma of cell biology. Elsevier, Amsterdam
Whitaker MJ, Steinhardt RA (1985) Ionic signaling in the sea urchin egg at fertilization. In: Metz CM, Monroy A (eds) Biology of fertilization, vol 3. Academic Press, Orlando, pp 456–485
Wilding M, Dale B (1998) Soluble extracts from ascidian spermatozoa trigger intracellular calcium release independently of the activation of the ADP ribose channel. Zygote 6:149–154
Wilding M, Kyozuka K, Russo GL, Tosti E, Dale B (1997) A soluble extract from human spermatozoa activates ascidian oocytes. Dev Growth Differ 39(3):329–336
Witalínski W (1985) Structure of the spermatophore in the mite *Trombidium holosericeum* (Acari, Trombidiformes). Acarologia 26:289–294
Woollacott RM, Porter ME (1977) A synchronized multicellular movement initiated by light and mediated by microfilaments. Dev Biol 61:41–57
Wright EJ, Sommerville RI (1977) Movement of a non-flagellate spermatozoon: a study of the male gamete of *Nematospiroides dubius* (Nematoda). Int J Parasit 7:355–359
Wright EJ, Sommerville RI (1985a) Structure and development of the spermatozoon of the parasitic nematode, *Nematospiroides dubius*. Parasitology 90:179–192
Wright EJ, Sommerville RI (1985b) Spermatogenesis in a nematode *Nippostrongylus brasiliensis*. Int J Parasit 15:283–299
Wright SJ, Schatten G (1995) Protein tyrosine phosphorylation during sea urchin fertilization: microtubule dynamics require tyrosine kinase activity. Cell Motil Cytoskel 30(2):122–135
Yoshida M, Sensui N, Inoue T, Morisawa M, Mikoshiba K (1998) Role of two series of Ca^{2+} oscillations in activation of ascidian eggs. Dev Biol 203:122–133
Young CM, Braithwaite LF (1980) Larval behavior and post-settling morphology in the ascidian, *Chelyosoma productum* Stimpson. J Exp Biol Ecol 42:157–169
Zalokar M (1974) Effect of colchicine and cytochalasin B on ooplasmic segregation of ascidian eggs. Rouxs Arch Dev Biol 175:243–248
Zalokar M, Sardet C (1984) Tracing of cell lineage in embryonic development in *Phallusia mammillata* (Ascidia) by vital staining of mitochondria. Dev Biol 102:195–205
Zeh DW, Zeh JA, Smith RL (1989) Ovipositors, amnions and eggshell architectue in the diversification of terrestrial arthropods. Q Rev Biol 64:147–168

CHAPTER 3

Fertilization in Fishes

Takashi Iwamatsu[1]

1 Introduction

In teleosts, usually females have separated ovaries, but in some fishes the right and left ovaries are so completely fused as to give the appearance of a single organ. The ovary connects to the body wall by a short oviduct and opens posteriorly to the anus. Eggs ovulated into ovarian cavities are spawned with the ovarian fluid out of the genital pore through the short oviduct during copulation. Males with internal fertilization have the anal fin modified into a gonopodium (genitalium) to introduce spermatozoa into the female genital tract; they do not have a penis. However, such a copulation organ is not usually found in fishes with external fertilization.

There are several physiological types of teleostean eggs (Yamamoto 1958) as well as spermatozoa. In internally fertilizing fishes, a spermatozoon is motile in the ovarian fluid and enters the ooplasm of an intrafollicular egg and induces egg activation (cortical reaction) in the ovary. Like these internally fertilized eggs, externally fertilized eggs of freshwater teleost fishes, such as the medaka, *Oryzias*, and the stickleback, *Pungitius*, do not auto-activate in ovarian fluid, isotonic saline or tap water unless they are penetrated by a spermatozoon (*Oryzias* type). In the process of fish evolution, eggs of the *Oryzias* type have appeared to conserve the ability to initiate development upon sperm penetration in ovarian fluid, isotonic saline or tap water, but are not activated by immersion only. Eggs of the seawater teleost fish *Alcichthys* can undergo neither sperm penetration nor activation in ovarian fluid or hypotonic saline, although they come into contact with spermatozoa before spawning; they can undergo activation by sperm penetration only after exposure to seawater following spawning (*Alcichthys* type). In contrast, salmonid eggs can incorporate a spermatozoon in isotonic saline, but a cortical reaction does not result under these conditions. They can undergo activation without sperm penetration if they are simply immersed in hypotonic saline or tap water (*Salmon* type). Furthermore, eggs of freshwater teleost fishes such as *Carassius auratus*, *Gnathopogon elongatus elongatus*, *Plecoglossus altivelis*, and *Tribolodon hakonensis* are activated spontaneously by exposure to tap water

[1] Department of Biology, Aichi University of Education, Igaya-cho, Kariya 448–8542, Japan

Juan J. Tarín and Antonio Cano (Eds.)
Fertilization in Protozoa and Metazoan Animals

at spawning, and also by artificial immersion in isotonic and hypotonic saline without sperm penetration (*Carassius* type). In eggs of *Salmon* and *Carassius*, the function of activating the egg shifts from the spermatozoon to the aqueous medium, where the gametes encounter each other in nature. It is very interesting to elucidate the mechanisms of diversity in various physiological types of gametes that allow them to adapt to environmental changes in the process of evolution.

This review focuses on external fertilization in teleosts. Our knowledge of the morphological and physiological events during external fertilization of fishes has gradually accumulated during the last half century. Recently, oocyte maturation and fertilization have become implicated as steps in the cycle of cell division. Investigations on fertilization have been focused on clarifying the mechanisms of cell division. From the viewpoint of understanding fertilization as the transitional step to the initiation of the somatic cell cycle, fertilization may be tentatively divided into seven events: (1) the functional arrangement of the cell structure for embryonic development, (2) the acquisition of developmental capacity, (3) the interaction and membrane fusion of gametes, (4) the response of the egg to membrane fusion, (5) polyspermy-preventing mechanisms, (6) formation of pronuclei, and (7) migration and karyogamy of pronuclei. During these events, the structure and distribution of egg components responsible for later development are fixed. However, we still do not understand fertilization, particularly the internal fertilization of teleosts, although several excellent reviews of fish fertilization have been published (Yamamoto 1961; Nakano 1969; Laale 1980; Gilkey 1981; Hart 1990).

2
Structural Preparation of Oocyte and Spermatozoon for Fertilization

2.1
Egg

To accomplish successful fertilization, the fundamental structures of mature oocytes are constructed during oogenesis. A hypothesis has been offered that egg polarity may depend on qualitative differences in mRNA at the animal and vegetal poles of the almost spherical previtellogenetic oocytes (Monroy 1985). In very young oocytes of most teleosts, the RNA-containing yolk nucleus appears in a restricted cytoplasmic region adjacent to the nuclear envelope (Guraya 1986). Although the function of the yolk nucleus is still unclear, it gives rise to asymmetry of the spherical egg. In *Oryzias* oocytes, the side containing the yolk nucleus may differentiate into the vegetal pole of the oocyte by interaction with the follicle cells (Iwamatsu and Nakashima 1996; Fig. 1). The animal pole is secondarily determined by oocyte rotation around the axis connecting

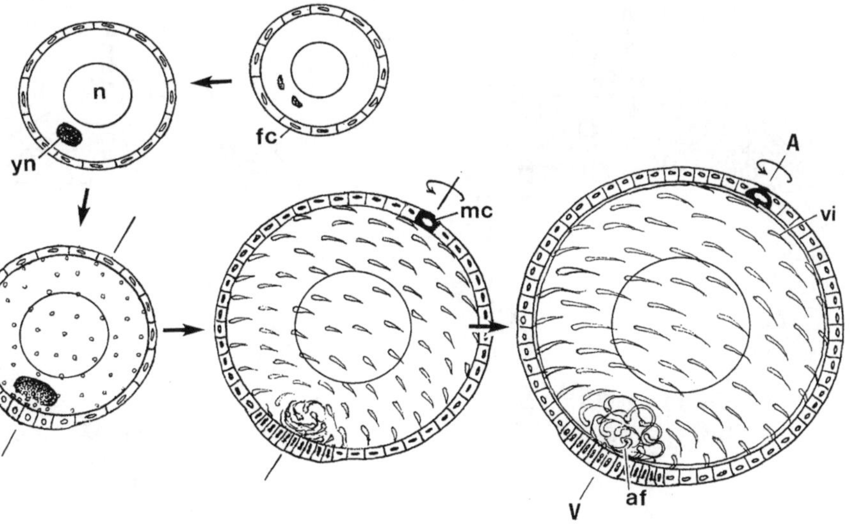

Fig. 1. Formation process of oocyte axis in previtellogenetic stage of oogenesis. Shortly after oogenesis begins, the yolk nucleus is recognized in small *Oryzias* oocytes. The follicle cell (*fc*) adjacent to yolk nucleus (*yn*) becomes somewhat columnar in morphology, fixing vegetal pole (*V*) with attaching filaments (*af*) of oocyte. Short villi-like filaments (*vi*) on chorion and follicle cells provide morphological evidence for rotation of immature oocytes. The micropylar cell (*mc*) that forms micropyle appears from a granulosa cell only at the animal pole (*A*) opposite vegetal pole

the centers of the oocyte nucleus and the yolk nucleus before the initiation of vitellogenesis (Iwamatsu et al. 1993b; Iwamatsu 1994). A micropyle is formed by a micropylar cell differentiated from a granulosa cell at the animal pole. Thus, the micropyle is located at the side opposite to the vegetal pole before the initiation of vitellogenesis (Nakashima and Iwamatsu 1989). The germinal vesicle migrates into the vicinity of the micropyle as yolk formation progresses. However, the mechanisms for nuclear translocation and for determination of the position of the germinal vesicle are still unknown.

2.1.1 Chorion

The egg envelope (*egg shell, egg membrane, vitelline membrane*, and *zona radiata*) surrounding the egg of teleost fishes is commonly called the chorion. The chorion with its variously specialized patterns and filamentous elaborations plays a key role in sperm–egg recognition, determination of the sperm entry point near the egg nucleus, the block to polyspermy, and protection of the developing embryo from various mechanical shocks, bacteria, and other factors (Laale 1980; Folz 1995). During oogenesis, the chorion displays zona

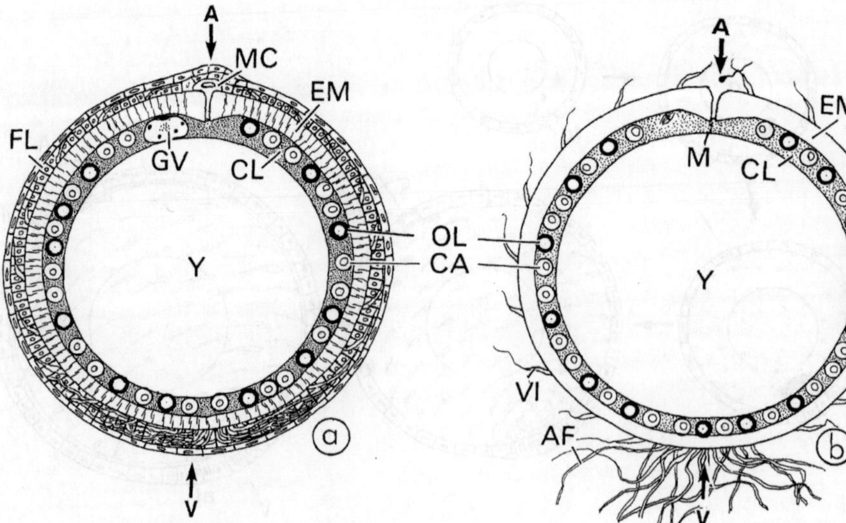

Fig. 2. Immature (**a**) and mature (**b**) *Oryzias* oocytes. The germinal vesicle (*GV*) is located in cortical cytoplasm (*CL*) at animal pole region (*arrow A*) of immature oocyte. The micropyle (*M*) in the chorion (*EM*) forms at the animal pole by a micropylar cell (*MC*) which differentiates from granulosa cells of follicle layer (*FL*). Cortical alveoli (*CA*) and oil droplets (*OL*) are embedded in cortical cytoplasm. Bulk of oocyte is occupied by a membrane-bounded yolk mass (*Y*). Long attaching filaments (*AF*) appear on chorion at vegetal pole region (*arrow V*) by ovulation, while nonattaching filaments (*VI*) are seen on whole chorion surface except for vegetal pole region

radiata by a radial insert of numerous long, slender, straight microvilli (cytoplasmic protrusions) from the vitelline surface and granulosa cells (Fig. 2). These microvilli withdraw during the oocyte maturation period. The matrix and the jelly coat surrounding the chorion vary in different species of animals and generally play an important role in assisting spermatozoa to detect the egg. There are excellent reviews on the chorion of teleost fishes (Busson-Mabillot 1973; Guraya 1978; Dumont and Brummett 1980, 1985; Laale 1980; Riehl 1991).

Morphology. The chorion is a thick acellular envelope composed mainly of proteins and glycoproteins, and its structural complexity and macromolecular composition differ from species to species. The morphology of the chorion surface in most teleosts is constructed in order to guide spermatozoa toward the micropyle (Fig. 3; Riehl 1978; *Barbus*, Amanze and Iyengar 1990; *Sturisoma*, Riehl and Patzner 1991; *Luciocephalus*, Riehl and Kokoscha 1993). It is generally thicker at the animal pole side than at the vegetal pole side. Morphological studies reveal that the chorion is a complex of several concentric lamellae

Fig. 3. Chorion surface around micropyle with grooves arranged to guide spermatozoa toward micropyle. A–C Egg of *Luciocephalus* sp.; D and E egg of *Sturisoma aureum*. A and D *Bar* 500 μm; B 10 μm; C 100 μm; E 50 μm. *Arrow* indicates micropyle. (Photographs by Dr. R. Riehl)

with different electron densities (Arndt 1960; Anderson 1967; Cherr and Clark 1982; Kudo 1982a; Hosokawa 1985). With the exception of the salmonid chorion, which is monolaminar (Brivio et al. 1991), this structure is normally multilayered. Both the thickness and number of layers vary in different species which spawn eggs in various places or bear the embryos. Despite these differences, structurally all chorions consist of a wave-shaped fibrillar component embedded in an amorphous matrix (for a review see Guraya 1986). Three main layers (ZI, ZII, ZIII) of the chorion are observed in the medaka *Oryzias latipes* (Tesoriero 1977a), *Carassius auratus* (Cotelli et al. 1988), and in the pipefish *Syngnathus fuscus* (Anderson 1967) and *S. scovelli* (Begovac and Wallace 1989). In contrast, nine layers are distinguishable in the killifish *Fundulus heteroclitus* (Dumont and Brummett 1980). Thus, the chorion is composed of the inner layers (zona radiata interna, ZI) and the outer layers or the outermost single layer (zona radiata externa, ZO). In most species, the outermost layer usually is a thin structure that serves as protection against bacteria and adhesiveness (Busson-Mabillot 1977; Ohta 1984; Koya et al. 1995). In the common carp egg, a bactericidal substance is detectable in this layer (Kudo and Inoue 1986, 1989, 1991).

In the chorions of viviparous teleosts with intraluminal and intrafollicular gestation, reduction in thickness is common. The reduced chorion thickness is correlated with the mode of reproduction. For example, the chorions of viviparous goodeids vary from only 0.5 µm (*Xenoophorus captivus*) to 1.5 µm (*Ameca splendens*) with short filaments, and are composed of a homogeneous electron-dense zona radiata (interna and externa) perforated mainly by oocyte microvilli (Riehl and Greven 1993). On the external surface of the outermost layer of the chorion in viviparous teleosts, such as *Ameca* (Riehl and Greven 1990), *Serrasalmus* (Wirz-Hlavacek and Riehl 1990), and *Chiondraco* (Riehl 1993), extremely diluted or adhesive structures can be found. These adhesive structures may be vestiges of structural characteristics of the chorion in oviparous fishes.

Chemical Components. The synthesis and assembly of chorion components are strictly related to oocyte development (Wourms 1976; Dumont and Brummett 1980; Selman et al. 1993). Autoradiographic techniques suggest that the oocyte itself is involved in choriogenesis (Tesoriero 1978). More recent data also implicate either the whole follicle (Begovac and Wallace 1989) or extra-ovarian organs such as the liver in *Oryzias latipes* (Hamazaki et al. 1989), *Scophthalms maximus* and *Gadus morhua* (Hyllner et al. 1991; Oppen-Berntsen et al. 1992a,b), and *Sparus aurata* and *Dicentrarchus labrax* (Hyllner et al. 1995).

The inner layer of the chorion of fishes has been considered as a part of the primary egg envelope. The precursors of major components of the inner layer of the chorion are derived from the liver of spawning females. The intrahepatic synthesis of precursor proteins, termed spawning female-specific (SF) substances (Hamazaki et al. 1987a,b) occurs in response to the action of oestro-

Table 1. Chorion proteins (molecular weight in kDa) of fishes

Species	Major protein	Minor protein
Oncorhynchus mykiss[a]	110, 64, 56, 50	
Oncorhynchus mykiss[b]	129, 62, 54, 47	
Oryzias latipes[c]	83, 78, 51	150
Oryzias latipes[d]	76~74, 49	
Oryzias latipes[e]	77~73, 49	150
Gadus morhua[f]	78~74, 54, 47	
Salmo salar[g]	60, 55, 50	
Dicentrarchus labrax[h]	170, 47	55~54
Branchydanio rerio[i]	116, 97, 50, 43	

[a] Iuchi et al. (1991).
[b] Brivio et al. (1991).
[c] Masuda et al. (1991).
[d] Murata et al. (1993).
[e] Iwamatsu et al. (1995).
[f] Oppen-Berntsen et al. (1990, 1992b).
[g] Oppen-Berntsen et al. (1994).
[h] Scapigliati et al. (1994).
[i] Bonsignorio et al. (1996).

gen (Murata et al. 1994; Yamagami et al. 1994) in *Oryzias latipes*. Some proteins similar to the chorion proteins with respect to molecular size and immunogenicity have also been found in the blood plasma of spawning and estradiol (E_2)-treated females of the Atlantic salmon *Salmo salar*, the cod *Gadus morhua*, the Sakhalin teimen *Hucho perryi*, and the rainbow trout *Oncorhynchus mykiss* (Hyllner et al. 1991; Hyllner and Haux 1992; Oppen-Berntsen et al. 1992b, 1994; Cotelli et al. 1993; Shimizu et al. 1998a,b). In *Oryzias latipes*, high and low molecular weight spawning female-specific (H-SF and L-SF) substances are liver-derived putative precursors of the inner layer (ZI) of the chorion (Hamazaki et al. 1989; Murata et al. 1993, 1995). The inner layer of the chorion of *Oryzias latipes* is separable into three major subunits: ZI-1, ZI-2 (74–76 kDa), and ZI-3 (49 kDa) (Hamazaki et al. 1987b; Murata et al. 1991, 1993). The SDS-PAGE patterns of the proteins and glycoproteins of the chorion have been reported for some fishes (Table 1). The inner layers are constructed of three or four major subunit proteins (Begovac and Wallace 1989; Oppen-Berntsen et al. 1990; Brivio et al. 1991; Hyllner et al. 1991; Scapigliati et al. 1994). The outermost layer of the chorion in *Oryzias latipes* is chemically similar to tubulin (Hart et al. 1984). Thus, the glycoproteins (ZI-1, ZI-2, and ZI-3) of the inner layers of the *Oryzias* chorion are produced as two female-specific (SF) substances by the liver and carried into the ovary via the bloodstream during vitellogenesis (Hamazaki et al. 1984, 1985, 1987a,b, 1989; Murata et al. 1991; Yamagami et al. 1992). These substances may be modified by structural components derived from both the oocyte itself and the follicle cells (*Oryzias*: Tesoriero 1978; *Pomatoschistus*: Riehl 1984).

2.1.2 Micropyle

The micropyle(s) is formed in the egg chorion during oogenesis. In most teleost fishes that have been studied, there is a single micropyle at the animal pole of the chorion. Two micropyles in the animal pole region of an egg are very rarely found in eggs of *Oryzias latipes*. In sturgeon (Ginsburg 1972; Cherr and Clark 1982) and paddlefish (Linhart and Kudo 1997) eggs, there are several micropyles, also restricted to the animal pole region.

The morphology of the micropyle in teleosts with external fertilization, but not those with internal fertilization, has been extensively investigated by light and electron microscopy (Riehl and Götting 1974, 1975; Szöllösi and Billard 1974; Riehl 1977, 1978, 1980; Riehl and Schulte 1977, 1978). The micropyle consists of a vestibule and a micropylar canal, which is generally funnel-shaped (Ginsburg 1963; Kudo 1980, 1982b; Laale 1980; Kobayashi and Yamamoto 1981). A single micropyle is always located in the vicinity of the egg nucleus at the animal pole. The position and the structure of the micropyle are determined by a single large, specialized follicle (granulosa) cell, the so-called micropylar cell (Eigenmann 1890 quoted in Laale 1980) or plug cell (Riehl 1978), which characteristically contains an increased amount of rough endoplasmic reticulum, Golgi complexes, microtubules, and tonofilaments around the flattened nucleus (Ohta and Takano 1982; Ohta and Teranishi 1982; Takano and Ohta 1982; Kobayashi and Yamamoto 1985; Nakashima and Iwamatsu 1989). The fine structural changes have been reported in the micropylar cell during oogenesis (pond smelt: Takano and Ohta 1982; medaka: Nakashima and Iwamatsu 1989) and during oocyte maturation (chum salmon: Kobayashi and Yamamoto 1985; medaka: Nakashima and Iwamatsu 1994). The structure of the micropylar vestibule and micropylar canal not only depends on the formation pattern and the thickness of the chorion but also reflects the shape of the micropylar cell associated with the arrangement of its cytoskeleton. Actually, in mature oocytes the outermost layer of the chorion surrounding the micropylar opening and the shape and size of the micropylar opening and canal are specialized among species (Riehl and Götting 1974, 1975; Szöllösi and Billard 1974; Kuchow and Scott 1977; Osanai 1977; Riehl 1977, 1978, 1980; Riehl and Schulte 1977; Stehr and Hawkes 1979; Dumont and Brummett 1980; Hart and Donovan 1983; Groot and Alderdice 1985). Several different types of micropyles have been described by Ginsburg (1972), Riehl and Götting (1974), Riehl and Schulte (1977), and Riehl (1991, 1999).

Yaron (1971) suggested that the micropylar cell dissolves or digests a channel through the fully formed chorion to form the micropyle. This appears to be supported by data on the blue gourami, *Trichogaster trichopterus* (Jackson et al. 1994). In the chum salmon (Kobayashi and Yamamoto 1985), the micropylar canal is formed by deposition of the chorionic inner layer material around the cytoplasmic process of the micropylar cell, which is intimately associated with the ooplasm. In *Oryzias latipes*, electron microscopic observations suggest that the broad cytoplasmic process of the micropylar cell attaches to the oocyte surface during or just after formation of the outermost layer of the chorion, since this attachment is detected only after formation of the outermost and inner layers of the chorion (Nakashima and Iwamatsu 1989). Similar observations have been reported for the pond smelt (Takano and Ohta 1982). The huge cytoplasmic process of the micropylar cell mechanically pushes through the chorion to come into contact with the oocyte surface as microfilaments and microtubules in the process grow. In *Oryzias* oocytes, evidence of this process is observed in the structural distortion of the inner layers of the chorion below the ridged outermost layer lining the wall of the micropylar canal (Nakashima and Iwamatsu 1989). However, the interaction of the oocyte and the micropylar cell during formation of the micropyle is still unclear.

2.1.3
Cortical Layer

The morphology of eggs of the loach, the zebrafish, and the common carp is quite different from that of the *Oryzias* egg in which the cortical cytoplasm is separated from the central yolk mass by a yolk membrane. The cortex at the animal pole region is depressed by an eminence of the chorion toward the vitelline surface caused by the protrusion of the micropylar cell toward the oocyte surface. The surface specialization of the plasma membrane of mature fish eggs shows a distinct polarity along the animal–vegetal axis (Hart and Donovan 1983; Hart et al. 1992). The microvilli on the oolemma of unfertilized eggs differ slightly in density between the animal and vegetal pole areas (Iwamatsu and Keino 1978). The polarity in cortical distribution establishes a distinctive gradient in the structural organization of the egg cortex from the site of sperm entry at the animal pole to the vegetal pole of *Brachydanio* eggs (Hart and Donovan 1983). Large cortical alveoli (10–40 μm in diameter) are absent from the cytoplasm around the micropyle in *Oryzias* eggs (Yamamoto 1944a). Within an approximate radius of 50 μm of the sperm entry site facing the inner end of micropyle, small alveoli (granules or vesicles: 2.7–2.8 μm in *Brachydanio*, Hart and Donovan 1983; 2–10 μm in *Oryzias*, Iwamatsu et al. 1991) are observable with the electron microscope. The cortical cytoplasm has a similar distribution in *Fundulus* eggs (Brummett and Dumont 1979) and *Oncorhynchus* eggs (Kobayashi and Yamamoto 1981).

In addition, bundles of cortical microfilaments are observed in *Brachydanio* eggs, with the meshwork being thickest at the sperm entry site of the cortex (Hart et al. 1992). A Ca^{2+}/Mg^{2+}-dependent actomyosin-like ATPase is detected at the egg plasma membrane in *Epiplatys* (Jørgensen 1972). Actin with a molecular weight of 43 kDa is a major component of the cortex (Ivanenkov et al. 1990; Hart et al. 1992; Becker and Hart 1993, 1994, 1996). F-actin is also distributed in the cortex and subcortex (Beams et al. 1985; Wolenski and Hart 1987; Chang 1991) of the zebrafish egg. Recently, Hart and Fluck (1995) reported that the distribution of F-actin is restricted to the plasma membrane, and that the nonfilamentous actin localizes to the cytoplasm surrounding the cortical alveoli and other organelles of the cortex. Becker and Hart (1996) also found that nonmuscle myosin-II (215–225 kDa) colocalizes with nonfilamentous actin at a deeper region to filamentous actin at the plasma membrane where it contributes to the submembranous cytoskeleton in the zebrafish egg. In this connection, myosin-II appears to be richly distributed over the surface of cortical alveoli. The cortical distribution of F-actin, myosin, and spectrin is important in maintaining the structural integrity of the plasma membrane (Wolenski and Hart 1988a).

In the ooplasm prior to the initiation of vitellogenesis, the cortical alveoli appear as small vesicles, the so-called yolk vesicles (*cortical vacuoles, vacuolar yolk, vesicular yolk, endogener Dotter, carbohydrate yolk*, and *intravesicular yolk*) (K. Yamamoto 1956; Nosek 1984; Guraya 1986; Masuda et al. 1986; Selman et al. 1988) which do not contain yolk per se (Wallace and Selman 1981; Selman et al. 1986). They move throughout the whole cortical layer except for the sperm entry site as yolk formation proceeds. Electron microscopic studies have demonstrated that the alveolar contents consist of a colloidal fluid and a spherical body in *Oryzias* (Fig. 4), and a dense body or a fine filamentous component in *Syngnathus* (Anderson 1968). The microfilamentous cortex in unfertilized loach eggs binds with cortical alveoli and the plasma membrane and prevents direct contact between cortical alveoli near the sites of fusion (Ivanenkov et al. 1990).

The alveolar contents were demonstrated early to contain polysaccharides that are perhaps combined with a protein in the eggs of the stickleback *Pungitius pungitius* (Kusa 1953b), *Gasterosteus aculeatus* (Thomopoulos 1953), the medaka *Oryzias latipes* (Aketa 1954), the herring *Clupea pallasii* (K. Yamamoto 1955), and sturgeons Acipenseridae (Dettlaff 1957). Other early studies also reported that cortical alveoli in various teleostean eggs contain polysaccharides (Devillers et al. 1953; Kusa 1954, 1956; Nakano 1956; K. Yamamoto 1956; Sakun 1960). Judging from the analysis of the carbohydrate components of the cortical alveoli and the perivitelline fluid by paper chromatography, Ohtsuka (1958) concluded that in *Oryzias* eggs, the polysaccharide in the perivitelline fluid originates from the contents of the cortical alveoli that are secreted at the time of egg activation because it contains such monosaccharides as galactose, mannose, and a sugar, presumed to be methylpentose, with the Gal predominating. Later, polysialoglycoprotein was found to be a major glycoprotein component of cortical alveoli of the rainbow trout egg (Inoue and Iwasaki 1978). A novel type of sialoglycoprotein (hyosophorin) with molecular mass ranging from 15 to 100 kDa has also been found to be a major component of cortical alveoli in unfertilized *Oryzias* eggs (Kitajima et al. 1989).

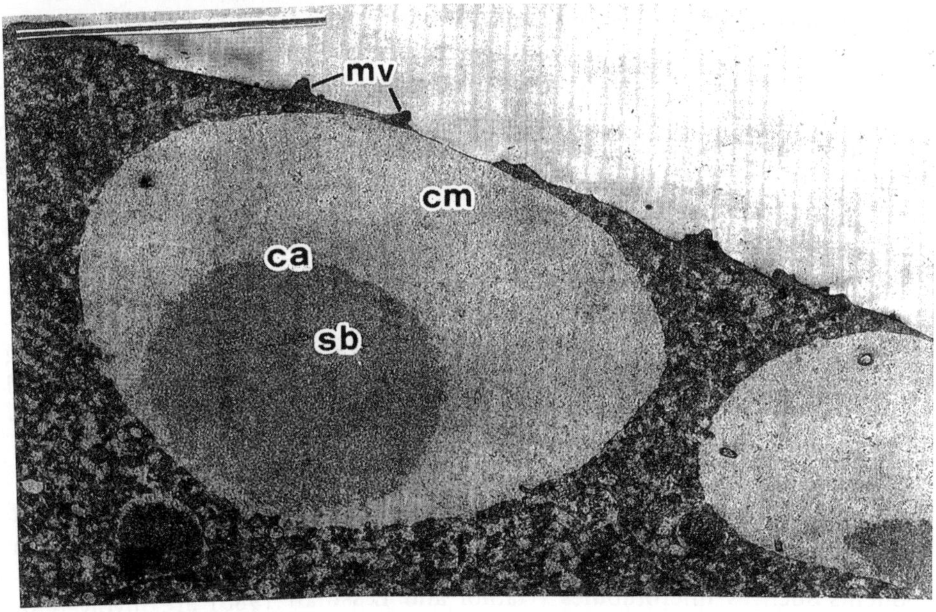

Fig. 4. Cortical alveoli in egg cortex of *Oryzias latipes*. *ca* cortical alveolus; *cm* colloidal material; *mv* microvilli; *sb* spherical body. *Bar* 10 µm. (Iwamatsu and Ohta, unpubl.)

2.1.4 Sperm Entry Site

The site on the egg surface where sperm penetration occurs is restricted by the position of the micropyle, through which the egg and spermatozoon interact, as determined by ultrastructural investigations (Brummett and Dumont 1979; Kudo 1980, 1982b; Iwamatsu and Ohta 1981; Kobayashi and Yamamoto 1981; Hart and Donovan 1983; Kessel et al. 1983; Ohta and Iwamatsu 1983; Ohta 1985; Wolenski and Hart 1988a,b). The "sperm entry site" at which attachment and fusion of the egg with the spermatozoon occurs is morphologically identifiable as an area with elongated microvilli in the central part of a critical depression facing the inner aperture of the micropylar canal in *Brachydanio* (Hart and Donovan 1983). The plasma membrane of the fertilization locus is generally specialized in *Fundulus* (Brummett and Dumont 1979), *Cyprinus* (Kudo 1980; Linhart et al. 1995), *Brachydanio* (Hart and Donovan 1983), and *Tribolodon* (Kudo 1988) eggs. However, spermatozoa can enter all over the egg surface, resulting in polyspermic fertilization when the chorion is removed (Kanoh and Yamamoto 1957; Yanagimachi 1957a; Yamamoto 1958; Gamo et al. 1960; Sakai 1961). Therefore, the significance of the morphological specializations at the sperm entry site on the egg surface is still unclear. Although sperm–egg fusion can occur anywhere on the egg surface, the frequency of sperm entry varies in different regions. Spermatozoa are more frequently incorporated in the animal hemisphere than in the vegetal hemisphere of the dechorionated *Oryzias* egg (Iwamatsu et al. 1992b). Nevertheless, it is not known whether or not these significant differences in sperm binding and incorporation between the animal and vegetal hemispheres are due to different distributions of receptor proteins (sea urchin egg: Aketa et al. 1972; Tsuzuki et al. 1977; Glabe and Vacquier 1978) on the vitelline surface.

2.1.5 Nucleus

Oogonia proliferate and differentiate into oocytes that are arrested for a long period at prophase of the first meiosis. During the hormonally regulated growth period, the nucleus (germinal vesicle) as well as the cytoplasm accumulate actively synthesized materials.

The germinal vesicle migrates to the animal pole area, where the micropyle is located, when the oocytes proceed to mature due to the action of maturation-inducing steroid (MIS: 17α, 20β-dihydroxy-4-pregnen-3-one in salmonids; Young et al. 1983). It has been reported that cytoskeletal components such as microtubules (Habibi and Lessman 1986) are involved in migration of the germinal vesicle which precedes its breakdown (Lessman and Kavumpurath 1984) in goldfish oocytes. Following the action of MIS, the early response of oocytes increases in free calcium levels and decreases in intracellular cAMP levels, followed by formation of maturation (metaphase)-promoting factor (MPF). The envelope of the germinal vesicle then breaks down due to phosphorylation of lamins and subsequent disassembly of the lamin-complex network by MPF (Ward and Kirschner 1990) and possibly other kinases (cf. Liu et al. 1998). Germinal vesicle material disperses and intermingles with the surrounding cytoplasm. The process from the end of the germinal vesicle stage (the first meiotic prophase) to metaphase of the second meiotic division (MII) proceeds within the ovarian follicle. Ovulated eggs are arrested at MII in the teleost species examined so far. In just ovulated *Oryzias* eggs, the MII spindle is located about 150 μm from the micropyle.

The crucian carp (*Carassius auratus langsdorfii*) reproduces by gynogenesis (Kobayashi 1971; Ojima and Asano 1977; Nakakuni et al. 1984), which is maintained by two mechanisms: failure of the reduction of egg ploidy during meiosis and inhibition of male pronucleus formation during fertilization (Yamashita et al. 1990). The meiotic cycle progresses normally with a transient drop in MPF between meiosis I and II. However, the gynogenetic crucian carp oocyte maintains somatic triploidy by failure of the first polar body extrusion via the formation of a tripolar spindle at meiosis I. An ordinary spindle is then formed and the genomes are reduced by extrusion of the second polar body. Consequently, in the crucian carp, triploid eggs are produced. In the triploid loach *Misgurnus anguillicaudatus*, two types of small haploid and large triploid eggs are simultaneously produced. Triploid eggs are formed by a quasinormal meiosis of large hexaploid oocytes, which are generated by premeiotic endomitosis. The haploid eggs are produced from bivalents after eliminating univalents in the meiosis of small triploid oocytes (Zhang et al. 1998). Chromosome elimination during oogenesis has been reported in unisexual *Poeciliopsis* (Cimino 1972). Moreover, production of diploid eggs has also been reported in the hybrids between *Poecilia formosa* and *P. sphenops* (Schueltz and Kallman 1968), *Oryzias latipes* and *O. celebensis* (Iwamatsu et al. 1984), and *O. latipes* and *O. curvinotus* (Hamaguchi and Sakaizumi 1992; Sakaizumi et al. 1993). However, it has not been determined how these hybrid diploid eggs are formed during meiotic division of the oocyte.

2.2 Spermatozoon

To restore the diploid number of chromosomes for the species by fertilization, the spermatozoon generally has a haploid number of chromosomes. The sperm nucleus is also unique in having a poreless envelope. It is more compact than the somatic cell nucleus and contains arginine-rich histones and protamines. There is no record of natural androgenetic reproduction in which diploid or polyploid spermatozoa are produced by inhibition of meiotic division of spermatocytes during spermatogenesis, although diploid spermatozoa can be produced by artificial interspecific hybridization (Shimizu et al. 1997). Why has such an androgenetic system of reproduction not been acquired in animals during evolution?

The sperm flagellum in teleosts varies from the biflagellate type found in *Porichthys notatus* (Stanley 1965), *Lepadogaster* sp. (Mattei and Mattei 1974), *Opsanus tau* (Casas et al. 1981), *Ictalurus punctatus* (Poirier and Nicholson 1982), *Zoarces elongatus* (Koya et al. 1993b; Hara and Okiyama 1998), and *Apogon semilineatus* (Hara and Okiyama 1998) to the aflagellate type in *Gymnarchus niloticus* (Mattei et al. 1967). The eels, *Anguilla australis schmidtii, A. dieffenbachii* (Todd 1976), *Anguilla anguilla* (Gibbons et al. 1983), and *A. japonica* (Hara and Okiyama 1998), have both a pseudoflagellum and a true 9+0 flagellum developed from the distal centriole. However, spermatozoa of most teleosts have the simplest morphology among vertebrate spermatozoa (Ginsburg 1972; Grier 1976; Sakai 1976; Hara and Okiyama 1998). The absence of an acrosome in the sperm head is the common feature of teleostean spermatozoa (Fig. 5). The sperm head of most externally fertilizing species is usually spherical (Burns et al. 1995) and 2–3 μm narrower than the inner diameter of the micropylar canal, whereas in the internally fertilizing species it is elongated (Porte and Follenius 1960; Billard 1970; Grier 1973, 1975; Mattei 1991). Elongation of the sperm head may be correlated with the presence of sperm bundles required for sperm movement through the testicular ducts (Atwood and Chia 1974; Burns et al. 1995), or with sperm passage through the female genital tract (Gardiner 1978; Jamieson 1987). Exceptions to nuclear and midpiece elongation of the spermatozoon are observed in goodeids (Grier 1981) and *Labideshes sicculus* (Grier et al. 1990) in which spermatozoa are not

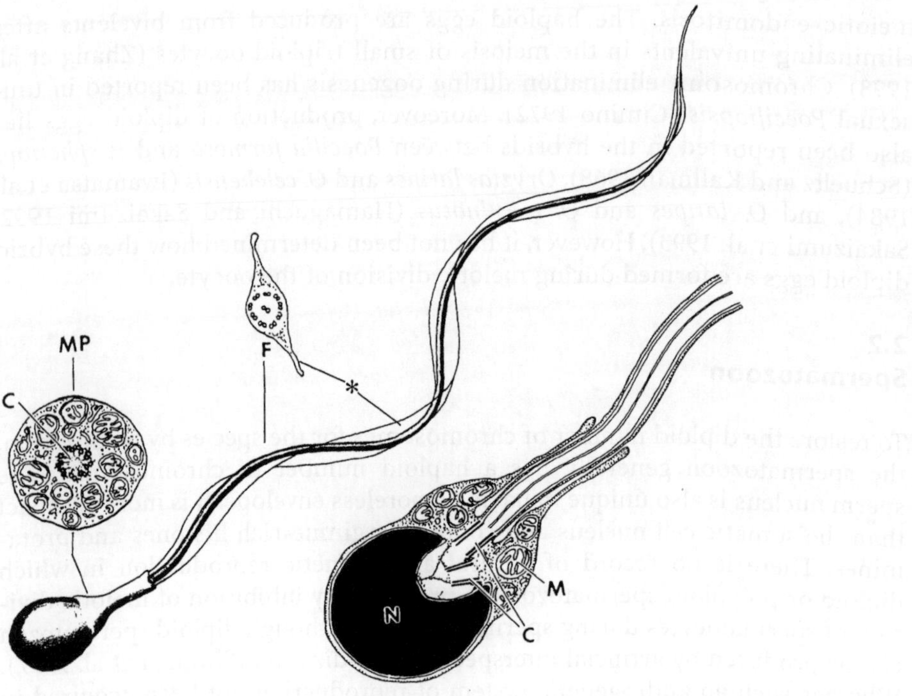

Fig. 5. Morphology of spermatozoon of *Oryzias latipes*. *C* centriole; *F* flagellum; *M* mitochondria; *MP* midpiece; *N* nucleus; * undulating membrane (lateral ridge)

packaged as spermatozeugmata or spermatophores, as is common in other internally fertilizing fishes (Grier 1981).

According to Miura et al. (1992), the acquisition of sperm motility in the salmon may be promoted by the testis-produced 17α, 20β-dihydroxy-4-pregnen-3-one, via increased pH in the testicular duct and elevated cAMP content of the spermatozoa. The duration of sperm motility varies among fish species but is generally very short in oviparous fishes (Scotte and Baynes 1980; Stoss 1983). Ishijima et al. (1998) have considered the following possible reasons for the short duration of motility (see Ginsburg 1972): (1) damage by osmotic shock, (2) rapid exhaustion of intracellular ATP, (3) decrease in intracellular pH, and (4) inadequate cAMP to sustain a high ATP concentration. The speed of movement is not very high (ca. 50–250 µm/s). *Oryzias latipes* spermatozoa with two lateral ridges (flat "wings") formed by the plasma membrane in the flagellum display a straight, forward movement along the chorion surface with an average velocity of 77–100 µm/s (Ishijima et al. 1993; Iwamatsu et al. 1993a). Similar ridges in the sperm tail have been observed in *Oligocottus maculosus* (Stanley 1969), *Neoceratodus forsteri* (Jespersen 1971), *Salmo gairdneri* (Billard 1983), and *Oreochromis niloticus* (Lou and Takahashi 1989). Swimming *Oryzias* spermatozoa rotate about their long flagellar axis with an average frequency of 8 Hz, and their flagella beat in waves with a small amplitude of 2.5 μm and a wavelength of 5–7 μm. The amplitude of flagellar beating approximates to the diameter of the micropylar canal.

The beating pattern of the sperm flagellum is closely related to the structure of the micropyle. Several types of motility patterns among fish spermatozoa have been reported by Ishijima et al. (1998). One type of motility pattern exhibited by spermatozoa of many teleost species is a fairly straight progression, in which spermatozoa rotate around the longitudinal axis, as in *A. anguilla* (Woolley 1998). This type of sperm motility is also observed in cartilaginous fishes (*Deania calcea; Chimera phantasma*) with internal fertilization, the Japanese eel (*Anguilla japonica*), and the Perciformes (*Entomacrodus striatus, Istiblennius bilitonesis, Apogon semilineatus, Arctoscopus japonica*). In the other type of motility pattern, seen in teleosts such as the sardine (*Sardinops melanostrictuss, Engraulis japonicus*), sleeper (*Eviota prasina*), and the eelpout (*Zoarces elongatus*) with biflagella (Ishijima et al. 1998), the spermatozoa swim in a circular path, the radius of curvature of which varies considerably among different species.

3
Oocyte Maturation for Acquisition of Fertilizability

Precocious fertilization of fully grown, preovulatory *Oryzias* oocytes has been used as an experimental tool for analysis of the mechanisms by which oocytes acquire fertilizability, i.e., arrange the factors necessary for fertilization. The fertilizability (developmental competence) of preovulatory fish oocytes increases gradually after germinal vesicle breakdown (GVBD), although the cortical reaction is slower than in ovulated eggs (Iwamatsu 1965a). Masui and Markert (1971) by their excellent experiments using maturing preovulatory frog oocytes were the first to identify an important MPF that appears in the cytoplasm upon treatment with MIS (progesterone) and induces GVBD. MPF was subsequently characterized biochemically by several investigators (Dunphy et al. 1988; Gautier et al. 1988; Lohka et al. 1988; Dorée et al. 1989) and found to consist of a 34 kDa (cdc2) protein and its partner cyclins (Jessus and Ozon 1993; Taieb et al. 1997; Handel and Eppig 1998).

MPF (metaphase-specific histone H1 kinase) consists of cdc2 kinase complexed with cyclin B in the goldfish, the carp (Yamashita et al. 1992a,b), and the zebrafish (Kondo et al. 1997). Cyclin B is absent in immature fish oocytes and is newly synthesized from stored (masked) mRNA just before GVBD (Hirai et al. 1992; Kajiura et al. 1993; Kondo et al. 1997), coinciding exactly with the appearance of 34 kDa active cdc2 kinase. In the goldfish, MPF activity appears upon formation of the complex between pre-existing monomeric cdc2 kinase and newly synthesized cyclin B after stimulation by MIS (17α, 20β-dihydroxy-4-pregnen-3-one) (Katsu et al. 1993). The synthesized cyclin B protein binds to the pre-existing cdc2, enabling an active cdk7–cyclin H complex to activate MPF by phosphorylating cyclin B-cdc2 on Thr 161. Therefore, if cyclin B protein is microinjected into growing oocytes, MPF activation and GVBD can be induced (Kondo et al. 1997). The activation of MPF is associated with phosphorylation of its Thr 161 and a mobility shift on SDS-PAGE from 35 to 34 kDa after binding of cdc2 to cyclin B (Yamashita et al. 1995). The mechanism of MPF activation in goldfish oocytes differs from that of *Xenopus* oocytes (Gautier and Maller 1991; Kobayashi et al. 1991) and starfish oocytes (Dorée et al. 1983). In *Xenopus* oocytes, the cdc2–cyclin B complex (pre-MPF) pre-exists with cdc2 phosphorylated on both Tyr 15 and Thr 161, thereby requiring Tyr 15 dephosphorylation catalyzed by cdc25 phosphatase for MPF activation (Yamashita et al. 1995). Inhibitory phosphorylation on Thr 14/Tyr 15 by wee1 and its dephosphorylation by cdc25 after the formation of the cdc2–cyclin B complex do not occur during oocyte maturation of the goldfish (Kondo et al. 1997). Thus, fully grown fish oocytes blocked at prophase of the first meiotic division by the absence of MPF activity are stimulated to resume meiosis by activation of MPF after stimulation by MIS secreted from the gonadotropin-stimulated follicle (granulosa) cells (Nagahama 1987). The MIS-stimulated oocytes gradually acquire fertilizability throughout the process from the GVBD stage to the second meiotic division (MII) within the follicle (Iwamatsu et al. 1976; Iwamatsu 1997). There is a gap between the times at which oocytes acquire the ability to undergo cortical reaction and the ability to incorporate the spermatozoon. Thus, the competence of the oocyte to initiate normal development appears to be related to the increase in MPF activity during the maturation period. When immature oocytes during the first meiotic division fail to undergo the global exocytosis of the cortical alveoli upon sperm penetration, the sperm nucleus transforms to metaphase chromosomes. On the other hand, if exocytosis of cortical alveoli occurs throughout the whole oocyte cortex, the first meiotic metaphase chromosomes and the sperm nucleus transform to female and male pronuclei, respectively, after extrusion of the first polar body. These pronuclei undergo karyogamy by abbreviating the second meiotic division. In this case, rapid inactivation of MPF occurs as a result of egg activation.

4 Interaction and Membrane Fusion Between Gametes

4.1 Behavior and Fertilizability of Spermatozoa

Unlike mammalian spermatozoa, spermatozoa of oviparous teleosts have no acrosome, so that acrosomal lytic enzymes such as hyaluronidase and a trypsin-like enzyme are not involved in the interaction of gametes. Also motility hyperactivation of flagellum has not been observed in fish spermatozoa, as seen in capacitated mammalian spermatozoa which exhibit large-amplitude beats (Yanagimachi 1972, 1988; Cooper 1986). Spermatozoa of most oviparous teleosts with external fertilization do not show evidence of a physiological event like sperm capacitation seen in mammals with internal fertilization, but maintain their activity in physiological or saline media longer than in water. However, spermatozoa of *Lebistes reticulatus* with internal fertilization take a long time to penetrate into dechorionated *Oryzias* eggs, while other externally fertilizing foreign spermatozoa penetrate immediately (Iwamatsu 1981). This suggests the possibility that spermatozoa of teleosts with internal fertilization acquire fertilizability in the female genital organ.

Spermatozoa of salmonids, which are immobile in seminal plasma, acquire activity not only in freshwater but also in physiological solutions, in diluted seawater, and in coelomic fluid (see Ginsburg 1972). In salmonid spermatozoa, this potential for sperm motility is acquired by the increase in pH in seminal plasma and intracellular cyclic AMP level during passage of spermatozoa from the testis to the sperm duct (Morisawa et al. 1993). In fact, spermatozoa of most teleosts are generally quiescent within the male genital organ, and change from the quiescent state to a vigorously motile state when released into such aquatic environments as freshwater or seawater at the time of external fertilization, or when ejaculated into the female genital tract for internal fertilization. According to Ishijima et al. (1998), there is no difference in the duration of motility between freshwater and seawater fishes. The major signals that initiate the motility of released or ejaculated spermatozoa are changes in the concentrations of inorganic substances such as CO_2, ions including H^+ and various metal ions (K^+, Na^+, or Ca^{2+}), and organic compounds, as well as non-chemical factors such as osmolarity and temperature (for review see Billard 1978; Morisawa and Suzuki 1980; Cosson et al. 1989; Goodall et al. 1989; Billard and Cosson 1992; Morisawa 1994). In the seminal plasma of salmonids and other freshwater species, sodium concentration is the same as or lower than that in blood plasma, but potassium concentration is much higher than that in blood plasma (Morisawa 1985). When spermatozoa of salmonid fishes are diluted in a medium containing K^+ at physiologically high concentrations, motility remains quiescent, while they move actively when diluted in K^+-free medium (Morisawa and Suzuki 1980; Morisawa et al. 1983). Placing trout spermatozoa in K^+-free medium causes a hyperpolarization of the plasma membrane induced by K^+ efflux (Tanimoto and Morisawa 1988; Gatti et al. 1990; Boitano and Omoto 1991). Hyperpolarization of the plasma membrane induces an increase in intracellular Ca^{2+} (Boitano and Omoto 1992), stimulating adenylate cyclase to synthesize cAMP, resulting in the initiation of sperm motility. An increase in intracellular pH, as well as intracellular Ca^{2+}, plays an important role in the initiation of sperm motility in marine teleosts such as puffers and flounder (Oda and Morisawa 1993). In the trout, the initiation of sperm motility is associated with Ca^{2+} influx (Cosson 1990), but not with a rise in intracellular pH (Boitano and Omoto 1991). By increasing the cAMP level, cAMP-dependent phosphorylation of motility-initiating phosphoprotein (15 kDa: Jin et al. 1994) in salmonid spermatozoa appears to occur in the basal region of the flagellum, where the cell signaling systems are located (Morisawa 1994). Thus, in the spermatozoa that are activated by extracellular signals, the intracellular cascade initiating motility is triggered by second messengers such as cyclic nucleotides and Ca^{2+}.

4.2 Sperm–Egg Interaction

The situations and strategies for external or internal fertilization vary among teleosts, and the physiological properties of their spermatozoa are different. In the elkhorn sculpin, *Alcichthys alcicornis*, spermatozoa introduced into the ovarian cavity by copulation associate with several eggs in the isotonic ovarian fluid. Consequently, spermatozoa enter the micropyle prior to spawning and are attached to the egg plasma membrane at the site of the inner opening of the micropylar canal (Munehara et al. 1989; Koya et al. 1993c). Sperm penetration into the egg cytoplasm occurs after spawning into seawater. This unusual phenomenon in which the encounter of the spermatozoon and egg is completed in the female genital tract is termed "internal gametic association" (Munehara et al. 1989, 1991) in order to distinguish it from true oviparity. In *A. alciornis*, the ovarian fluid provides a suitable environment for spermatozoa (Koya et al. 1993a).

Herring spermatozoa are almost immotile or quiescent in seawater, but they begin to move around the micropyle of ripe, unfertilized eggs (Yanagimachi and Kanoh 1953; Yanagimachi 1957a). The sperm motility initiating factor (SMIF), a 105 kDa glycosylated peptide, of the herring egg resides within the micropyle area (Pillai et al. 1993). Spermatozoa require the SMIF bound to the micropyle region of the chorion for initiation of motility, and continued contact with the factor appears to be necessary for the maintenance of motility (Yanagimachi 1957a,b; Yanagimachi et al. 1992). Diffusible sperm-activating proteins (HSAPs: 81 kDa, PI 5.1) emanating from the egg of the Pacific herring, *Clupea pallasii* (Morisawa et al. 1992; Oda et al. 1995), also may participate in maintaining motility for fertilization. The water-soluble HSAPs with a strong homology to Kazal-type trypsin inhibitors are synthesized in follicle cells, secreted, and accumulated in the chorion during oocyte development (Oda et al. 1998), as postulated in the origin of a sperm-attracting component of low molecular weight in the bitterling egg (Suzuki 1961). In several species of bitterling, *Acheilognathus lancolata, A. tabira, Rhodeus ocellatus*, and *Sarcocheilichthys variegatus*, unfertilized eggs maintain a conspicuous capacity for sperm activation and aggregation around the micropyle region of the chorion for 30 min in distilled water (Suzuki 1959b). The factor responsible for sperm activation and aggregation is not species-specific.

4.3 Sperm–Egg Membrane Fusion

The movement of intramembranous components begins following the binding of spermatozoon to a sperm receptor (recognition site) and subsequent membrane fusion, probably due to induced membrane fluidity. The dynamic change in the intercellular membrane fusion gives rise to physiological and biochemical changes in the plasma membrane and the submembranous layer of the egg, leading to intracellular membrane fusion such as exocytosis of cortical alveoli. Consequently, shortly after fertilization, the sperm incorporation and exocytosis result in forming a mosaic membrane derived from the cortical vesicles (granules) and the spermatozoon with the original plasma membrane (Schroeder 1979; Longo 1981).

In spermatozoa of marine invertebrates, the lamprey (Nicander and Sjoden 1971), amphibian (Picheral and Charbonneau 1982), and bird (Okamura and Nishiyama 1978), sperm–egg fusion initiates between the inner acrosomal membrane and egg plasma membrane (Colwin and Colwin 1961), whereas in eutherian mammals the sperm plasma membrane above the equatorial segment fuses first with the egg plasma membrane (Bedford et al. 1979; Moore and Bedford 1983) but the inner acrosomal membrane does not. Like eutherian mammalian spermatozoa, the plasma membrane of teleostean spermatozoa lacking an acrosome fuses with the egg plasma membrane at the sperm entry site. Sperm–egg fusion in teleosts (*Oryzias*, Iwamatsu and Ohta 1978; *Rhodeus*, Ohta and Iwamatsu 1983; Ohta 1985, 1991; *Fundulus*, Brummett et al. 1985; *Cyprinus*, Kudo and Sato 1985; *Brachydanio*, Wolenski and Hart 1987; Kudo et al. 1994) begins between the microvilli of the sperm entry site on the egg surface and the plasma membrane of the apical or equatorial region of the spermatozoon, and gradually progresses toward the end of the sperm tail.

Gamete fusion requires the presence of calcium ions (Yamamoto 1944a, 1954, 1961; Yanagimachi and Kanoh 1953; Ginsburg 1963; Gilkey et al. 1978; Iwamatsu et al. 1985). In calcium-free media, sperm penetration is not induced by Ca^{2+}-microinjection to increase intracellular Ca^{2+} (Iwamatsu et al. 1985). This indicates that the presence of the extracellular calcium ion is required for intercellular membrane fusion between the gametes, while the increased intracellular calcium is not. Sperm–egg fusion is separable from the fusion between the plasma membrane and cortical alveolus membrane, which is induced only by increased intracellular calcium ions. In *Oryzias latipes*, the concentration of extracellular Ca^{2+} most effective in inducing sperm penetration is higher than 0.5 mM. Similarly, the threshold concentration of extracellular calcium ions is 0.57 $mmol\,kg^{-1}$ for fertilization in *A. alcicornis*, so that fertilization is triggered by a rise in the calcium concentration of the external medium (Munehara et al. 1994). The failure of internal fertilization in ovarian fluid (0.41 $mmol\,kg^{-1}$) is caused by a deficiency of calcium ions.

Actins that distribute restrictedly in the egg plasma membrane (Hart and Fluck 1995) may not participate in sperm–egg fusion, since treatment with an inhibitor of microfilaments (cytochalasin B) does not prevent membrane fusion of the spermatozoon with the egg (Wolenski and Hart 1988a; Iwamatsu et al. 1992b). However, the presence of actin-filaments in the cortex might play a role in sperm incorporation into the deep region of the cytoplasm.

Sperm penetration is also affected by the pH of external media. In natural conditions, freshly spawned eggs were surrounded by ovarian fluid, and spermatozoa ejaculated during copulation reach the egg surface through the micropyle that is filled with ovarian fluid. In *Oryzias* eggs, fresh ovarian fluid obtained from the ovarian cavity has a pH of 8.3. This pH value coincides with the range between 7 and 8.5 which allows the highest rate of sperm penetration (Iwamatsu 1984).

The size and morphology of the spermatozoon are very similar among teleosts, so that the spermatozoon can directly touch the plasma membrane of eggs of heterologous species through the micropyle. Some early investigators succeeded in artificial hybridization among several species and obtained many adult specimens (Suzuki and Fukuda 1971). Dechorionated *Oryzias* eggs incorporate various heterologous spermatozoa and exhibit exocytosis. Therefore, the species specificity does not seem to reside distinctly in the egg plasma membrane. However, under natural conditions, membrane fusion between gametes of heterologous species is rare due to selective mating of homologous species.

5 Responses of Egg to Membrane Fusion with Spermatozoon

5.1 Change in Membrane Potential

According to Nuccitelli (1980a,b), the resting potential of the unfertilized egg of *Oryzias latipes* averages −47 mV in 10% Yamamoto's Ringer and a small depolarization of about 4 mV in 10% Yamamoto's Ringer which is triggered within 1–5 s after sperm entry into the micropyle (Nuccitelli 1977) continues for about 20 s, independent of such extracellular ions as $[Na^+]_o$, $[Ca^{2+}]_o$, and $[Cl^-]_o$. Depolarization of the egg plasma membrane is detected prior to the increase in intracellular Ca^{2+}. Exocytosis of the cortical alveoli follows this increase and occurs concomitantly with a negative-going shift of membrane potential (hyperpolarization). There are three possible mechanisms for the initiation of these dynamic cortical changes upon sperm–egg membrane fusion. One cause is the integration of the activated sperm plasma membrane into the egg plasma membrane by fusion. The open Ca channels of the sperm plasma membrane possibly induce the increase in ooplasmic Ca^{2+}. Another cause may be cytoplasmic factors derived from the spermatozoon, such as a small amount of Ca^{2+}, inositol 1,4,5-trisphosphate (IP_3), or truncated *c-kit* gene product (tr-kit), which are enough to induce the release of free Ca^{2+} from microsomal components into the restricted submembranous cytoplasm of the egg. The possibility of such diffusible activators has been discussed with regard to sea urchin eggs (Whitaker and Crossley 1989) and mouse eggs (Sette et al. 1997). The final possibility is that the binding of a spermatozoon to its receptor induces a series of reactions producing metabolites [diacylglycerol (DG) and

IP_3] of phosphatidylinositol-4,5-diphosphate (PIP_2). When unfertilized and activated *Oryzias* eggs are dechorionated and treated with wheatgerm agglutinin, these eggs sometimes fuse with each other. However, this fusion does not cause the unfertilized egg to undergo exocytosis. Unfertilized eggs may undergo cortical reaction upon fusion with another activating egg by the way of propagation of a Ca^{2+} increase.

The depolarization is replaced by a large hyperpolarizing phase (about 155 s) with an average amplitude of about 31 mV (Nuccitelli 1980b). During this hyperpolarization phase, K^+ permeability increases (Ito 1962, 1963). There are at least two possible mechanisms for generating this increase in K^+ permeability (Nuccitelli 1980a): (1) the opening of K^+ channels due to a rise in internal free $[Ca^{2+}]_i$ for about 2 min after sperm entry (Ridgway et al. 1977; Gilkey et al. 1978; Yoshimoto et al. 1986), and (2) the addition of K^+ channels from the envelopes of the cortical alveoli to the plasma membrane by fusion. Nuccitelli (1980b) proposed the second possibility based on the experimental data of Kiyohara and Ito (1968) who observed that, in centrifuged eggs with cortical alveoli accumulated on one side of the egg cortex, the hyperpolarization phase does not begin until the exocytosis of the accumulated alveoli (envelope fusion of cortical alveolus to plasma membrane) has begun. The addition of new membrane resulting from cortical alveolus fusion brings about an increase in K^+ permeability and the beginning of hyperpolarization. At present, unfortunately, there is no evidence for a high density of K^+ channels in the envelope of the cortical alveolus.

One possible explanation for the subsequent decrease in K^+ permeability at the end of the hyperpolarization phase is the removal of some K^+ channels from the plasma membrane. Electron microscopic observations on exocytosis of cortical alveoli have revealed that membrane components move endocytotically into the submembranous cortical cytoplasm (Donovan and Hart 1982, 1986; Hart and Collins 1991). Another possibility is that many newly inserted K^+ channels are closed by the alveolar contents secreted into the perivitelline space.

5.2 Fertilization Wave and Wave of Increase in Cytoplasmic Ca^{2+}

It has been postulated in early investigations that the impulse "fertilization wave" elicited by sperm attachment is propagated throughout the cortical layer by a chain reaction in fish (Yamamoto 1944a,b). It became clear that the wave of a single increase in cytoplasmic free Ca^{2+} is propagated from the point of sperm entry toward the antipode of the egg (Fig. 6), preceding the exocytosis wave. Thus, the Ca^{2+} wave as an invisible cortical reaction was considered to correspond to the fertilization wave (Gilkey et al. 1978). The increase in intracellular Ca^{2+} can also start at the animal pole region following Ca-ionophore treatment in the absence of external free Ca^{2+}. Furthermore, no distinct increase in uptake of external $^{45}Ca^{2+}$ has been measured during the increase

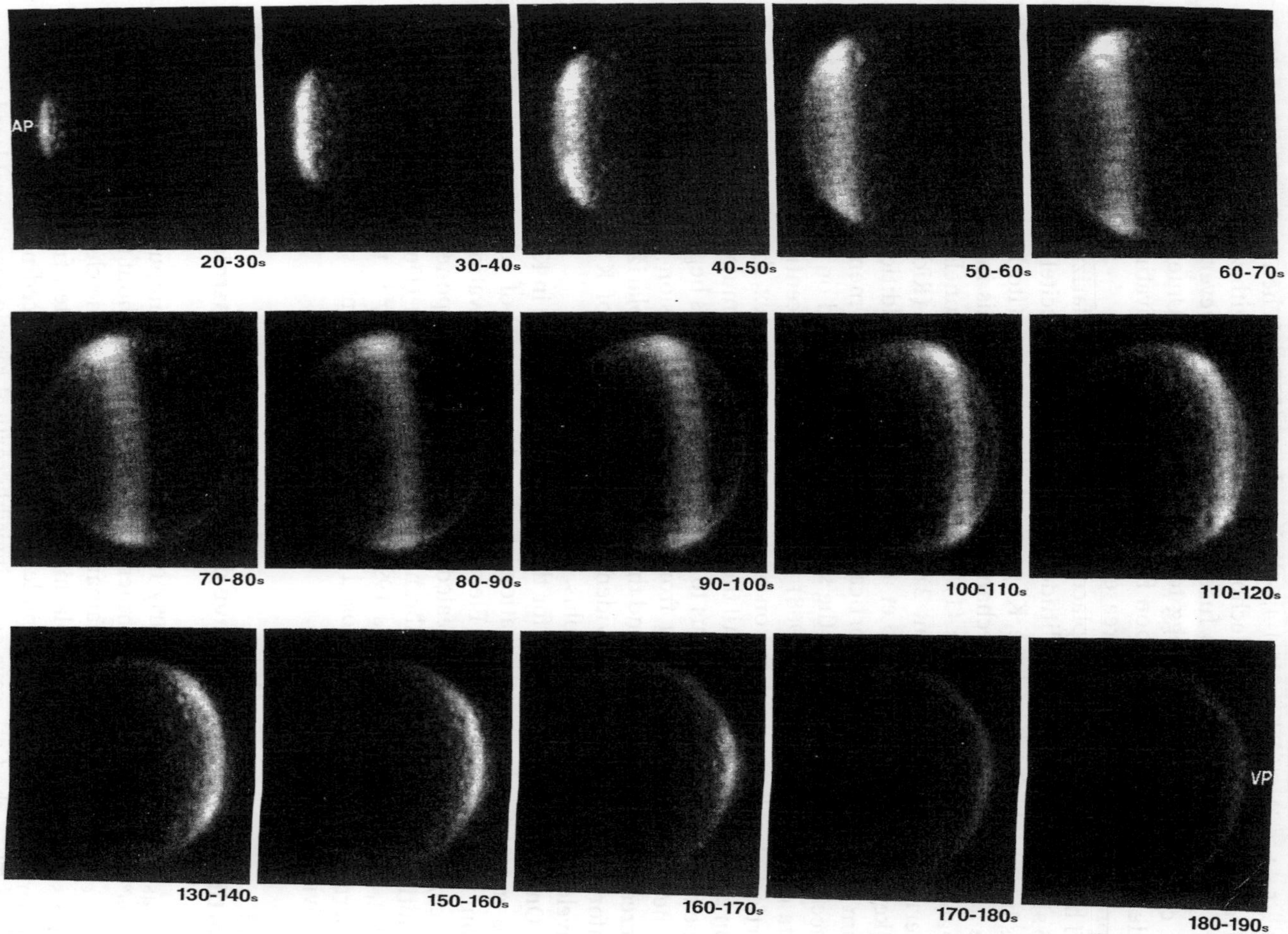

Fig. 6. A Ca^{2+} wave induced by fertilization and visualized by luminescence of aequorin previously loaded into cortical cytoplasm of a *Oryzias* egg. Photographs were taken at 10s intervals beginning 40s after insemination. Each photograph is a 5s exposure of videotape

in cytoplasmic Ca^{2+} (Iwamatsu et al. 1985), although a possible Ca^{2+} influx has been suggested by Nuccitelli (1987). Moreover, the amount of $^{45}Ca^{2+}$ previously incorporated in the microsomal fraction decreases after fertilization (Iwamatsu, unpubl. data). These facts support the idea (Gilkey et al. 1978; Jaffe 1983) that Ca^{2+} is stored in the microsomal fraction. Thus, sperm stimulation seems to induce a release of free Ca^{2+} from microsomes into the cytosol, and the wave of increase in cytoplasmic free Ca^{2+} is propagated to the whole egg surface prior to the wave of exocytosis (see review of Jaffe 1985).

5.3 Exocytosis of Cortical Alveoli

Breakdown (exocytosis) of cortical alveoli in a wave-like fashion starts at the animal pole and ends at the vegetal pole in *Oryzias* eggs (Yamamoto 1939, 1944a,b), and alveolar contents (colloidal material and spherical bodies) are secreted into the perivitelline space. The secretion of the alveolar contents is followed by the formation of the widening perivitelline space (Yamamoto 1939; Devillers et al. 1953; Kusa 1956; Rothschild 1958). Yamamoto (1944b) performed a physiological analysis of the fertilization reaction using centrifuged *Oryzias* eggs in which cortical alveoli were forced to accumulate in the centrifugal side, equatorial region or vegetal hemisphere. The eggs that had no alveoli in the animal hemisphere could be fertilized, and the cortical alveoli embedded in the cortical cytoplasm of the equatorial region or vegetal hemisphere underwent exocytosis.From this observation, it was postulated that the cause of the exocytosis of the cortical alveoli is an "impulse" provoked in the cortical layer by the spermatozoon and that the invisible impulse is propagated to the vegetal pole through the cortical layer which has been depleted of cortical alveoli. This impulse was termed the "fertilization wave" by Yamamoto (1944b).

The process of discharge of the cortical alveolar contents in teleostean eggs has been studied morphologically by Kusa (1956) with the stickleback *Pungitius pungitius* and by Yamamoto (1962) with *Oryzias latipes* using light microscopy. In the illustration of Yamamoto (1958), the cortical alveolar contents are released into the perivitelline space through a very small aperture in the cortical alveolus. Following release of the contents, the cortical alveolus membrane shrinks; the membranes of large alveoli sometimes remain as patches on the activated egg surface after forming many microvilli. According to observations with electron microscopy (*Oryzias latipes*, Iwamatsu and Ohta 1976; *Brachydanio*, Hart et al. 1977; Hart and Yu 1980; *Fundulus*, Brummett and Dumont 1981; *Oncorhynchus keta*, Kobayashi 1985), fusion between the plasma and cortical alveolus membranes forms a large aperture in the cortical alveolus. Fusion appears to occur at multiple sites between the limiting membrane of the cortical alveolus and the overlying plasma membrane (Iwamatsu and Keino 1978; Brummett and Dumont 1981; Kobayashi 1985). The formation of a large aperture opening onto the egg surface during exocytosis has been observed in eggs of the stickleback (Kusa 1956) and codfish *Gadus morrhua* (Hagström and Lönning 1968).

Exocytosis of cortical alveoli is triggered by microinjection of Ca^{2+} into the cytoplasm of *Oryzias* eggs (Gilkey 1983; Iwamatsu et al. 1985, 1988a,b; Iwamatsu and Ito 1986; Iwamatsu 1989) and loach eggs (Ivanenkov et al. 1990). Calcium chelating agents inhibit exocytosis. Thus, the increase in intracellular Ca^{2+} appears to lead to reorganization or partial disintegration of the microfilamentous cortex between the cortical alveoli and the plasma membrane. The effects of the calmodulin inhibitors W-5, W-7, and W-13 on exocytosis were examined by microinjecting the inhibitors into the cortical cytoplasm of unfertilized eggs (Iwamatsu 1992). This treatment failed to inhibit exocytosis, suggesting that a Ca-dependent protein other than calmodulin is involved in exocytosis. Ivanenkov et al. (1990) postulated that the process of cortical exocytosis is as follows: the increase in cytoplasmic Ca^{2+} might activate calcium-dependent actin-regulatory proteins resulting in the partial disintegration of the microfilamentous cortex. Consequently, the membranes of the cortical alveoli come into contact with the plasma membrane. Concomitant with this contact between cortical alveoli and the plasma membrane, the increased Ca^{2+} activates the contraction of the microfilamentous cortex and induces the membranes to fuse, with subsequent secretion of the alveolar contents. The special site of secretion of the contents seems to be the cortical alveolus surface, since isolated cortical alveoli undergo breakdown from different sides regardless of which surface is exposed to Ca^{2+} (T. Iwamatsu, unpubl. data). However, the existence of intramembranous fusion proteins in the cortical alveolus has still not been demonstrated, and the exact mechanism of intracellular fusion between the plasma membrane and the membrane of the cortical alveolus is unclear.

During exocytosis the alveolar membrane fuses with the plasma membrane and becomes part of it. After the alveolus membrane fuses with the plasma membrane and the contents are discharged, the inner surface of the alveolar membrane, which is probably masked by the alveolar contents, becomes located on the outer surface of the plasma membrane and might be activated by being unmasked. Consequently, the change in the function and temporary increase in volume of the egg plasma membrane are brought about just after exocytosis.

Kusa (1956) suggested that the cortical alveolus contents are probably forced out by regional contractions of the surface of stickleback eggs. In the process of cortical alveolar breakdown in *Oryzias* eggs, each cortical alveolus forms an empty pocket upon discharge of its contents into the perivitelline space, and the pocket then shrinks as illustrated diagramatically in Fig. 7 (Iwamatsu and Keino 1978). Numerous microvilli appear on the inner surface of the pocket. In zebrafish eggs, F-actin accumulates around the rim of the fusion pore and adjacent walls of exocytotic profiles (Hart and Becker 1992; Hart and Fluck 1995) and may be related to the stabilization of the limiting membrane between the plasma membrane and the alveolar membrane and to formation of new microvilli. The microvilli or extensive folds bend like the crest of a wave in the

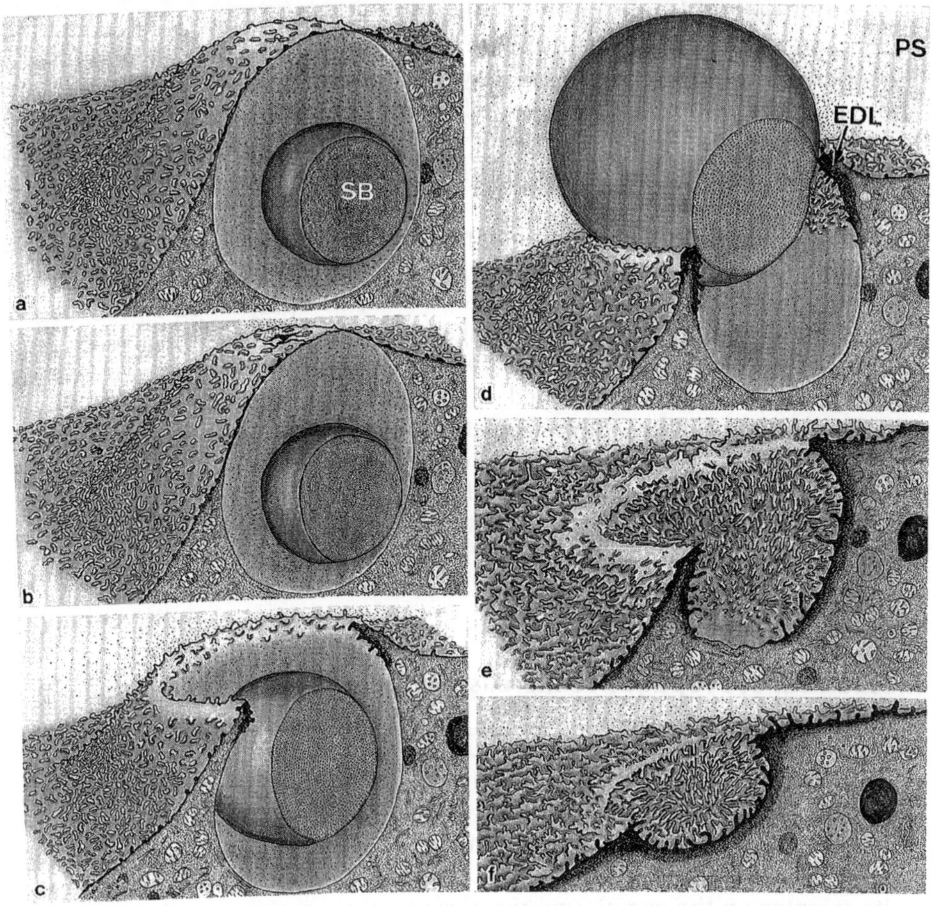

Fig. 7. Exocytosis of cortical alveolus (*CA*) in *Oryzias* egg. Exocytosis begins with formation of holes or gaps (**b**) at apical region of CA. Release of colloidal material and a swelling spherical body (*SB*, **a–d**) into perivitelline space (*PS*) follows formation of a large aperture (**c**). CA envelope shrinks and transforms into an electron-dense layer (*EDL*) with many microvilli (**c–f**). (From Iwamatsu and Keino 1978)

direction of the centrifugal side of the pocket. The shrinkage of the cortical alveolar pockets following the release of their contents into the perivitelline space is accompanied by transformation of the membrane into new microvilli and formation of an electron-dense layer just beneath the egg plasma membrane (Fig. 7). The whole plasma membrane of the fertilized egg must be increased by the addition of the membranes of the broken cortical alveoli. The volume of the vitellus itself, on the other hand, decreases as a result of the discharge of the cortical alveolar contents into the perivitelline space (Yamamoto 1940). Therefore, redundancy in the egg surface should accompany the exocytosis of cortical alveolar contents, and the tension at the egg surface should reduce transiently during the cortical reaction (Sakai 1961). The redundant membrane material may then decrease as it is used to form new microvilli and the submembranous electron-dense layer (Iwamatsu and Ohta 1976; Iwamatsu and Keino 1978) as well as by internalization of membrane through endocytosis (Donovan and Hart 1982, 1986; Hart and Collins 1991). Consequently, the tension of the egg surface recovers with the development of cytoskeletal networks.

Small cortical alveoli at the animal pole region, within 15 μm of the sperm-binding site, begin simultaneously to undergo exocytosis about 9 s (24.5 °C) after sperm attachment (Iwamatsu et al. 1991). If sperm–egg fusion begins concomitantly with the cessation of flagellar movement (4 s), the alveolar exocytosis starts after a short delay in *Oryzias*. In eggs of *Brachydanio*, cortical alveolus exocytosis is not initiated at the site of sperm binding, and occurs simultaneously over the egg surface after a delay of about 30 s (Hart and Yu 1980; Schalkoff and Hart 1986). Moreover, small cortical alveoli in the vicinity of the micropyle do not begin to break down until about 60 s (Wolenski and Hart 1987). Extremely delayed initiation of cortical alveolar exocytosis is observed in *Rhodeus* (Suzuki 1959a; Ohta et al. 1990). In this case, eggs require 20 min for the completion of exocytosis.

5.4 Mechanism of Propagative Increase in Cytoplasmic Ca^{2+} and pH

Sperm–egg interaction in the sea urchin stimulates the metabolism of polyphosphoinositides to IP_3 and DG (Ciapa and Whitaker 1986). IP_3 then stimulates the release of Ca^{2+} from intracellular stores (Clapper and Lee 1985; Oberdorf et al. 1986; Swann and Whitaker 1986). Our results suggest that the same sequence of events occurs in *Oryzias* eggs. Binding of the spermatozoon to its receptor on the egg plasma membrane causes activation of phosphodiesterase [phospholipase C (PLC)] to catalyse PIP_2 via a G-protein (Gp). PLC cleaves inositol-diphospholipids into IP_3 and DG. The release of Ca^{2+} from cytoplasmic stores is then triggered by IP_3. It has been found in the sea urchin egg that DG derived from polyphosphoinositides stimulates Na^+–H^+ exchange (Lau et al. 1986; Shen and Burgart 1986). A propagated rise in intracellular pH is also observed in the cytoplasm of activated *Oryzias* eggs (Iwamatsu 1984). In addition, propagation of the Ca^{2+} wave is slowed by the injection of buffers with pH 6.9 or below (Gilkey 1983), and progressive exocytosis is prevented by holding the egg in acidic saline (Iwamatsu 1984). The prevention of exocytosis may indicate (1) that Na^+–H^+ exchange stimulated by DG is inhibited in acidic medium, and/or (2) that the threshold for the induction of Ca^{2+} release by IP_3 rises. Increased cytoplasmic Ca^{2+} brings about the exocytosis of cortical alveoli concurrently with inactivation of MPF (histone H1 kinase) and MAP kinase.

5.5 Relationship Between Wave of Increase in Cytoplasmic Ca^{2+} and Exocytosis Wave

As described above, exocytosis is propagated from the sperm stimulation point toward the antipode of the egg. Some batches of eggs that undergo an incomplete breakdown of cortical alveoli upon sperm stimulation continue breakdown when the animal pole is pricked with a glass needle 15 μm in diameter (Yamamoto 1956, 1958). A similar phenomenon has been observed in maturing oocytes (Iwamatsu 1965a). Yamamoto (1944b) proposed that the fertilization wave or impulse is propagated with diminishing intensity and velocity. In incompletely anesthetized eggs, cortical alveoli break down in the animal hemisphere (AE) upon insemination, while those in the vegetal hemisphere (EV) remain unchanged. According to this idea, if the wave of increase in cytoplasmic Ca^{2+} corresponds to the fertilization wave, the former should diminish as it is propagated. However, the wave of increase in cytoplasmic Ca^{2+} has the same velocity in the AE area as in the EV area and is propagated without diminishing in intensity or velocity (Yoshimoto et al. 1986). This suggests that the incomplete exocytosis of cortical alveoli results from a process subsequent to the fertilization wave, rather than from a diminishing of the intensity and velocity of the fertilization wave.

In order to verify diminishment of exocytotic velocity, we re-examined the velocity in *Oryzias* eggs (Iwamatsu et al. 1992b). The velocity of the exocytosis wave does not diminish between the starting point of exocytosis and the end point in the egg cortex. When Ca^{2+} or IP_3 is microinjected into different regions of the cortical cytoplasm of unfertilized eggs, exocytosis begins earlier at the animal pole than at other regions. Since the Ca^{2+} wave is propagated from the starting point to the antipode of the egg without a delay, the propagation velocity of exocytosis seems to result from a difference in the time (seconds) required for the beginning of IP_3- or Ca^{2+}-induced exocytosis in different areas. The difference in the propagation velocity of exocytosis in AE and EV areas indicates that the exocytosis reaction begins at different intervals in response to stimulation by increased intracellular Ca^{2+} or IP_3 in different regions of the egg.

5.6 Perivitelline Space Formation and Change in Chorion

Elevation of the chorion and formation of the perivitelline space are considered to result from both a decrease in the volume of the ooplasm and an increase in the osmotic pressure of the fluid in the perivitelline space due to the colloidal contents of the cortical alveoli (cf. Yamamoto 1962). In addition, the thickness of the chorion is decreased with reconstruction of chorionic components by the exudate from the cortex. Sakai (1961) noted first that spherical bodies (cores, exudate in *Cyprinus*: Kudo 1982a) are released from the cortical alveoli of the dechorionated *Oryzias* egg that has been activated by spermatozoa. The swollen spherical bodies of cortical alveolar origin, however, persist in the perivitelline space during development of the egg (Iwamatsu and Ohta 1976). The spherical bodies eventually appear to keep the vitellus away from the chorion.

A major glycoprotein component (high molecular weight polysialoglycoprotein, H-PSGP: rainbow trout 200 kDa, Inoue and Iwasaki 1978; flounder 100–120 kDa, Seko et al. 1989; *Oryzias* 15–100 kDa, Kitajima et al. 1989) of the cortical alveolus is made up of tandem repeats of glycodecapeptides or glycotridecapeptides. It undergoes proteolytic depolymerization to the least repeating units (low molecular weight mass: rainbow trout 9 kDa, Inoue and Iwasaki 1978, Kitajima and Inoue 1988; flounder 6 kDa, Seko et al. 1989; medaka 7 kDa, Kitajima et al. 1989, Taguchi et al. 1993, 1994) by the action of a unique polysialoglycoproteinase (named hyosophorinase or PSGPase: Inoue and Inoue 1986; Kitajima and Inoue 1988; Kitajima et al. 1988a,b) at or during exocytosis upon fertilization or parthenogenetic egg activation (Inoue and Inoue 1986, 1987; Kitajima et al. 1986, 1988a,b; Inoue et al. 1987; Kitajima and Inoue 1988). N-linked glycan chains are detected among the low molecular weight glycoproteins resulting from hydrolysis with peptide:N-glycosidase (PNGase) (Seko et al. 1991), probably in the perivitelline fluid.

After fertilization or parthenogenetic activation, the chorion undergoes a calcium-dependent hardening process, which induces a tenfold increase in the mechanical strength of the chorion (Zotin 1958; Iwamatsu 1969; Davenport et al. 1981; Lönning et al. 1984). The process of hardening induced by nonphysiological concentration of calcium ions (Kanoh 1952; Kusa 1953a; Yamamoto 1957; Masuda et al. 1991) or by drying is different from that occurring at fertilization, although Ca^{2+} in the external medium is indispensable for the hardening process: during exocytosis of the cortical alveoli, $^{45}Ca^{2+}$ is rapidly absorbed from the external medium into the chorion but not into the vitellus (Iwamatsu et al. 1985). However, Ca^{2+} detectable with aequorin is not released from the cortex into the perivitelline space during exocytosis (Iwamatsu et al. 1988a). The Ca^{2+}-hardening of isolated chorions seems to be due to some hardening mechanism that functions with Ca^{2+} (Masuda et al. 1992). The hardening process of the chorion seems to progress in two steps in *Oryzias* eggs: (1) the involvement of egg exudate secreted from the cortical layer (cortical alveoli); (2) the involvement of a hardening mechanism within the chorion. The egg exudate contains a proteolytic enzyme of cortical alveoli (the optimum pH is 7: Iwamatsu et al. 1995; Shibata and Iwamatsu 1996), and the hardening mechanism (Masuda et al. 1991, 1992) includes transglutaminase (the optimum pH is 5–6: Iuchi et al. 1991). The content of γ-glutamyl-ε-lysine (γ-Glu-ε-Lys) crosslink in the chorion increases after hardening (Iuchi et al. 1996). Thus, fertilized eggs proceed with the chorion hardening reaction by a multiple enzyme system (Iuchi et al. 1995).

In other teleost fish eggs, polymerization of constituent proteins through participation of transglutaminase results in an insoluble matrix (rainbow trout, Hagenmaier et al. 1976; cod, Oppen-Berntsen et al. 1990). When present during the egg activation process, monodansyl cadaverine (MDC) inhibits chorion hardening and at the same time becomes covalently incorporated into the chorion (Oppen-Berntsen et al. 1990). The mechanism of this process is unknown, but Hagenmaier et al. (1976) and Oppen-Berntsen et al. (1990) found isopeptide bonds (between the γ-carbonyl group of glutamine and the ε-amino group of lysine) by transglutaminases in the chorion proteins of the rainbow trout only after fertilization, suggesting that the increase in mechanical strength may be due to the formation of iso-peptide bonds during the hardening process.

5.7 Fertilization Cone and Sperm Incorporation

Following fusion with a spermatozoon, the egg surface responds by a contraction that appears to be based on an actomyosin complex (Gilkey et al. 1978; Abraham et al. 1993a,b). A small blister-like swelling surrounding the sperm head, called an early fertilization cone (Kudo 1983; Kudo and Sato 1985), is formed following fusion of the sperm plasma membrane with the egg plasma membrane. It also pushes a second spermatozoon out of the micropylar canal (Iwamatsu et al. 1992b; Iwamatsu 1998). The swollen apex of the early fertilization cone reaches the micropylar vestibule, and the bulbar- or dome-shaped top of the fertilization cone is seen at the outside of the micropylar vestibule (Fig. 8; micropylar plug: Yamamoto 1952a,b, 1953; Sakai 1961; Kudo and Sato 1985; Iwamatsu et al. 1991). The morphology and the time required for appearance and disappearance of the fertilization cone vary among teleosts (Hart 1990).

After the swollen apex of the early fertilization cone withdraws or collapses with time, a cytoplasmic eminence forms around the incorporated spermatozoon (Iwamatsu and Ohta 1981) as a "later fertilization cone" (Kudo and Sato 1985; Linhart et al. 1995). In the top of the fertilization cone, the sperm tail is sometimes detected. The fertilization cone appears 30–60 s after insemination and disappears by 180 s in *Oryzias* eggs (Iwamatsu and Ohta 1981) and by 300 s in *Oncorhynchus* eggs (Kobayashi and Yamamoto 1987). Thickening and rearrangement of the actin meshwork associated with the sperm nucleus occur between 30 and 60 s after insemination (Hart et al. 1992). Hart and Fluck (1995) have pointed out that reorganization of the F-actin cytoskeleton, which correlates with sperm incorporation, appears to take place independently of any cytoskeletal alteration required for growth of the fertilization cone, since movement of the sperm nucleus into the egg cytoplasm can take place under experimental conditions in the absence of either egg activation or fertilization cone formation (Wolenski and Hart 1988b). However, the mechanism of sperm incorporation following membrane fusion still needs to be clarified, although

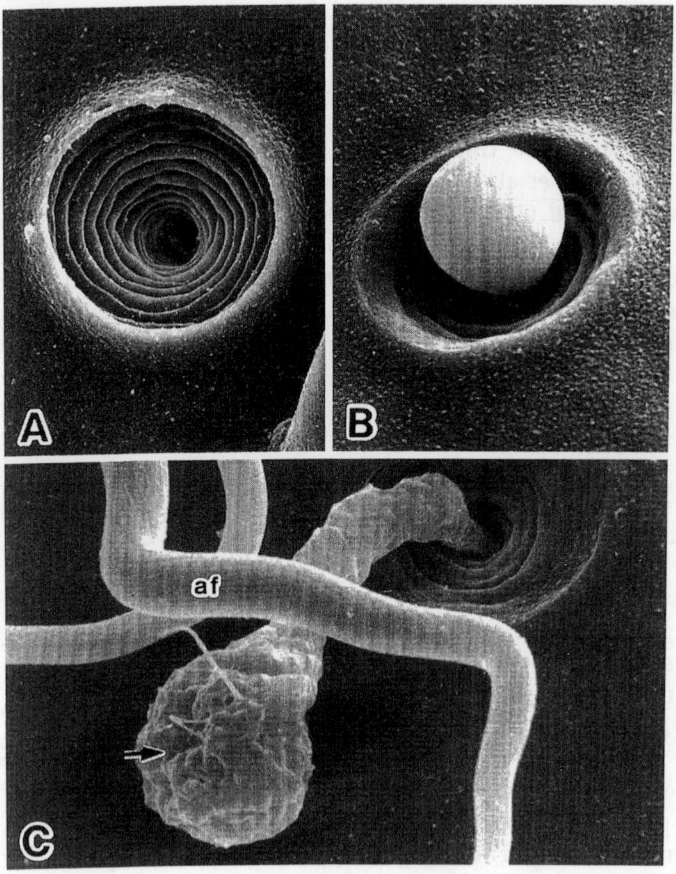

Fig. 8. Scanning electron micrographs of micropylar aperture in *Oryzias* eggs before and after insemination. **A** Micropyle of an unfertilized egg. **B** and **C** Exudation of perivitelline fluid or ooplasmic protrusion (fertilization cone) (arrow in **C**) out of micropyle 15 s after sperm attachment to egg surface. *af* attaching filaments. Bar 10 μm. (Iwamatsu et al. 1991)

the spermatozoon is completely incorporated concomitant with the appearance of the fertilization cone.

The male pronucleus migrates towards a deep inner region of the cortical cytoplasm as the fertilization cone surrounding it shrinks. The female pronucleus that formed at the end of the second meiotic division also moves toward the same depth as that of the male pronucleus. The female pronucleus migrates toward the male pronucleus from approximately 20 to 40 min after sperm incorporation in *Oryzias*. The fully developed male and female pronuclei come into close approximation at the center of the blastodisc in fertilizing eggs. The relationship between breakdown of the pronuclear envelopes and the mingling of gamete chromosomes is unclear. The direction of their encounter depends on the cortical pattern that was constructed during oocyte growth, and that also determines the direction of female pronucleus migration (Iwamatsu 1998).

6 Polyspermy-Preventing Mechanism

In monospermic eggs, the attachment of a fertilizing spermatozoon to the egg surface provokes a series of protective reactions that prevent all supernumerary spermatozoa from entering the egg. This phenomenon is called the block to polyspermy, and it is very important to ensure normal development of monospermic eggs (Rothschild 1956; Ginsburg 1961, 1972, 1987; Yamamoto 1961; Monroy 1965; Dale and Monry 1981; Jaffe and Gould 1985; Longo 1987; Kudo 1988).

The physiologically polyspermic eggs of arachnids, bryozoa, some insects, elasmobranchs, urodeles, reptiles, and birds incorporate several or many haploid spermatozoa, but only one sperm nucleus fuses with the female nucleus. Such an intracellular block to polyspermy has not been defined in teleost fishes, although it has been reported that only a limited number of sperm nuclei that have penetrated one egg can develop into pronuclei (Iwamatsu et al. 1982). This implies that one factor (Yamashita et al. 1990) to break down the nuclear envelope, another to decondense sperm nuclei and the male pronucleus growth factor (MPGF) (Thibault et al. 1975) exist in the egg cytoplasm, and that the amounts of these factors in the egg cytoplasm seem to be limited, as suggested in mammalian eggs (Hunter 1976).

The fertilization action potential has been originally proposed to cause the fast incomplete block to polyspermy at the level of the plasma membrane (Jaffe 1976). However, the fast block to polyspermy is not recognized in *Oryzias* eggs (Nuccitelli 1980b). In fish eggs with a single micropyle, the fast propagating cortical reaction covering the whole surface of the egg with respect to sperm-blockage seems to be unnecessary (Aketa 1966). When the perivitelline fluid is removed from around the dechorionated egg after activation, many spermatozoa penetrate into the ooplasm (Kanoh and Yamamoto 1957; Yanagimachi 1957c; T.S. Yamamoto 1958; Gamo et al. 1960; Ginsburg 1961; Sakai 1961; Iwamatsu and Ohta 1978; Iwamatsu et al. 1992b). These experiments may support the absence of the fast incomplete block to polyspermy in fish eggs.

The internal aperture of the micropyle in most species is only wide enough to allow one spermatozoon to pass through (Ginsburg 1972; Kuchow and Scott 1977; Brummett and Dumont 1979; Gilkey 1981; Iwamatsu and Ohta 1981; Kobayashi and Yamamoto 1981; Hart and Donovan 1983; Kudo et al. 1994; Linhart et al. 1995). In contrast, the inner opening (5 μm in diameter) of the micropylar canal in common carp eggs is wide enough to admit several spermatozoa (head 2.4 μm in diameter) at once (Fig. 9; Kudo 1980). In these common carp eggs, the size of the micropylar canal does not ensure

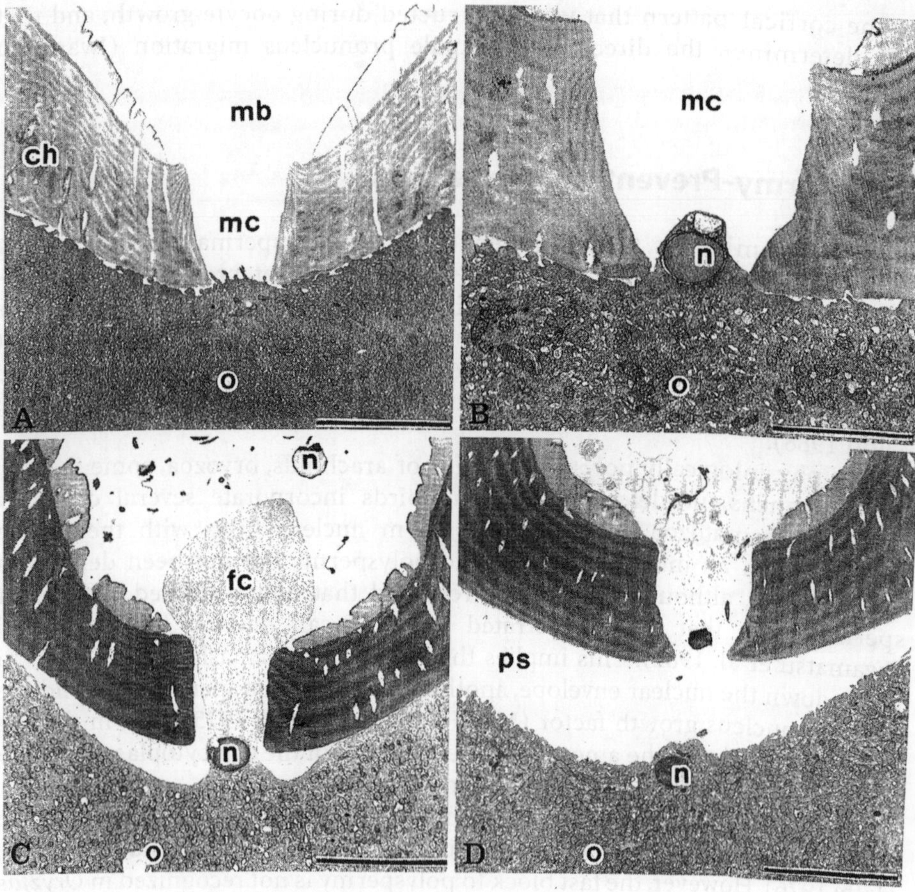

Fig. 9. Electron micrographs of sperm penetration in common carp eggs. **A** Micropyle and its cytoplasmic region in a mature egg before immersion in fresh water. The micropyle consists of a micropylar vestibule (*mb*) and micropylar canal (*mc*). A few microvilli at "sperm entry site" protrude from egg cortex toward the canal. **B** Spermatozoon which has just entered the ooplasm (*o*) prior to growth of fertilization cone. Sperm nucleus (*n*) remains enclosed by a double membrane. **C** An inseminated egg fixed about 15 s after immersion in fresh water. A slight elevation of the chorion forms a narrow perivitelline space. First sperm nucleus (*n*) is located at base of early fertilization cone (*fc*). **D** Inseminated egg fixed about 30 s after immersion in fresh water. The typical early fertilization cone over the first sperm nucleus (*n*) is not seen. At this time, a true plug structure is not seen in the micropylar canal. *ch* chorion; *fc* fertilization cone; *mb* micropylar vestibule; *mc* micropylar canal; *o* ooplasm. *Bar* 10 µm in **A**, **C**, and **D**; 5 µm in **B**. (Kudo 1980)

monospermic fertilization, and yet no polyspermy is found. The binding affinity and fusion ability of the restricted membrane of the early fertilization cone, which includes the sperm plasma membrane, seem to be changed, as inferred by Aketa (1966).

The slow burst of perivitelline fluid through the micropylar canal pushes supernumerary spermatozoa out of the micropyle, and frequently forms a "fertilization plug" as it is called by Hart and Donovan (1983). In the flounder, *Liopsetta obscura* and *L. schrenki*, a ball-like plug found at the micropylar vestibule (Yamamoto 1952b) may be formed by a surface precipitation reaction of the colloidal substance eliminated from the cortical alveoli (Yamamoto 1953). Thus, before the closure of the micropylar canal, the micropyle is plugged by a colloidal substance (Yamamoto 1952b, 1953; Sakai 1961; Iwamatsu et al. 1991; see Fig. 8) or by an ooplasmic outgrowth (early fertilization cone) (Kudo 1980; Kobayashi and Yamamoto 1981) that prevents supernumerary spermatozoa from penetrating the ooplasm.

In *Oryzias* eggs, the outflow of perivitelline fluid continuously expels supplementary spermatozoa that enter the micropylar canal before the occlusion (shrinkage) of its inner one-third portion (covered with no outermost chorion layer) by a reduction in thickness of the inner layers of the chorion. As soon as supplementary spermatozoa touch the perivitelline fluid containing sperm-agglutinating substances derived from cortical alveoli, they stop moving and their flagella straighten. Similar observations on the sperm-agglutinating activity of perivitelline fluid have been reported in eggs of the trout (Ginsburg 1961, 1972) and the chum salmon (Jaana and Yamamoto 1984). Thus, *Oryzias* eggs reject supplementary spermatozoa that enter the open micropyle for an average of 2.5 min (at 25 °C) after the first spermatozoon attaches to the egg surface. Finally, a complete block to polyspermy is established by occlusion of the micropyle, which is induced by shrinkage of the inner layer of the chorion with polymerization of the chorion glycoproteins (Iwamatsu et al. 1997b). The occlusion of the micropylar canal occurs gradually during 3–5 min after fertilization in chum salmon eggs (Kobayashi and Yamamoto 1981, 1993) and *Oryzias* eggs (Iwamatsu et al. 1992b). According to some investigations (flounder, Yamamoto 1952b; chum salmon, Kobayashi and Yamamoto 1981), the micropyle is narrowed and occluded after complete elevation of the chorion. Yamamoto and Kobayashi (1992) postulated that the closure of the micropylar canal is a result of the stretching of the envelope in eggs of *Paralichthys*, *Theragra*, and *Branchiostegus*. Bitterling eggs retain the cortical cytoplasm adhering to the inner surface of the chorion at the animal pole region, even after eggs have almost completed the cortical reaction (Suzuki 1959a). About 20 min after insemination, a part of the cytoplasm adhering around the micropyle tears away from the cortical cytoplasm and plugs the micropyle from the inside (Ohta and Iwamatsu 1983).

The frequency of sperm entry into the micropyle diminishes upon fertilization even if the outflow of perivitelline fluid ceases. This implies that some factors participate in guidance of spermatozoa into the micropylar canal of unfertilized *Oryzias* eggs and that they are lost upon fertilization (Iwamatsu et al. 1993a). Recently, chorion glycoproteins (ZI-3: 49 kDa), homologous to mammalian zona pellucida proteins (Epifano et al. 1995) with a high affinity for spermatozoa, have been found as a broadly diluted mucous area distributed on the whole outer surface of the chorion, including the micropyle in unfertilized *Oryzias* eggs (Iwamatsu et al. 1997a,b). Similar morphological observations on an outer adhesive (flocculent) coating over the chorion surface in eggs of some salmonids have been reported by Schmehl and Graham (1987). The diminished incidence of sperm entry into the micropyle of fertilized eggs appears to be caused by a gradual modification of chorion glycoproteins in the diluted mucous area in *Oryzias*. This area transforms to a thin, compact layer on the chorion surface in response to the contents of the cortical alveoli during the cortical reaction. The canal occludes completely with shrinkage (with insolubilization by polymerization of chorion proteins) of the inner layer of the chorion about 3.5 min after sperm attachment. This is a slow change to block polyspermy. On the other hand, supplementary spermatozoa actually often remain in the perivitelline space and on the egg plasma membrane in activated eggs where they could not be incorporated (Iwamatsu and Ohta 1981). Therefore, it is considered that factors such as cortical alveolus exudate (possibly containing a sperm-coating substance, or sperm agglutinin), reconstruction of the egg plasma membrane by addition of sperm plasma membrane and alveolar membrane components, or a combination of these factors may act to prevent sperm incorporation.

In brief, the mechanisms for the block to polyspermy in the fish egg are considered to be as follows: (1) the diameter of the micropylar canal restricts admission of spermatozoa; (2) spermatozoa are extruded by outgrowth of the early fertilization cone or by formation of a colloidal fluid plug in the micropylar canal; the fertilization cone membrane (sperm plasma membrane fused with the egg plasma membrane or a limiting membrane newly formed by perivitelline fluid) is not able to fuse with the sperm plasma membrane; (3) perivitelline fluid containing sperm agglutinin flows out through the micropylar canal caused by the increased osmotic pressure; (4) sperm guidance substance and attractant to the micropyle from the chorion surface are lost; (5) a decrease in diameter or complete closure of the micropylar canal caused by shrinkage of the inner layer of the chorion occurs.

7 Formation of Male and Female Pronuclei

7.1 Progression of Second Meiotic Metaphase to Interphase

MPF activity is at a high level in maturing oocytes during the first and second meiotic metaphases, and is sustained at this high level until egg activation occurs at fertilization (Fig. 10), as in starfish (Dorée et al. 1983), *Xenopus* (Gerhart et al. 1984), and mouse (Hashimoto and Kishimoto 1988) oocytes.

Mature teleost oocytes with high MPF and MAP kinase activity are arrested at the second meiotic metaphase (MII) and ovulate. Following rapid inactivation of MPF upon sperm penetration, the egg begins to exit MII (Fig. 10). The second polar body is extruded within 3 min (Brummett et al. 1985) or 5 min (Huver 1960) in *Fundulus*, 5 min in *Alcichthys* (Koya et al. 1993c), 5–10 min in

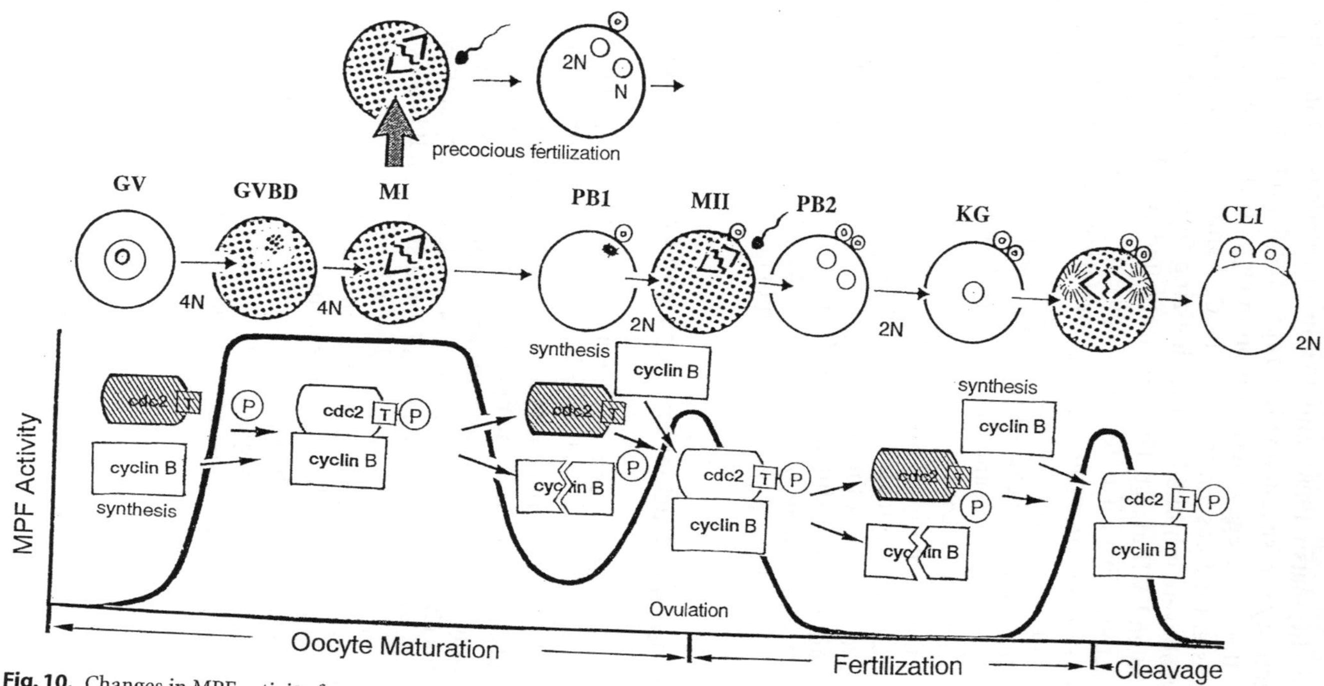

Fig. 10. Changes in MPF activity from maturation to cleavage in the fish egg. *Top of figure* illustrates precocious fertilization at MI. *KG* karyogamy stage; *CL1* two-cell stage; *T* threonine 161

Oryzias (Iwamatsu 1998), and 15 min in *Rhodeus* (Ohta 1986). Rapid loss of MPF activity is associated with a rapid disintegration of cyclin B upon activation of the whole egg at fertilization in *Oryzias* (Iwamatsu et al. 1999b). When the transient increase in cytoplasmic Ca^{2+} and subsequent exocytosis of cortical alveoli were inhibited by anesthetizing *Oryzias* eggs, penetration by sperm failed to end MII arrest. These nonactivated eggs had high levels of MPF and MAP activity. Even if unfertilized eggs are treated with only an inhibitor of protein synthesis or protein phosphorylation, MPF and MAP kinase activities remain high, and the nucleus remains arrested at MII. However, following sperm penetration, the same treated nonactivated eggs induce transformation of the sperm nucleus, and they exit MII to form an interphase nucleus but do not extrude the second polar body. In this case, MPF activity is maintained at a high level, but MAP kinase activity is lost in spite of there being no transient increase in cytoplasmic Ca^{2+} (Iwamatsu et al. 1999a). These observations appear to reveal that the activity of MAP kinase rather than MPF may play an important role in maintaining metaphase chromosomes, and that inactivation of MAP kinase may be involved in the exit of MII to interphase. Similar findings have been reported in amphibian eggs (Haccard et al. 1995).

7.2 Breakdown of Nuclear Envelope and Chromatin Decondensation

The incorporated sperm nucleus transforms into a pronucleus via a series of successive processes including breakdown of the poreless nuclear envelope, decondensation of the chromatin, and reformation of a new nuclear envelope (Fig. 11). Breakdown of the nuclear envelope starts within 20 s of sperm incorporation in *Fundulus* eggs (Brummett and Dumont 1981), or after 3 min in *Oryzias* (Iwamatsu and Ohta 1978) and *Alcichthys* eggs (Koya et al. 1993c), and 5 min in *Rhodeus* eggs (Ohta and Iwamatsu 1983). It begins at the anterior and lateral region of the sperm head, which was first to fuse with the egg plasma membrane, and finishes at the region near the centriole. Before or at the same time as the vesiculation of the sperm nucleus envelope, inactivation of MPF activity begins rapidly with the increase in cytoplasmic Ca^{2+}. Thus, the breakdown process of the sperm nuclear envelope progresses in the cytoplasm simultaneously with loss of MPF activity in normal fertilization. In contrast, somatic cells require the presence of active MPF. When compact sperm nuclei are injected into the cytoplasm of activated eggs just after inactivation of MPF, they transform into pronuclei 20 min before reappearance of MPF activity (Iwamatsu et al. 1999), as also seen in mammalian fertilization (Usui et al. 1997). These findings may reflect the difference between the nuclear envelopes of the spermatozoon and that of the somatic cell. Decondensation of the chromatin starts from the vesiculated region of the nuclear envelope, and the sperm nucleus swells greatly within about 30 min, when it encounters the female pronucleus in the absence of MPF activity. MPF activity appears again with de novo synthesis of cyclin B. The new nuclear envelope of the swelling

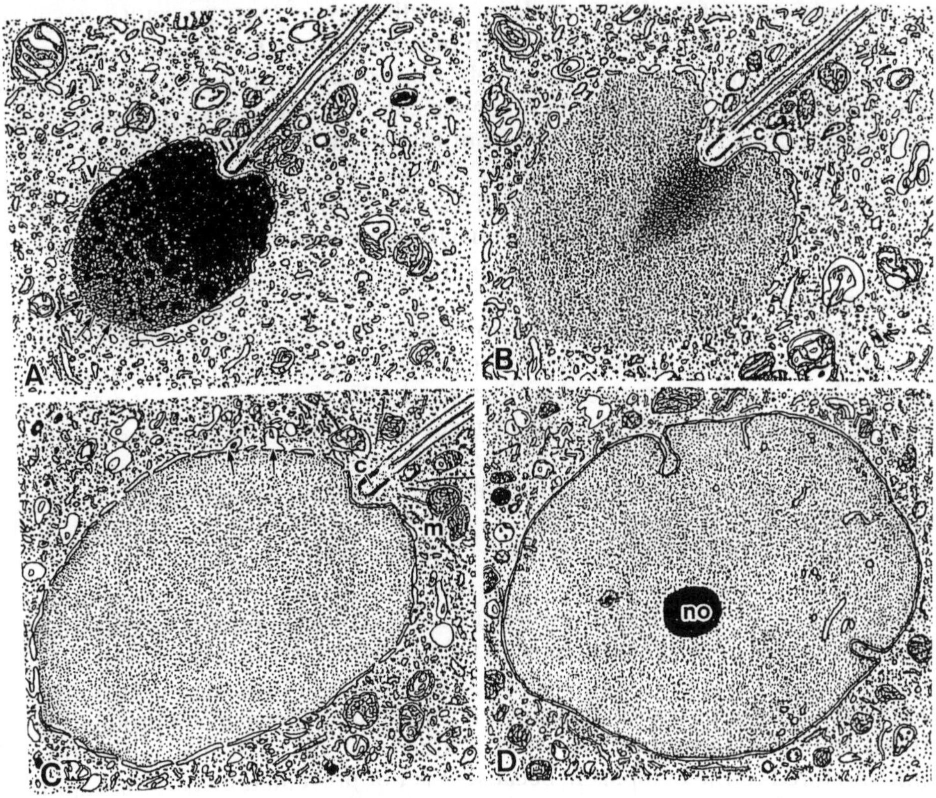

Fig. 11. Male pronucleus formation in *Oryzias* eggs. **A** An early stage in dispersal of sperm chromatin 3 min after insemination. Vesiculation (*v*) of nuclear envelope and chromatin dispersion (*arrows*) begin at apical region of sperm head. **B** Sperm nucleus at stage of dispersing chromatin still has an intact nuclear envelope surrounding the centriole (*c*). **C** The early male pronucleus 5 min after insemination. Long or flattened vesicles containing minute vesicles (*arrows*) have accumulated along the surface of dispersed chromatin with centriole and sperm mitochondria (*m*) excluded. **D** Fully-grown male pronucleus with a large nucleolus (*no*) 25 min after insemination

nucleus begins to be reformed about 3 min after sperm incorporation, as described above.

7.3 Pronucleus Formation and MPF Activity

When oocytes at first meiotic metaphase are precociously penetrated by spermatozoa and exhibit propagative exocytosis of cortical alveoli, they form a female pronucleus and a male pronucleus, skipping the second meiotic division (Fig. 10; Iwamatsu 1997). In this case, the male pronucleus is larger than the female pronucleus. It is unclear whether or not the difference in size is due to the presence of male pronucleus growth factor (MPGF; Thibault et al. 1975). The difference in size might also occur because the nuclear envelope of the female pronucleus is derived solely from ooplasmic vesicles such as the endoplasmic reticulum (ER) and other vesicular components, while that of the male pronucleus is derived in part from the sperm nucleus envelope in addition to the ooplasm-derived vesicles (Longo 1987). The fate of ribosomes present on the chromatin side of rough ER as a component of the nuclear envelope remains unclear, although nuclear lamins contribute to the formation of the inner layer of the nuclear envelope (cf. Schatten et al. 1985; Prather and Schatten 1992).

In the mouse (Ecklund and Levine 1975; Kopecny and Pavlok 1975), it has been found that during the decondensation the sperm chromatin loses protamines, which are replaced by histones in the egg cytoplasm. In the frog, the nuclear proteins of the spermatozoon are also replaced by histones within 5 min of incorporation into the ooplasm (Ohsumi and Katagiri 1991). Similar findings have also been reported in mammals (Yanagimachi 1994) but not for teleosts.

The activity of MPF is lost by destruction of cyclin B in the cdc2–cyclin B complex concurrently with the increase in cytoplasmic Ca^{2+}. Approximately 30 min (at 26 °C) after sperm incorporation, cyclin B synthesis begins anew. MPF activity is then again induced synchronously with de novo synthesis of cyclin B. This is supported by the fact that neither an increase in the activity of MPF nor the appearance of cyclin B occurs in the presence of a protein synthesis inhibitor. Moreover, the rapid synthesis of cyclin B may result from mRNA stored in the cytoplasm, because it is not prevented by actinomycin D. The rise in MPF activity seems to relate to phosphorylation of the cdc2–cyclin B complex (Fig. 10), since MPF activity cannot be detected in the presence of de novo synthesized cyclin B if a phosphorylation inhibitor (6-DMAP) is present in the medium.

7.4 DNA Synthesis and Pronucleus Formation

In fertilization of the sand dollar (Simmel and Karnofsky 1961) and *Xenopus* (Graham 1966), DNA synthesis occurs during formation of the female and male pronuclei. However, in teleostean eggs, the timing of DNA synthesis has not yet been reported. When eggs are fertilized in the presence of aphidicolin (α-DNA polymerase inhibitor), the fertilization process from sperm entry to karyogamy progresses normally, followed by only a slight delay in subsequent cytokinesis, and formation of a thread-like chromatin stretched between blastomeres. Judging from the effect of aphidicolin on the fertilization process, DNA synthesis does not seem to be required for pronucleus formation.

8
Migration and Karyogamy of Female and Male Pronuclei

The cytoskeletal network (microfilaments, microtubules, and intermediate filaments) generally conducts the dynamic responses such as ooplasmic segregation, pronuclear migration, and polar body extrusion in fertilized and parthenogenetically activated eggs (Abraham et al. 1993a,b; Webb and Fluck 1995). The *Oryzias* egg has two independent microtubule networks that are present during the first cell cycle. One network is near the animal pole and is associated with the pronuclei, while the other is the parallel array of microtubules near the vegetal pole (Abraham et al. 1995).

During ooplasmic segregation, the fertilized egg assembles and disassembles a complex network of microtubules, by which the female pronucleus moves to encounter the male pronucleus away from the second polar body. If the formation of networks of microtubules in the cortical cytoplasm is blocked by treatment of *Oryzias* eggs with demecolcine, the growth of the blastodisc consequently slows and the movements of the male and female pronuclei are inhibited (Webb et al. 1995).

In general, the major force responsible for the streaming of the ooplasm containing pronuclei into the forming blastodisc may be provided by the filamentous components of the cytoskeleton including both microfilaments and microtubules (Beams et al. 1985). In *Brachydanio* (Katow 1983), *Misgurnus* (Ivanenkov et al. 1987), and *Oryzias* (Webb and Fluck 1993; Webb et al. 1995) eggs, blastodisc formation resulting from ooplasmic segregation is severely obstructed by treatment with or microinjection of cytochalasin B or D, which disorganizes the microfilamentous cortex in the protoplasmic coat. It has been suggested that actin-based contraction machinery might be located in the cortical cytoplasm during blastodisc formation in *Brachydanio* eggs (Hart and Fluck 1995). Intracellular Ca^{2+} may play a role in the contraction of a cortically located actin filament network, possibly providing the driving force for segregation in the zebrafish (Leung et al. 1998).

9
Fertilization Without an Increase in Cytoplasmic Ca^{2+} and Exocytosis of Cortical Alveoli

Eggs of a salmonid fish retain fertilizability for a long time in an isotonic solution (Runström 1920), and are penetrated by spermatozoa without egg activation upon insemination either in an isotonic solution (Kusa 1950) or at high concentrations of external calcium (Hamano 1949; Ginsburg 1963). The salmonid egg is able to initiate fertilization without breakdown of cortical alveoli (Kanoh 1952; Devillers et al. 1953; Kusa 1953a; Yamamoto 1962, 1967). Therefore, the visible cortical change in eggs at the time of fertilization and parthenogenetic activation may not be a necessary factor for the initiation of the developmental activity (Kusa 1953b).

When unfertilized *Oryzias* eggs are inseminated in an anesthetized state induced by anesthetics such as chloretone, phenylurethane, or MS222, exocytosis of cortical alveoli is not observed (Yamamoto 1949; Aketa 1966). Furthermore, no propagative increase in intracellular Ca^{2+} is induced in these eggs despite sperm incorporation. Such a nonactivated egg retains the high MPF activity after sperm incorporation, and displays metaphase chromosomes (MIII) from the sperm nucleus in addition to its own MII chromosomes. However, acetone-treated eggs begin to develop without exocytosis of cortical alveoli upon sperm stimulation (Sakai 1964; Iwamatsu 1965b). In acetone-treated eggs that have been loaded with aequorin, no increase in intracellular Ca^{2+} is detectable (Iwamatsu et al. 1992a). Immersion alone in acetone-saline without insemination causes eggs to undergo no exocytosis of cortical alveoli and shows parthenogenetic transformation of MII chromosomes into a pronucleus after extrusion of the second polar body. In the exit of MII to interphase caused by the acetone treatment, the decreases in MPF and MAP kinase activities take place concurrently. Furthermore, it has been observed that in *Oryzias* eggs inseminated in the anesthetized state, pronucleus formation occurs without the distinct increase in the cytoplasmic Ca^{2+} normally initiated by sperm incorporation if a protein synthesis inhibitor (cycloheximide) or phosphorylation inhibitor (6-DMAP) is present. In the case of nonactivated eggs with incorporated spermatozoa, the activity of MPF remains high in the presence of these inhibitors. These facts suggest that in fish eggs, transformation of MII chromosomes and the sperm nucleus into pronuclei may not be caused by loss of MPF activity but instead by the loss of activity of a short-lived protein kinase(s) sensitive to 6-DMAP. However, high MPF activity is necessary to induce chromosome condensation. The conspicuous increase in intracellular Ca^{2+} and subsequent exocytosis of cortical alveoli are considered to be dispensable for fertilization since normal formation, migration, and karyogamy of the pronuclei occur regardless of whether they occur. It is plausible that the transient increase in intracellular Ca^{2+} and exocytosis are not only separable from the factors responsible for nuclear behavior, but also unnecessary for initiation of development. However, a small amount of Ca^{2+} may be required for a protein involved in metabolic changes in the egg, because all fertilization reactions are completely blocked when the Ca-chelator EGTA is microinjected.

Acknowledgements. The author is grateful to the following colleagues who provided valuable advice and information: Drs. K. Aketa, R.A. Fluck, N.H. Hart, Y. Koya, M. Kusa, M. Yamashita, and R. Yanagimachi. The author is also indebted to Dr. C.A. Brown and Y. Shibata for their help in the preparation of the manuscript. A portion of our work cited here was supported by grants from the Narishige Zoological Science Award.

References

Abraham VC, Gupta S, Fluck RA (1993a) Ooplasmic segregation in medaka (*Oryzias latipes*) egg. Biol Bull 184:115–124

Abraham VC, Miller AL, Fluck RA (1995) Microtubule arrays during ooplasmic segregation in the medaka fish egg (*Oryzias latipes*). Biol Bull 188:136–145

Abraham VC, Miller AL, Jaffe LF, Fluck RA (1993b) Cytoplasmic microtubule arrays in *Oryzias latipes* (medaka) eggs during ooplasmic segregation. Biol Bull 185:305–306

Aketa K (1954) The chemical nature and the origin of the cortical alveoli in the egg of the medaka, *Oryzias latipes*. Embryologia 2:63–66

Aketa K (1966) A study of the block mechanism against polyspermy in the medaka, *Oryzias latipes*. Annot Zool Jpn 39:149–155

Aketa K, Onitake K, Tsuzuki H (1972) Tryptic disruption of sperm-binding site of sea urchin egg surface. Exp Cell Res 71:27–32

Amanze D, Iyengar A (1990) The micropyle: a sperm guidance system in teleost fertilization. Development 109:495–500

Anderson E (1967) The formation of the primary envelope during oocyte differentiation in teleosts. J Cell Biol 35:193–212

Anderson E (1968) Cortical alveoli formation and vitellogenesis during oocyte differentiation in the pipefish, *Syngnathus fuscus*, and the killifish, *Fundulus heteroclitus*. J Morphol 125:23–60

Arndt EA (1960) Untersuchungen über die Eihällen von Cyprinidien. Z Zellforsch Mikrosk Anat 52:315–327

Atwood DL, Chia F-S (1974) Fine structure of an unusual spermatozoon of a brooding sea cucumber, *Cucumaria lubrica*. Can J Zool 52:519–523

Beams HW, Kessel RG, Shih CY, Tung HN (1985) Scanning electron microscopy studies on blastodisc formation in the zebrafish, *Brachydanio rerio*. J Morphol 184:41–49

Becker KA, Hart NH (1993) Actin and myosin-II in zebrafish eggs: reorganization of actin during cortical granule exocytosis. Mol Biol Cell 4:141a

Becker KA, Hart NH (1994) Spectrin and the actin cytoskeleton in the zebrafish egg. Mol Biol Cell 5:463a

Becker KA, Hart NH (1996) The cortical actin cytoskeleton of unactivated zebrafish eggs: spatial organization and distribution of filamentous actin, nonfilamentous actin, and myosin II. Mol Reprod Dev 43:536–547

Bedford JM, Moore HDM, Flanklin LE (1979) Significance of equatorial segment of the acrosome of the spermatozoa in eutherian mammals. Exp Cell Res 119:119–126

Begovac PC, Wallace RA (1989) Major vitelline envelope proteins in pipefish oocytes originate within the follicle and are associated with the Z3 layer. J Exp Zool 251:56–73

Billard R (1970) La spermatogenèse de *Poecilia reticulata*. IV. La spermiogenèse. Étude ultrastructurale. Ann Biol Anim Biochem Biophys 10:493–510

Billard R (1978) Changes in structure and fertilizing ability of marine and fresh fish spermatozoa diluted in media of various salinities. Aquaculture 14:187–198

Billard R (1983) Spermiogenesis in the rainbow trout (*Salmo gairdneri*). An ultrastructural study. Cell Tissue Res 233:265–284

Billard R, Cosson MP (1992) Some problems related to the assessment of sperm motility in freshwater fish. J Exp Zool 261:122–131

Boitano S, Omoto CK (1991) Membrane hyperpolarization activates trout sperm without an increase in intracellular pH. J Cell Sci 98:343–349

Boitano S, Omoto CK (1992) Trout sperm swimming patterns and role of intracellular Ca^{++}. Cell Motil Cytoskel 21:74–82

Bonsignorio D, Perego L, Giacco LD, Cotelli F (1996) Structure and macromolecular composition of the zebrafish egg chorion. Zygote 4:101–108

Brivio MF, Bassi R, Cotelli F (1991) Identification and characterization of the major components of the *Oncorhynchus mykiss* egg chorion. Mol Reprod Dev 28:85–93

Brummett AR, Dumont JN (1979) Initial stages of sperm penetration into the egg of *Fundulus heteroclitus*. J Exp Zool 210:417–434

Brummett AR, Dumont JN (1981) Cortical vesicle breakdown in fertilized eggs of *Fundulus heteroclitus*. J Exp Zool 216:63–79

Brummett AR, Dumont JN, Richter CS (1985) Later stages of sperm penetration and second polar body and blastodisc formation in the egg of *Fundulus heteroclitus*. J Exp Zool 234:423–439

Burns JR, Weitzman SH, Grier HJ, Menezes NA (1995) Internal fertilization, testis and sperm morphology in glandulocaudine fishes (Teleostei: Characidae: Glandulocaudinae). J Morphol 224:131–145

Busson-Mabillot S (1973) Évolution des enveloppes de l'ovocyte et de l'oeuf chez un poisson téléostéen. J Microsc 18:23–44

Busson-Mabillot S (1977) Un type particulier de sécrétion exocrine: celui de l'a ppareil adhésif de l'oeuf d'un poisson Téléostéen. Biol Cell 30:233–244

Casas MT, Munoz-Guerra S, Subirana JA (1981) Preliminary report on the ultrastructure of chromatin in the histone containing spermatozoa of a teleost fish. Biol Cell 40:87–92

Chang DC (1991) Structures of actin in zebra fish embryos studied by confocal microscopy. Biophys J 59:55a

Cherr GN, Clark WH Jr (1982) Fine structure of the envelope and micropyles in the eggs of the white sturgeon, *Acipenser transmontanus* Richardson. Dev Growth Differ 24:341–352

Cherr GN, Clark WH Jr (1984) Jelly release in the eggs of the white sturgeon, *Acipenser transmontanus*: an enzymatically mediated event. J Exp Zool 230:145–149

Ciapa B, Whitaker MJ (1986) Two phases of inositol polyphosphate and diacylglycerol production at fertilization. FEBS Lett 195:347–351

Cimino MC (1972) Egg-production, polyploidization and evolution in a diploid all-female fish of the genus *Poeciliopsis*. Evolution 26:294–306

Clapper D, Lee H-C (1985) Inositol trisphosphate induces calcium release from non-mitochondrial stores in sea urchin egg homogenates. J Biol Chem 260:13947–13954

Colwin AL, Colwin LH (1961) Changes in the spermatozoon during fertilization in *Hydroides hexagonus* (Annelida). II. Incorporation with the egg. J Biophys Biochem Cytol 10:255–274

Cooper TG (1986) The epididymis, sperm maturation and fertilization. Springer, Berlin Heidelberg New York

Cosson MP (1990) Sperm chemotaxis. In: Gagnon C (ed) Control of sperm motility: biological and clinical aspect. CRC Press, Boca Raton, pp 103–135

Cosson MP, Billard R, Letellier L (1989) Rise of internal Ca^{2+} accompanies the initiation of trout sperm motility. Cell Motil Cytoskel 14:424–434

Cotelli F, Andronico F, Lora Lamia C (1988) Structure and composition of the fish egg chorion (*Carassius auratus*). J Ultrastruct Mol Struct Res 99:70–78

Cotelli F, Bonsignorio D, Lora Lamia C (1993) Structural and biochemical comparative analysis of teleostean egg chorion. J Reprod Dev 39 (Suppl)114

Dale B, Monroy A (1981) How is polyspermy prevented? Gamete Res 4:151–169

Davenport J, Lonning S, Kjorsvik E (1981) Osmotic and structural changes during early development of eggs and larvae of the cod, *Gadus morhua*. I. J Fish Biol 19:317–331

Dettlaff TA (1957) Cortical granules and substances secreted from the animal portion of the egg during the period of activation in *Acipenseridae* (in Russian). Daklay Acad Nauk SSSR 116:341–344

Devillers C, Thomopoulos A, Colas J (1953) Différenciation bipolaire et formation de l'espace périvitelline dans l'oeuf de *Salmo irideus*. Bull Soc Zool France 78:462–470

Donovan M, Hart NH (1982) Uptake of ferritin by the mosaic egg surface of *Brachydanio*. J Exp Zool 223:299–304

Donovan M, Hart NH (1986) Cortical granule exocytosis is coupled with membrane retrieval in eggs of *Brachydanio*. J Exp Zool 237:391–405

Dorée M, Peaucellier G, Picard A (1983) Activity of the maturation-promoting factor and the extent of protein phosphorylation oscillate simultaneously during meiotic maturation of starfish oocytes. Dev Biol 99:489–501

orée M, Labbe JC, Picard A (1989) M phase promoting factor: its identification as the M phase specific histone H1 kinase and its activation by phosphorylation. J Cell Sci 12 (Suppl):39–51
ımont JN, Brummett AR (1980) The vitelline envelope, chorion, and micropyle of *Fundulus heteroclitus* eggs. Gamete Res 3:25–44
ımont JN, Brummett AR (1985) Egg envelopes in vertebrates. In: Browder LW (ed) Developmental biology, vol 1. Plenum Press, New York, pp 235–288
unphy WG, Brizuela L, Beach D, Newport J (1988) *Xenopus* cdc2 protein is a component of MPF, a cytoplasmic regulator of mitosis. Cell 54:423–431
cklund PS, Levine L (1975) Mouse sperm basic nuclear protein. Electrophoretic characterization and fate after fertilization. J Cell Biol 66:251–262
pifano O, Liang L-F, Dean J (1995) Mouse *Zp1* encodes a zona pellucida protein homologous to egg envelope proteins in mammals and fish. J Biol Chem 270:27254–27258
luck RA, Miller AL, Abraham VC, Jaffe LC (1994) Calcium buffer injections inhibit ooplasmic segregation in medaka eggs. Biol Bull 186:254–262
olz KR (1995) Sperm binding proteins. Int Rev Cytol 163:249–303
Gamo H, Yamauchi E, Suzuki R (1960) The fertilization of dechorionated egg in two species of fish. Bull Aichi Gakugei Univ 9:117–126
Gardiner DM (1978) Cyclic changes in fine structure of the epithelium lining the ovary of the viviparous teleost, *Cymatogaster aggregata* (Perciformes: Embiotocidae). J Morphol 156:367–380
Gatti JL, Billard R, Christen R (1990) Ionic regulation of the plasma membrane potential of rainbow trout (*Salmo gairdneri*) spermatozoa: role in the initiation of sperm motility. J Cell Physiol 143:546–554
Gautier J, Maller JL (1991) Cyclin B in *Xenopus* oocytes: implications for the mechanism of pre-MPF activation. EMBO J 10:177–182
Gautier J, Minshull J, Lohka M, Glotzer M, Hunt T, Maller JL (1990) Cyclin is a component of maturation-promoting factor from *Xenopus*. Cell 60:487–494
Gautier J, Norbury C, Lohka M, Maller J (1988) Purified maturation-promoting factor contains the product of a *Xenopus* homolog of the fission yeast cell cycle gene cdc^{2+}. Cell 54:433–439
Gerhart JC, Wu M, Kirschner M (1984) Cell cycle dynamics of an M-phase-specific cytoplasmic factor in *Xenopus laevis* oocyte and eggs. J Cell Biol 98:1247–1255
Gibbons BH, Gibbons IR, Baccetti B (1983) Structure and motility of the 9+0 flagellum of eel spermatozoa. J Submicrosc Cytol 15:15–20
Gilkey JC (1981) Mechanisms of fertilization in fishes. Am Zool 21:359–375
Gilkey JC (1983) Roles of calcium and pH in activation of eggs of the medaka fish, *Oryzias latipes*. J Cell Biol 97:669–678
Gilkey JC, Jaffe LF, Ridgway EB, Reynolds GT (1978) A free calcium wave traverses the activating egg of the medaka, *Oryzias latipes*. J Cell Biol 76:448–466
Ginsburg AS (1961) The block to polyspermy in sturgeon and trout with special reference to the role of cortical granules (alveoli). J Embryol Exp Morphol 9:173–190
Ginsburg AS (1963) Sperm–egg association and its relationship to the activation of the egg in salmonid fishes. J Embryol Exp Morphol 11:13–33
Ginsburg AS (1972) Fertilization in fishes and the problem of polyspermy. US Dep Comm Nat Tech Inf Ser, Springfield, 366 pp
Ginsburg AS (1987) Egg cortical reaction during fertilization and its role in block to polyspermy. Soc Sci Rev F Physiol Gen Biol 1:307–375
Glabe CG, Vacquier VD (1978) Egg surface glycoprotein receptor for sea urchin bindin. Proc Natl Acad Sci USA 75:881–885
Goodall JA, Blackshaw AW, Capra MF (1989) Factors affecting the activation and duration of motility of the spermatozoa of the summer whiting (*Sillago ciliata*). Aquaculture 77:243–250
Graham CF (1966) The regulation of DNA synthesis and mitosis in multinucleate from eggs. J Cell Sci 1:363–374
Grier HJ (1973) Ultrastructure of the testis in the teleost *Poecilia latipinna*, spermatogenesis with reference to the intercentriolar lamellated body. J Ultrastruct Res 45:82–92

Grier HJ (1975) Spermiogenesis in the teleost *Gambusia affinis* with particular reference to the role played by microtubules. Cell Tissue Res 165:89–102

Grier HJ (1976) Sperm development in the teleost *Oryzias latipes*. Cell Tissue Res 168:419–431

Grier HJ (1981) Cellular organization of the testis and spermatogenesis in fishes. Am Zool 21:345–357

Grier HJ, Moody DP, Cowell CB (1990) Internal fertilization and sperm morphology in the brook silverside *Labidesthes sicculus* (Cope). Copeia 1990(1):221–226

Groot EP, Alderdice DF (1985) Fine structure of the external egg membrane of five species of Pacific salmon and steelhead trout. Can J Zool 63:552–566

Guraya SS (1978) Maturation of the follicular wall of nonmammalian vertebrates. In: Jones RE (ed) The vertebrate ovary. Plenum Press, New York, pp 261–329

Guraya SS (1986) The cell and molecular biology of fish oogenesis. Monographs in developmental biology, vol 18. Sauer/Karger, Basel, pp 1–262

Habibi HR, Lessman CA (1986) Effect of demecolcine (colcemid) on goldfish oocyte meiosis in vitro. Gamete Res 13:103–114

Haccard O, Lewellyn A, Hartley RS, Elinson E, Maller JL (1995) Induction of *Xenopus* oocyte meiotic maturation by MAP kinase. Dev Biol 168:677–682

Hagenmaier HE, Schmitz I, Föhles J (1976) Zum Vorkommen von Isopeptidbindungen in der Eihülle der Regenbogenforelle (*Salmo gairdneri* Rich). Hoppe-Seylers Z Physiol Chem 357:1435–1438

Hagström BE, Lönning S (1968) Electron microscopic studies of unfertilized and fertilized eggs from marine teleosts. Sarsia 33:73–80

Hamaguchi S, Sakaizumi M (1992) Sexually differentiated mechanisms of sterility in interspecific hybrids between *Oryzias latipes* and *O. curvinotus*. J Exp Zool 263:323–329

Hamano S (1949) On the swelling layer of egg membrane of salmon, *Oncorhynchus keta* (in Japanese with English Summary). Physiol Ecol 3:43–46

Hamazaki TS, Iuchi I, Yamagami K (1984) Chorion glycoprotein-like immunoreactivity in some tissues of adult female medaka. Zool Sci 1:148–150

Hamazaki TS, Iuchi I, Yamagami K (1985) A spawning female-specific substance reactive to anti-chorion (egg envelope) glycoprotein antibody in the teleost, *Oryzias latipes*. J Exp Zool 235:269–279

Hamazaki TS, Iuchi I, Yamagami K (1987a) Production of a "spawning female-specific substance" in hepatic cells and its accumulation in the ascites of the estrogen-treated adult fish, *Oryzias latipes*. J Exp Zool 242:325–332

Hamazaki TS, Iuchi I, Yamagami K (1987b) Isolation and partial characterization of a "spawning female-specific substance" in the teleost, *Oryzias latipes*. J Exp Zool 242:343–349

Hamazaki TS, Nagahama Y, Iuchi I, Yamagami K (1989) A glycoprotein from the liver constitutes the inner layer of the egg envelope (zona pellucida interna) of the fish, *Oryzias latipes*. Dev Biol 133:101–110

Handel MA, Eppig JJ (1998) Sexual dimorphism in the regulation of mammalian meiosis. Curr Top Dev Biol 37:333–358

Hara M, Okiyama M (1998) An ultrastructural review of the spermatozoa of Japanese fishes. Bull Ocean Res Inst Univ Tokyo 33:1–138

Hart NH (1990) Fertilization in teleost fish: mechanisms of sperm–egg interactions. Int Rev Cytol 121:1–66

Hart NH, Becker KA (1992) Actin and myosin-II in unfertilized zebrafish eggs: filamentous and nonfilamentous actin domains in the cortex. Mol Biol Cell 3:7a

Hart NH, Becker KA, Wolenski JS (1992) The sperm entry site during fertilization of the zebrafish egg: localization of actin. Mol Reprod Dev 32:217–228

Hart NH, Collins GC (1991) An electron-microscopic and freeze-fracture study of the egg cortex of *Brachydanio rerio*. Cell Tissue Res 265:317–328

Hart NH, Donovan M (1983) Fine structure of the chorion and site of sperm entry in the egg of *Brachydanio*. J Exp Zool 227:277–296

Hart NH, Fluck RA (1995) Cytoskeleton in teleost eggs and early embryos: contributions to cytoarchitecture and motile events. Curr Top Dev Biol 31:343–381

Hart NH, Pietri R, Donovan M (1984) The structure of the chorion and associated surface fragments in *Oryzias* – evidence for the presence of extracellular tubules. J Exp Zool 230:273–296

Hart NH, Yu S-F (1980) Cortical granule exocytosis and cell surface reorganization in eggs of *Brachydanio*. J Exp Zool 213:137–159

Hart NH, Yu S-F, Greenhut VA (1977) Observations on the cortical reaction in eggs of *Brachydanio rerio* as seen with the scanning electron microscope. J Exp Zool 201:325–331

Hashimoto N, Kishimoto T (1988) Regulation of meiotic metaphase by a cytoplasmic maturation-promoting factor during mouse oocyte maturation. Dev Biol 126:242–252

Hirai T, Yamashita M, Yoshikuni M, Lou Y-H, Nagahama Y (1992) Cyclin B in fish oocytes: its cDNA and amino acid sequences, appearance during maturation, induction of p34cdc2 activation. Mol Reprod Dev 33:131–140

Hosokawa K (1985) Electron microscopic observation of chorion formation in the teleost, *Navodon modestus*. Zool Sci 2:513–522

Hunter RHF (1976) Sperm–egg interactions in the pig: monospermy, extensive polyspermy, and the formation of chromatin aggregates. J Anat 122:43–59

Hurley D, Fisher K (1966) The structure and development of the external membrane in young eggs of the brook trout, *Salvelinus fontinalis* (Mitchill). Can J Zool 44:173–190

Huver CW (1960) The stages at fertilization of the egg of *Fundulus heteroclitus*. Biol Bull 119: 320

Hyllner SJ, Fernandez-Palacios Barber H, Larsson DGJ, Haux C (1995) Amino acid composition and endocrine control of vitelline envelope proteins in European sea bass (*Dicentrarchus labrax*) and gilthead sea bream (*Sparus aurata*). Mol Reprod Dev 41:339–347

Hyllner SJ, Haux C (1992) Immunochemical detection of the major vitelline envelope proteins in the plasma and oocytes of the female rainbow trout, *Oncorhynchus mykiss*. J Endocrinol 135:303–309

Hyllner SJ, Oppen-Berntsen DO, Helvik JV, Walther BT, Haux C (1991) Oestradiol-17β induces the major vitelline envelope proteins in both sexes in teleosts. J Endocrinol 131:229–236

Inoue S, Inoue Y (1986) Fertilization (activation)-induced 200- to 9-kDa depolymerization of glycoprotein, a distinct component of cortical alveoli of rainbow trout eggs. J Biol Chem 261:5256–5261

Inoue S, Inoue Y (1987) Structures and possible functions of fish egg polysialoglycoproteins: molecular exploitation in glycoconjugate research. Seikagaku 59:327–346

Inoue S, Iwasaki M (1978) Isolation of a novel glycoprotein from the eggs of rainbow trout: occurrence of disialosyl groups on all carbohydrate chains. Biochem Biophys Res Commun 83:1018–1023

Inoue S, Kitajima K, Inoue Y, Kudo S (1987) Localization of polysialoglycoprotein as a major glycoprotein component in cortical alveoli of the unfertilized eggs of *Salmo gairdneri*. Dev Biol 123:442–454

Ishijima S, Hamaguchi Y, Iwamatsu T (1993) Sperm behavior in the micropyle of the medaka egg. Zool Sci 10:179–182

Ishijima S, Hara M, Okiyama M (1998) Comparative studies on the spermatozoon motility of Japanese fishes. Bull Ocean Res Inst Univ Tokyo 33:139–152

Ito S (1962) Resting potential and activation potential of the *Oryzias* egg. II. Change of membrane potential and resistance during fertilization. Embryologia 7:47–55

Ito S (1963) Resting potential and activation potential of the *Oryzias* egg. III. The effect of monovalent cations. Embryologia 7:344–354

Iuchi I, Ha C-R, Masuda K (1995) Chorion hardening in medaka (*Oryzias latipes*) egg. Fish Biol J Medaka 7:15–20

Iuchi I, Ha C-R, Sugiyama H, Nomura K (1996) Analysis of chorion hardening of eggs of rainbow trout, *Oncorhynchus mykiss*. Dev Growth Differ 38:299–306

Iuchi I, Masuda K, Yamagami K (1991) Change in component proteins of the egg envelope (chorion) of rainbow trout during hardening. Dev Growth Differ 33:85–92

Ivanenkov VV, Minin AA, Ozerova SG (1990) Phalloidin inhibits cortical granule exocytosis and ooplasmic segregation in loach egg. Cell Differ Dev 29:21–36

Ivanenkov VV, Minin AA, Meshcheyakov VN, Martynova LE (1987) The effect of local microfilament disorganization on ooplasmic segregation in the loach (*Misgurnus fossilis*) egg. Cell Differ 22:19–28

Iwamatsu T (1965a) On fertilizability of pre-ovulation eggs in the medaka, *Oryzias latipes*. Embryologia 8:327–336

Iwamatsu T (1965b) Effect of acetone on the cortical changes at fertilization of the egg of the medaka, *Oryzias latipes*. Embryologia 9:1–12

Iwamatsu T (1969) Changes of the chorion upon fertilization in the medaka, *Oryzias latipes*. Bull Aichi Univ Educ 18:43–56

Iwamatsu T (1981) Development of *Oryzias latipes* eggs inseminated with several foreign spermatozoa (in Japanese with English Resumé). Bull Aichi Univ Educ 30 (Nat Sci):141–151

Iwamatsu T (1984) Effect of pH on the fertilization response of the medaka egg. Dev Growth Differ 26:533–544

Iwamatsu T (1989) Exocytosis of cortical alveoli and its initiation time in medaka eggs induced by microinjection of various agents. Dev Growth Differ 31:39–44

Iwamatsu T (1992) Egg activation. Fish Biol J Medaka 4:1–9

Iwamatsu T (1994) Medaka oocytes rotate within the ovarian follicles during oogenesis. Dev Growth Differ 36:177–186

Iwamatsu T (1997) Abbreviation of the second meiotic division by precocious fertilization in fish oocytes. J Exp Zool 277:450–459

Iwamatsu T (1998) Studies on fertilization in the teleost. I. Dynamic responses of fertilized medaka eggs. Dev Growth Differ 40:475–483

Iwamatsu T, Ishijima S, Nakashima S (1993a) Movement of spermatozoa and changes in micropyles during fertilization in medaka eggs. J Exp Zool 266:57–64

Iwamatsu T, Ito S (1986) Effects of microinjected cations on the early event of fertilization in the medaka egg. Dev Growth Differ 28:303–310

Iwamatsu T, Keino H (1978) Scanning electron microscopic study on the surface change of eggs of the teleost, *Oryzias latipes*, at the time of fertilization. Dev Growth Differ 20:237–250

Iwamatsu T, Kikuyama M, Hiramoto Y (1992a) Fertilization reaction without changes in intracellular Ca^{2+} in medaka eggs – an experiment with acetone-treated eggs. Dev Growth Differ 34:709–717

Iwamatsu T, Nakashima S (1996) Dynamic growth of oocytes of the medaka, *Oryzias latipes*. I. A relationship between establishment of the animal–vegetal axis of the oocyte and its surrounding granulosa cells. Zool Sci 13:873–882

Iwamatsu T, Ohta T (1976) Breakdown of the cortical alveoli of medaka eggs at the time of fertilization, with a particular reference to the possible role of spherical bodies in the alveoli. Wilhelm Rouxs Arch 180:297–309

Iwamatsu T, Ohta T (1978) Electron microscopic observation on sperm penetration and pronuclear formation in the fish egg. J Exp Zool 205:157–180

Iwamatsu T, Ohta T (1981) Scanning electron microscopic observations on sperm penetration in teleostean fish. J Exp Zool 218:261–277

Iwamatsu T, Ohta T, Nakayama N, Shoji H (1976) Studies of oocyte maturation of the medaka, *Oryzias latipes*. III. Cytoplasmic and nuclear changes during in vitro maturation. Annot Zool Jpn 49:28–37

Iwamatsu T, Ohta T, Oshima E, Sugiura T (1985) Requirement of extracellular calcium ions for the early fertilization events in the medaka egg. Dev Growth Differ 27:751–762

Iwamatsu T, Onitake K, Yoshimoto Y, Hiramoto Y (1991) Time sequence of early events in fertilization in the medaka egg. Dev Growth Differ 33:479–490

Iwamatsu T, Onitake K, Matsuyama K, Satoh M, Yukawa S (1997a) Effect of micropylar morphology and size on rapid sperm entry into the eggs of the medaka. Zool Sci 14:623–628

Iwamatsu T, Onitake K, Nakashima S (1992b) Polarity of responsiveness to sperm and artificial stimuli in medaka eggs. J Exp Zool 264:351–358

Iwamatsu T, Onitake K, Nakashina S (1993b) Spiral patterns in the micropylar wall and filaments on the chorion in eggs of the medaka, *Oryzias latipes*. J Exp Zool 267:225–232

Iwamatsu T, Shibata Y, Kanie T (1995) Changes in chorion proteins induced by the exudate released from the egg cortex at the time of fertilization in the teleost, *Oryzias latipes*. Dev Growth Differ 37:747–759

Iwamatsu T, Shibata Y, Yamashita M (1999a) Studies on fertilization of the teleost. II. Nuclear behavior and changes in histone H1 kinase. Dev Growth Differ 41:473–482

Iwamatsu T, Tada Y, Nishiyama Y (1982) On sperm nuclei and development in polyspermic eggs of the teleost, *Oryzias latipes*. Annot Zool Jpn 55:91–99

Iwamatsu T, Uwa H, Inden A, Hirata K (1984) Experiments on interspecific hybridization between *Oryzias latipes* and *Oryzias celebensis*. Zool Sci 1:653–663

Iwamatsu T, Shibata Y, Yamashita M (1999b) Studies on fertilization of the teleost. III. The relationship between nuclear behavior and the histone H1 kinase activity in anesthetized medaka eggs. Dev Genet 25:137–145

Iwamatsu T, Yoshimoto Y, Hiramoto Y (1988a) Cytoplasmic Ca^{2+} release induced by microinjection of Ca^{2+} and effects of microinjected divalent cations on Ca^{2+} sequestration and exocytosis of cortical alveoli in the medaka egg. Dev Biol 125:451–457

Iwamatsu T, Yoshimoto Y, Hiramoto Y (1988b) Mechanism of Ca^{2+} release in medaka eggs microinjected with inositol 1,4,5-trisphosphate and Ca^{2+}. Dev Biol 129:191–197

Iwamatsu T, Yoshizaki N, Shibata Y (1997b) Changes in the chorion and sperm entry into the micropyle during fertilization in the teleostean fish, *Oryzias latipes*. Dev Growth Differ 39:33–41

Jaana H, Yamamoto TS (1984) Local variation of the cell surface in chum salmon sperm as revealed by their agglutination reaction. J Exp Zool 230:449–463

Jackson K, Abraham M, Degani G (1994) Oocyte maturation triggered by the presence of male in the blue gourami, *Trichogaster trichopterus*. J Morphol 220:1–9

Jaffe LA (1976) Fast block to polyspermy in sea urchin eggs is electrically mediated. Nature 261:68–71

Jaffe LF (1983) Sources of calcium in egg activation: a review and hypothesis. Dev Biol 99:265–276

Jaffe LF (1985) The role of calcium explosions, waves, and pulses in activating eggs. In: Metz CB, Monroy A (eds) Biology of fertilization, vol 3. Academic Press, New York, pp 127–165

Jaffe LA, Gould M (1985) Polyspermy-preventing mechanisms. In: Metz CB, Monroy A (eds) Biology of fertilization, vol 3. Academic Press, New York, pp 223–250

Jamieson BGM (1987) A biological classification of sperm types, with special reference to annelids and molluscs, and an example of spermiocladistics. In: Mohri H (ed) New horizons in sperm cell research. Gordon and Breach Scientific Publishing, New York, pp 311–332

Jespersen A (1971) Fine structure of the spermatozoon of the Australian lungfish *Neoceratodus forsteri* (Krefft). J Ultrastruct Res 37:178–185

Jessus C, Ozon R (1993) Regulation of cell divisions during oogenesis of vertebrates: the *Xenopus* oocyte paradigm. Comp Biochem Physiol A 106:431–448

Jin ZX, Nakajima T, Morisawa M, Hayashi H (1994) Isolation and properties of a protein complex containing flagellar movement – initiating phosphoprotein from testes of a white salmon. J Biochem 115:552–556

Jørgensen N-C (1972) Actomyosin-like ATPase at the surface of fish eggs. Exp Cell Res 71:460–464

Kajiura H, Yamashita M, Katsu Y, Nagahama Y (1993) Isolation and characterization of goldfish cdc2, a catalytic component of maturation-promoting factor. Dev Growth Differ 35:647–654

Kanoh Y (1952) Über die Beziehung zwischen dem Zerfallen der Kortikalalveoli und der Entwicklung bei Fischeiern. Jpn J Ichthyol 2:99–103

Kanoh Y, Yamamoto TS (1957) Removal of the membrane of the dog salmon egg by means of proteolytic enzymes. Bull Jpn Soc Sci Fish 23:166–172

Katow H (1983) Obstruction of blastodisc formation by cytochalasin B in the zebrafish, *Brachydanio rerio*. Dev Growth Differ 25:477–484

Katsu Y, Yamashita M, Kajiura H, Nagahama Y (1993) Behavior of the components of maturation-promoting factor, cdc2 and cyclin B, during oocyte maturation of goldfish. Dev Biol 160:99–107

Kessel RG, Beams HW, Tung HN, Roberts R (1983) Unusual particle arrays in the plasma membrane of zebrafish spermatozoa. J Ultrastruct Res 84:268–274

Kitajima K, Inoue S (1988) A proteinase associated with cortices of rainbow trout eggs involved in fertilization-induced depolymerization of polysialoglycoproteins. Dev Biol 129:270–274

Kitajima K, Inoue Y, Inoue S (1986) Polysialoglycoproteins of *Salmon*idae fish eggs. Complete structure of 200-kDa polysialoglycoprotein from the unfertilized eggs of rainbow trout (*Salmo gairdneri*). J Biol Chem 261:5262–5269

Kitajima K, Inoue S, Inoue Y, Troy FA (1988a) Use of a bacteriophage-derived endo-N-acetylneuraminidase and an equine anti-polysialyl antibody to characterize the polysialyl residues in salmonid fish egg polysialoglycoproteins. Substrate and immunospecificity studies. J Biol Chem 263:18269–18276

Kitajima K, Inoue S, Inoue Y (1989) Isolation and characterization of a novel type of sialoglycoproteins (hyosophorin) from the eggs of medaka, *Oryzias latipes*: nonapeptide with a large N-linked glycan chain as a tandem repeat unit. Dev Biol 132:544–553

Kitajima K, Sorimachi H, Inoue S, Inoue Y (1988b) Comparative structures of the apo-polysialoglycoproteins from the unfertilized and fertilized eggs of salmonid fishes. Biochemistry 27:7141–7145

Kiyohara T, Ito S (1968) Activation potential observed in the centrifused egg of the Japanese killifish (*Oryzias latipes*). Kumamoto J Sci Ser B Sect 2 (9):23–27

Kobayashi H (1971) A cytological study on gynogenesis of the triploid ginbuna (*Carassius auratus langsdorfii*). Zool Mag 80:316–322

Kobayashi H, Minshull J, Ford C, Golsteyn R, Poon R, Hunt T (1991) On the synthesis and destruction of A- and B-type cyclins during oogenesis and meiotic maturation in *Xenopus laevis*. J Cell Biol 114:755–765

Kobayashi W (1985) Electron microscopic observation of the breakdown of cortical vesicles in the chum salmon egg. J Fac Sci Hokkaido Univ Ser VI Zool 24:87–102

Kobayshi W, Yamamoto TS (1981) Fine structure of the micropylar apparatus of the chum salmon egg, with a discussion of the mechanism for blocking polyspermy. J Exp Zool 217:265–275

Kobayashi W, Yamamoto TS (1985) Fine structure of the micropylar cell and its changes during oocyte maturation in the chum salmon, *Oncorhynchus keta*. J Morphol 184:263–276

Kobayashi W, Yamamoto TS (1987) Light and electron microscopic observations of sperm entry in the chum salmon egg. J Exp Zool 243:311–322

Kobayashi W, Yamamoto TS (1993) Factors inducing closure of the micropylar canal in the chum salmon egg. J Fish Biol 42:385–394

Kondo T, Yamagata T, Yoshida N, Yamashita M (1997) Introduction of cyclin B induces activation of the maturation promoting factor and breakdown of germinal vesicle in growing zebrafish unresponsive to maturation-inducing hormone. Dev Biol 190:142–152

Kopecny V, Pavlok A (1975) Autoradiographic study of mouse spermatozoan arginine-rich nuclear protein in fertilization. J Exp Zool 191:85–96

Koya Y, Munehara H, Takano K (1995) Formation of egg adhesive material in masked greenling, *Hexagrammos octogrammus*. Jpn J Ichthyol 42:45–52

Koya Y, Munehara H, Takano K, Iwasaki H (1993a) Effects of extracellular environments on the motility of spermatozoa in several marine sculpins with internal gametic association. Comp Biochem Physiol 106A:25–29

Koya Y, Ohara S, Ikeuchi T, Adachi S, Matsubara T, Yamauchi K (1993b) Testicular development and sperm morphology in the viviparous teleost, *Zoarces elongatus* (in Japanese with English Abstract). Bull Hokkaido Natl Fish Res Inst 57:21–31

Koya Y, Takano K, Takahashi H (1993c) Ultrastructural observations on sperm penetration in the egg of elkhorn sculpin, *Alcichthys alcicornis*, showing internal gametic association. Zool Sci 10:93–101

Kuchow KP, Scott JR (1977) Ultrastructure of the chorion and its micropylar apparatus in the mature *Fundulus heteroclitus* (Walbaum) ovum. J. Fish Biol 10:197–201
Kudo S (1980) Sperm penetration and the formation of a fertilization cone in the common carp egg. Dev Growth Differ 22:403–414
Kudo S (1982a) Ultrastructure and ultracytochemistry of fertilization envelope formation in the carp egg. Dev Growth Differ 24:327–339
Kudo S (1982b) Ultrastructure of a sperm entry site beneath the micropylar canal in fish eggs. Zool Mag 91:213–220
Kudo S (1983) Response to sperm penetration of the cortex of eggs of the fish, *Plecoglossus altivelis*. Dev Growth Differ 25:163–170
Kudo S (1988) Polyspermy-preventing mechanisms in fish eggs. In: Inoue S, Shirai T, Egar MW, Aiyama S, Géraudie J, Nobunaga T, Sato NL (eds) Proc 6th Meet Singer Symp, Gunma, pp 395–410
Kudo S, Inoue M (1986) A bactericidal effect of fertilization envelope extract from fish eggs. Zool Sci 3:323–329
Kudo S, Inoue M (1989) Bactericidal action of fertilization envelope extract from eggs of the fish *Cyrinus carpio* and *Plecoglossus altivelis*. J Exp Zool 250:219–228
Kudo S, Inoue M (1991) Enzyme activities responsible for the defense mechanism of the fertilization envelope. Naturwissenschaften 78:172–173
Kudo S, Linhart O, Billard R (1994) Ultrastructural studies of sperm penetration in the egg of the European catfish, *Sillurus glanis*. Aquat Living Resour 7:93–98
Kudo S, Sato A (1985) Fertilization cone of carp eggs revealed by scanning electron microscopy. Dev Growth Differ 27:121–128
Kusa M (1950) Physiological analysis of fertilization in the egg of the salmon, *Oncorhynchus keta*. I. Why are the eggs not fertilized in isotonic Ringer solution? Annot Zool Jpn 24:22–28
Kusa M (1953a) Significance of cortical change in the initiation of development of the salmon egg (physiological analysis of fertilization in the egg of the salmon, *Oncorhynchus keta* II). Annot Zool Jpn 26:73–77
Kusa M (1953b) On some properties of the cortical alveoli in the egg of the stickleback. Annot Zool Jpn 26:138–144
Kusa M (1954) The cortical alveoli of salmon egg. Annot Zool Jpn 27:1–6
Kusa M (1956) Studies on cortical alveoli in some teleostean eggs. Embryologia 3:105–129
Laale HW (1980) The perivitelline space and egg envelopes of bony fishes: a review. Copeia 2:210–226
Lau AF, Royson RC, Humphreys T (1986) Tumor promoters and diacylglycerol activate the Na^+–H^+ antiporter of sea urchin eggs. Exp Cell Res 186:23–30
Lessman CA, Kavumpurath S (1984) Cytological analysis of nuclear migration and dissolution during steroid-induced meiotic maturation in vitro of follicle-enclosed oocytes of the goldfish (*Carassius auratus*). Gamete Res 10:21–29
Leung CF, Webb SE, Miller AL (1998) Calcium transients accompany ooplasmic segregation in zebrafish embryos. Dev Growth Differ 40:313–326
Linhart O, Kudo S (1997) Surface ultrastructure of paddlefish eggs before and after fertilization. J Fish Biol 51:573–582
Linhart O, Kudo S, Billard R, Slechta V, Mikodina EV (1995) Morphology, composition and fertilization of carp eggs: a review. Aquaculture 129:75–93
Liu L, Ju J-C, Yang X (1998) Parthenogenetic development and protein patterns of newly matured bovine oocytes after chemical activation. Mol Reprod Dev 49:298–307
Lohka MJ, Hayes MK, Maller JL (1988) Purification of maturation-promoting factor, an intracellular regulator of early mitotic events. Proc Natl Acad Sci USA 85:3009–3013
Longo FJ (1981) Morphological features of the surface of the sea urchin (*Arbacia punctulata*) egg: oolemma-cortical granule association. Dev Biol 54:173–182
Longo FJ (1987) Fertilization. Chapman and Hall, London, 183 pp
Lönning S, Kjørsvik E, Davenport J (1984) The hardening process of the egg chorion of the cod, *Gadus morhua* L., and lumpsucker, *Cycloperus lumpus* L. J Fish Biol 24:505–522.

Lou Y-H, Takahashi H (1989) Spermiogenesis in the Nile tilapia, *Oreochromis niloticus* with notes on a unique pattern of nuclear chromatin condensation. J Morphol 200:321–330
Masuda K, Iuchi I, Iwamori M, Nagai Y, Yamagami K (1986) Presence of a substance crossreacting with cortical alveolar material in "yolk vesicles" of growing oocytes of *Oryzias latipes.* J Exp Zool 238:261–265
Masuda K, Iuchi I, Yamagami K (1991) Analysis of hardening of the egg envelope (chorion) of the fish, *Oryzias latipes.* Dev Growth Differ 33:75–83
Masuda K, Murata K, Iuchi I, Yamagami K (1992) Some properties of the hardening process in chorions isolated from unfertilized eggs of medaka, *Oryzias latipes.* Dev Growth Differ 34:545–551
Masui Y, Markert CL (1971) Cytoplasmic control of nuclear behavior during meiotic maturation of frog oocyte. J Exp Zool 177:129–146
Mattei X (1991) Spermatozoon ultrastructure and its systematic implications in fishes. Can J Zool 69:3038–3055
Mattei X, Boisson C, Mattei C, Reizer C (1967) Spermatozoïdes aflagellés chez un poisson: *Gymnarchus niloticus* (Téléostéen, Gymnachidae). CR Acad Sci (Paris) 365D:2010–2012
Mattei X, Mattei C (1974) La spermiogénèse de *Lepadogaster* sp. (Téléostéen Gobiesocidae). J Microsc 20:68a–69a
Miura T, Yamauchi K, Takahashi H, Nagahama Y (1992) The role of hormones in the acquisition of sperm motility in salmonid fish. J Exp Zool 261:359–363
Monroy A (1965) Chemistry and physiology of fertilization. Holt, Rinehart and Winston, New York, 150 pp
Monroy A (1985) Gametogenesis: general conclusions. In: Metz CB, Monroy A (eds) Biology of fertilization, vol 1. Academic Press, New York, pp 89–94
Moore HDM, Bedford JM (1983) The interaction of mammalian gametes in the female. In: Hartmann JF (ed) Mechanism and control of animal fertilization. Academic Press, New York, pp 453–497
Morisawa M (1985) Initiation mechanism of sperm motility at spawning in teleosts. Zool Sci 2:605–615
Morisawa M (1994) Cell signaling mechanisms for sperm motility. Zool Sci 11:647–662
Morisawa S, Ishida K, Okuno M, Morisawa M (1993) Roles of pH and cyclic adenosine monophosphate in the acquisition of potential for sperm motility during migration from the sea to the river in chum salmon. Mol Reprod Dev 34:420–426
Morisawa M, Suzuki K (1980) Osmolality and potassium ion: their roles in initiation of sperm motility in teleosts. Science 210:1145–1147
Morisawa M, Suzuki K, Morisawa S (1983) Effects of potassium and osmolality on spermatozoan motility of salmonid fishes. J Exp Biol 107:105–113
Morisawa M, Tanimoto S, Ohtake H (1992) Characterization and partial purification of sperm-activating substance from eggs of the herring, *Clupea palasii.* J Exp Zool 264:225–230
Munehara H, Koya Y, Takano K (1994) Conditions for initiation of fertilization of eggs in the copulating elkhorn sculpin. J Fish Biol 45:1105–1111
Munehara H, Takano K, Koya Y (1989) Internal gametic association and external fertilization in the elkhorn sculpin, *Acichthys alcicornis.* Copeia 1989:673–678
Munehara H, Takano K, Koya Y (1991) The little dragon sculpin *Blepsias cirrhosus*, another case of internal gamete association and external fertilization. Jpn J Ichthyol 37:391–394
Murata K, Hamazaki TS, Iuchi I, Yamagami K (1991) Spawning female-specific egg envelope glycoprotein-like substances in *Oryzias latipes.* Dev Growth Differ 33:553–562
Murata K, Iuchi I, Yamagami K (1993) Isolation of H-SF substances, the high-molecular-weight precursors of egg envelope proteins, from the ascites accumulated in the oestrogen-treated fish, *Oryzias latipes.* Zygote 1:315–324
Murata K, Iuchi I, Yamagami K (1994) Synchronous production of the low- and high-molecular-weight precursors of the egg envelope subunits, in response to estrogen administration in the teleost fish *Oryzias latipes.* Gen Comp Endocrinol 95:232–239

Murata K, Sasaki T, Yasumasu S, Iuchi I, Enami J, Yasumasu I, Yamagami K (1995) Cloning cDNAs for the precursor protein of a low-molecular-weight subunit of the inner layer of the egg envelope (chorion) of the fish *Oryzias latipes*. Dev Biol 167:9–17

Nagahama Y (1987) Endocrine control of oocyte maturation. In: Norris DO, Jones RE (eds) Hormones and reproduction in fishes, amphibians, and reptiles. Plenum Press, New York, pp 171–201

Nakakuni M, Toya H, Sawano K, Kajishima T (1984) On the fertilization of the triploid ginbuna. J Fac Sci Shinshu Univ 19:53–61

Nakano E (1956) Changes in the egg membrane of the fish egg during fertilization. Embryologia 3:89–103

Nakano E (1969) Fishes. In: Metz CB, Monroy A (eds) Fertilization: comparative morphology, biochemistry and immunology, vol 2. Academic Press, New York, pp 295–324

Nakashima S, Iwamatsu T (1989) Ultrastructural changes in micropylar cells and formation of the micropyle during oogenesis in the medaka *Oryzias latipes*. J Morphol 202:339–349

Nakashima S, Iwamatsu T (1994) Ultrastructural changes in micropylar and granulosa cells during in vitro oocyte maturation in the medaka, *Oryzias latipes*. J Exp Zool 270:547–556

Nicander L, Sjoden I (1971) An electron microscopical study of the acrosomal complex and its role in fertilization in the river lamprey, *Lampetra fluviatilis*. J Submicrosc Cytol 3:309:317

Nosek J (1984) Biogenesis of the cortical granules in fish oocytes. Histochem J 16:435–437

Nuccitelli R (1977) A study of the extracellular electrical currents and membrane potential changes generated by the medaka egg during activation. J Cell Biol 75:23a (Abstr)

Nuccitelli R (1980a) The electrical changes accompanying fertilization and cortical vesicle secretion in the medaka egg. Dev Biol 76:483–498

Nuccitelli R (1980b) The fertilization potential is not necessary for the block to polyspermy or the activation of development in the medaka egg. Dev Biol 76:499–504

Nuccitelli R (1987) The wave of activation current in the egg of the medaka fish. Dev Biol 122:522–534

Oberdorf JA, Head JF, Kaminer B (1986) Calcium uptake and release by isolated cortices and microsomes from the unfertilized egg of the sea urchin *Strongylocentrotus droebachiensis*. J Cell Biol 102:2205–2210

Oda S, Igarashi Y, Ohtake H, Sakai K, Shimizu N, Morisawa M (1995) Sperm-activating proteins from unfertilized eggs of the Pacific herring, *Clupea pallasii*. Dev Growth Differ 37:257–261

Oda S, Igarashi Y, Manaka K, Koibuchi N, Sakai-Sawada M, Sakai K, Morisawa M, Ohtake H, Shimizu N (1998) Sperm-activating proteins obtained from the herring eggs are homologous to trypsin inhibitors and synthesized in follicle cells. Dev Biol 204:55–63

Oda S, Morisawa M (1993) Rises of intracellular Ca^{2+} and pH mediate the initiation of sperm motility by hyperosmolality in marine teleosts. Cell Motil Cytoskel 25:171–178

Ohsumi K, Katagiri C (1991) Occurrence of H1 subtypes specific to pronuclei and cleavage-stage cell nuclei of anuran amphibians. Dev Biol 147:110–120

Ohta H (1984) Electron microscopic study on adhesive material of Pacific herring (*Clupea pallasi*) eggs. Jpn J Ichthyol 30:404–411

Ohta H, Takano K (1982) Ultrastructure of micropylar cells in the pre-ovulatory follicles of Pacific herring, *Clupea pallasi valennciennes*. Bull Fac Fish Hokkaido Univ 33:57–64

Ohta H, Teranishi T (1982) Ultrastructure and histochemistry of granulosa and micropylar cells in the ovary of the loach, *Misgurnus anguillicaudatus* (Cantor). Bull Fac Fish Hokkaido Univ 33:1–8

Ohta T (1985) Electron microscopic observations on sperm entry and pronuclear formation in naked eggs of the rose bitterling in polyspermic fertilization. J Exp Zool 234:273–281

Ohta T (1986) Electron microscopic observations on the process of the polar body formation in artificially activated eggs of the rose bitterling. J Exp Zool 237:263–270

Ohta T (1991) Initial stages of sperm–egg fusion in freshwater teleost, *Rhodeus ocellatus ocellatus*. Anat Rec 229:195–202

Ohta T, Iwamatsu T (1983) Electron microscopic observations on sperm entry into eggs of the rose bitterling, *Rhodeus ocellatus*. J Exp Zool 227:109–119

Ohta T, Iwamatsu T, Tanaka M, Yoshimoto Y (1990) Cortical alveolus breakdown in the eggs of the freshwater teleost *Rhodeus ocellatus ocellatus*. Anat Rec 227:486–496

Ohtsuka E (1958) Carbohydrate component of the perivitelline fluid and its origin in the egg of *Oryzias latipes* (in Japanese with English Abstract). Zool Mag 67:96–99

Ojima Y, Asano M (1977) A cytological evidence for gynogenetic development of the ginbuna (*Carassius auratus langsdorfii*). Proc Jpn Acad 53 (Ser B):138–142

Okamura F, Nishiyama H (1978) The passage of spermatozoa through the vitelline membrane in the domestic fowl, *Gallus gallus*. Cell Tissue Res 188:497–508

Oppen-Berntsen DO, Helvik JV, Walther BT (1990) The major structural proteins of cod (*Gadus morhua*) eggshells and protein crosslinking during teleost egg hardening. Dev Biol 137:258–265

Oppen-Berntsen DO, Hyllner SJ, Haux C Helvik JV, Walther BT (1992a) Zona radiata proteins are synthesized by rainbow trout (*Oncorhynchus mykiss*) hepatocytes in response to oestradiol-17β. J Endocrinol 135:293–302

Oppen-Berntsen DO, Hyllner SJ, Haux C, Helvik JV, Walther BT (1992b) Eggshell zona radiata-proteins from cod (*Gadus morhua*): extra-ovarian origin and induction by estradiol-17β. Int J Dev Biol 36:247–254

Oppen-Berntsen DO, Olsen SO, Rong CJ, Taranger GL, Swanson P, Walther BT (1994) Plasma levels of eggshell zr-proteins, estradiol-17β, and gonadotropins during an annual reproductive cycle of Atlantic salmon (*Salmo salar*). J Exp Zool 268:59–70

Osanai K (1977) Scanning electron microscopy of the envelopes surrounding the chum-salmon oocytes. Bull Mar Biol Stat Asamushi Tôhoku Univ 16:21–25

Picheral B, Charbonneau M (1982) Anuran fertilization: a morphological reinvestigation of some early events. J Ultrastruct Res 81:306–321

Pillai MC, Shields TS, Yanagimachi R, Cherr GN (1993) Isolation and partial characterization of the sperm motility initiation factor from eggs of the Pacific herring, *Clupea pallasi*. J Exp Zool 265:336–342

Poirier GR, Nicholson N (1982) Fine structure of the testicular spermatozoa from the channel catfish, *Ictalurus punctatus*. J Ultrastruct Res 80:104–110

Porte A, Follenius E (1960) La spermiogénèse chez *Lebistes reticulatus*. Etude au microscope électronique. Bull Soc Zool Fr 85:82–88

Prather RS, Schatten G (1992) Construction of the nuclear matrix at the transition from maternal to zygotic control of development in the mouse: an immunocytochemical study. Mol Reprod Dev 32:203–208

Ridgway EB, Gilkey JC, Jaffe LF (1977) Free calcium increases explosively in activating medaka eggs. Proc Natl Acad Sci USA 74:623–627

Riehl R (1977) Licht- und elektronenmikroskopische Untersuchungen zu Bau und Entwicklung der Mikropylen von *Noemacheilus barbatulus* (L.) und *Gobio gobio* (L.) (Pisces, Teleostei). Zool Anz 198:313–327

Riehl R (1978) Die Oocyten der Grundel Pomatoschistus minutus. II. Lichtmikroskopische Untersuchungen zur Kenntnis der Mikropyle. Microsc Acta 80:287–291

Riehl R (1980) Micropyle of some salmonins and coregonins. Environ Biol Fish 5:59–66

Riehl R (1984) The oocytes of the goby *Pomatoschistus minutus*. IV. Electron microscopic autoradiography of [^{3}H]-proline incorporation. Cytologia (Tokyo) 49:127–142

Riehl R (1991) Die Struktur der Oocyten und Eihüllen oviparer Knochenfische – eine Übersicht. Acta Biol Benrodis 3:27–65

Riehl R (1993) Surface morphology and micropyle as a tool for identifying fish eggs by scanning elecron microscopy. Microsc Anat (May):29–31

Riehl R (1999) The micropyle of teleost fish eggs: morphological and functional aspects. In: Séret B, Sire JY (eds) Proc 5th Indo-Pacific Fish Conf, Nouméa, 1997. Soc Fr Ichthyol (Paris) 1999, pp 589–599

Riehl R, Götting K (1974) Zu Structure und Vorkommen der Mikropyle an Eizellen und Eirn von Knochenfischen (Teleostei). Arch Hydrobiol 74:393–402

Riehl R, Götting K (1975) Bau und Entwicklung der Mikropylen in der Oocyten einiger Süsswasser-Teleosteer. Zool Anz 195:363–373

Riehl R, Greven H (1990) Electron microscopical studies on oogenesis and development of egg envelopes in two viviparous teleosts, *Heterandria formosa* (Poeciliidae) and *Ameca splendens* (Goodeidae). Zool Beitr NF 33:247–252

Riehl R, Greven H (1998) Fine structure of egg envelopes in some viviparous goodeid fishes, with comments on the relation of envelope thinness to viviparity. Can J Zool 71:91–97

Riehl R, Kokoscha M (1993) A unique surface pattern and micropylar apparatus in the eggs of *Luciocephalus* sp. (Perciformes, Luciocephalidae). J Fish Biol 43:617–620

Riehl R, Patzner RA (1991) Breeding, egg structure and larval morphology of the catfish *Sturisoma aureum* (Steindachner) (Teleostei, Loricariidae). J Aquaricult Aquat Sci 6:1–6

Riehl R, Schulte E (1977) Vergleichende rasterelektronenmikroskopische Untersuchungen an den Mikropylen ausgewählter Süßwasser-Teleosteer. Arch Fisch Wiss 28:95–107

Riehl R, Schulte E (1978) Bestimmugsschlüssel der wichtigsten deutschen Süßwasser A.-Teleosteer anhand ihrer Eistructrur. Arch Hydrobiol 83:200–212

Rothschild Lord (1956) Fertilization. Methuen, London, 170 pp

Rothschild Lord (1958) Fertilization in fish and lampreys. Biol Rev 33:372–392

Runström J (1920) Über osmotischen Druck und Eimembranfunktion bei den Lachsfischen. Acta Zool Stockh 1:321–336

Runström J, Kriazat G (1952) The cortical propagation of the activation impulse in the sea urchin egg. Exp Cell Res 3:419–426

Sakai YT (1961) Method for removal of chorion and fertilization of the naked egg in *Oryzias latipes*. Embryologia 5:357–368

Sakai YT (1964) Studies on the ooplasmic segregation in the egg of the fish *Oryzias latipes*. I. Ooplasmic segregation of the partially activated egg. Embryologia 8:135–145

Sakai YT (1976) Spermiogenesis of the teleost *Oryzias latipes* with special reference to formation of flagellar membrane. Dev Growth Differ 18:1–13

Sakaizumi M, Shimizu Y, Matsuzaki T, Hamaguchi S (1993) Unreduced diploid eggs produced by interspecific hybrids between *Oryzias latipes* and *O. curvinotus*. J Exp Zool 266:312–318

Sakun OF (1960) Khimicheskaya priroda i znachenie vklyuchenii v ootsitak kostistykh ryb (Chemical nature and the significance of inclusions in teleost oocytes). Arkhiv Anat Gistol Embriol 38(1):38–42 (quoted from Ginsburg 1972)

Scapigliati G, Carcupino M, Taddei AT, Mazzini M (1994) Characterization of the main egg envelope proteins of the sea bass *Dicentrarchus labrax* L. (Teleostea, Serranidae). Mol Reprod Dev 38:48–53

Schalkoff ME, Hart NH (1986) Effects of A23187 upon cortical granule exocytosis in eggs of *Brachydanio*. Rouxs Arch Dev Biol 195:39–48

Schatten G, Simerly C, Schatten H (1985) Microtubule configurations during fertilization, mitosis and early development in the mouse and the requirement for egg microtubule-mediated motility during mammalian fertilization. Proc Natl Acad Sci USA 82:4152–4156

Schmehl MK, Graham EF (1987) Comparative ultrastructure of the zona radiata from eggs of six species of salmonids. Cell Tissue Res 250:513–519

Schroeder T (1979) Surface area change at fertilization: resorption of the mosaic membrane. Dev Biol 70:306–326

Schueltz RJ, Kallman KD (1968) Triploid hybrids between the all-female teleost *Poecilia formosa* and *Poecilia sphenops*. Nature 319:280–282

Scotte AP, Baynes SN (1980) A review of the biology, handling and storage of salmonid spermatozoa. J Fish Biol 17:707–739

Seko A, Kitajima K, Inoue Y, Inoue S (1991) Peptide:N-glycosidase activity found in the early embryos of *Oryzias latipes* (medaka fish). J Biol Chem 266:22110–22114

Seko A, Kitajima K, Iwasaki M, Inoue S, Inoue Y (1989) Structural studies of fertilization-associated carbohydrate-rich glycoproteins (Hyosophorin) isolated from the fertilized and unfertilized eggs of flounder, *Paralichthys olivaceus*. J Biol Chem 264:15922–15929

Selman K, Wallace RA, Barr V (1986) Oogenesis in *Fundulus heteroclitus*. IV. Yolk-vesicle formation. J Exp Zool 239:277–288

Selman K, Wallace RA, Barr V (1988) Oogenesis in *Fundulus heteroclitus*. V. The relationship of yolk vesicles and cortical alveoli. J Exp Zool 426:42–56

Selman K, Wallace RA, Sarka A, Qi X (1993) Stages of oocyte development in the zebrafish, *Brachydanio rerio*. J Morph 218:203–224

Sette C, Bevitacqua A, Bianchini A, Mangia F, Geremia R, Rossi P (1997) Parthenogenetic activation of mouse eggs by microinjection of a truncated c-kit tyrosine kinase present in spermatozoa. Development 124:2267–2274

Shen SS, Burgart LJ (1986) 1,2-diacylglycerols mimic phobol 12-myristate 13-acetate activation of the sea urchin egg. J Cell Physiol 127:330–340

Shibata Y, Iwamatsu T (1996) Evidence for involvement of the exudate released from the egg cortex in the change in chorion proteins at the time of egg activation in *Oryzias latipes*. Zool Sci 13:271–275

Shimizu M, Fujita T, Hara A (1998a) Purification of the precursors to vitelline envelope proteins from serum of Sakhalin taimen, *Hucho perryi*. J Exp Zool 282:385–359

Shimizu M, Fujita T, Hiramatsu N, Hara A (1998b) Immnochemical detection, estrogen induction and occurrence in serum of vitelline envelope-related proteins of Sakhalin taimen *Hucho perryi*. Fish Sci 64:600–605

Shimizu Y, Shibata N, Yamashita M (1997) Spermiogenesis without preceding meiosis in the hybrid medaka between *Oryzias latipes* and *O. curvinotus*. J Exp Zool 279:102–112

Simmel EB, Karnofsky DA (1961) Observations on the uptake of tritiated thymidine in the pronuclei of fertilized sand dollar embryos. J Biophys Biochem Cytol 10:59–65

Stanley HP (1965) Electron microscopic observations on the biflagellated spermatids of the teleost fish *Porichthys notatus*. Anat Rec 151:477

Stanley HP (1969) An electron microscope study of spermiogenesis in the teleost fish *Oligocottus maculosus*. J Ultrastruct Res 27:230–243

Stehr CM, Hawkes JW (1979) The comparative ultrastructure of the egg membrane and associated pore structures in the starry flounder, *Platichthys stellatus* (Pallas), and pink salmon, *Oncorhynchus gorbuscha* (Walbaum). Cell Tissue Res 202:347–356

Stoss J (1983) Fish gamete preservation and spermatozoan physiology. In: Hoar WS, Randall DJ (eds) Fish physiology, vol 9B. Academic Press, New York, pp 305–350

Suzuki R (1958) Sperm activation and aggregation during fertilization in some fishes. I. Behavior of spermatozoa around the micropyle. Embryologia 4:93–102

Suzuki R (1959a) Sperm activation and aggregation during fertilization in some fishes. II. Effect of distilled water on the sperm-stimulating capacity and fertilizability of eggs. Embryologia 4:359–367

Suzuki R (1959b) Sperm activation and aggregation during fertilization in some fishes. III. Non specific-specificity of stimulating factor. Annot Zool Jpn 32:105–111

Suzuki R (1961) Sperm activation and aggregation during fertilization in some fishes. VI. The origin of the sperm-activating factor. Annot Zool Jpn 34:24–29

Suzuki R, Fukuda Y (1971) Survival potential of F_1 hybrids among salmonid fishes. Bull Freshwater Fish Res Lab 21:69–83

Swann KS, Whitaker MJ (1986) The part played by inositol trisphosphate and calcium in the propagation of the fertilization wave in sea urchin eggs. J Cell Biol 103:2333–2342

Swann KS, Whitaker MJ (1990) Second messengers at fertilization in sea-urchin eggs. J Reprod Fertil 42 (Suppl):141–153

Szöllösi D, Billard R (1974) The micropyle of trout eggs and its reaction to different incubation media. J Microsc 21:55–62

Taguchi T, Seko A, Kitajima K, Inoue S, Iwamatsu T, Khoo K-H, Morris HR, Dell A, Inoue Y (1993) Structural studies of a novel type of tetra-antennary sialoglycan unit in a carbohydrate-rich

glycopeptide isolated from the fertilized eggs of Indian medaka fish, *Oryzias melastigma*. J Biol Chem 268:2353–2362
Taguchi T, Seko A, Kitajima K, Muto Y, Inoue S, Khoo K-H, Morris HR, Dell A, Inoue Y (1994) Structural studies of a novel type of penta-antennary large glycan unit in the fertilization-associated carbohydrate-rich glycopeptide isolated from the fertilized eggs of *Oryzias latipes*. J Biol Chem 269:8762–8771
Taieb F, Thibier C, Jessus C (1997) On cyclins, oocytes, and eggs. Mol Reprod Dev 48:397–411
Takano K (1964) On the egg formation and the follicular changes in *Lebistes reticulatus*. Bull Fac Fish Hokkaido Univ 15:147–155
Takano K, Ohta H (1982) Ultrastructure of micropylar cells in the ovarian follicles of the pond smelt, *Hypomesus transpacificus nipponensis*. Bull Fac Fish Hokkaido Univ 33:65–78
Tanimoto S, Morisawa M (1988) Roles for potassium and calcium channels in the initiation of sperm motility in rainbow trout. Dev Growth Differ 30:117–124
Tesoriero JV (1977a) Formation of the chorion (zona pellucida) in the teleost, *Oryzias latipes*. I. Morphology of early oogenesis. J Ultrastruct Res 59:282–291
Tesoriero JV (1977b) Formation of the chorion (zona pellucida) in the teleost, *Oryzias latipes*. II. Polysaccharide cytochemistry of early oogenesis. J Histochem Cytochem 25:1376–1380
Tesoriero JV (1978) Formation of the chorion (zona pellucida) in the teleost, *Oryzias latipes*. III. Autoradiography of [^{3}H]-proline incorporation. J Ultrastruct Res 64:315–326
Thibault C, Gerard M, Menezo Y (1975) Acquisition par l'ovocyte de lapine et deveau du facteur de decondensation de noyan du spermatozoide fecondant (MPGF). Ann Biol Anim Bioch Biophys 15:705–715
Thomopoulos A (1953) Sur l'oeuf de l'epinoche (*Gasterosteus aculeatus* L.). Bull Soc Zool Fr 78:142–149
Todd PR (1976) Ultrastructure of the spermatozoa and spermiogenesis in New Zealand freshwater eels (Anguillidae). Cell Tissue Res 171:221–232
Tsuzuki H, Yoshida M, Onitake K, Aketa K (1977) Purification of the sperm-binding factor from the egg of the sea urchin, *Hemicentrotus pulcherrimus*. Biochem Biophys Res Commun 76:502–511
Usui N, Oguri A, Kimura Y, Yanagimachi R (1997) Sperm nuclear envelope: breakdown of intrinsic envelope and de novo formation in hamster oocytes and eggs. Zygote 5:35–46
Wallace RA, Selman K (1981) Cellular and dynamic aspects of oocyte growth in teleosts. Am Zool 21:325–343
Ward GE, Kirschner MW (1990) Identification of cell cycle-regulated phosphorylation sites on nuclear lamin C. Cell 61:561–577
Wassarman PM (1987) The biology and chemistry of fertilization. Science 235:553–560
Webb TA, Fluck RA (1993) Microfilament- and microtubule-based movements during ooplasmic segregation in the medaka fish egg (*Oryzias latipes*). Mol Biol Cell 4:274a
Webb TA, Fluck RA (1995) Spatiotemporal pattern of microtubules in parthenogenetically activated *Oryzias latipes* (medaka) eggs. Fish Biol J Medaka 7:25–35
Webb TA, Kowalski WJ, Fluck RA (1995) Microtubule-based movements during ooplasmic segregation in the medaka fish egg (*Oryzias latipes*). Biol Bull 188:146–156
Whitaker M, Crossley I (1989) How does a sperm activate a sea urchin egg? In: Dale B (ed) Mechanism of fertilization. NATO ASI Ser, vol H45. Springer, Berlin Heidelberg New York, pp 433–443
Wirz-Hlavacek G, Riehl R (1990) Fortpflanzungsverhalten und Eistruktur des Piranhas Serrasalmus nattereri (KNER, 1860). Acta Biol Benrodis 2:19–38
Wolenski JS, Hart NH (1987) Scanning electron microscope studies of sperm incorporation into the zebrafish (*Brachydanio*) egg. J Exp Zool 243:259–273
Wolenski JS, Hart NH (1988a) Effects of cytochalasin B and D on the fertilization of zebrafish (*Brachydanio*) eggs. J Exp Zool 246:202–215
Wolenski JS, Hart NH (1988b) Sperm incorporation independent of fertilization cone formation in the danio egg. Dev Growth Differ 30:619–628

Woolley DM (1998) Studies on the eel sperm flagellum. 2. The kinematics of normal motility. Cell Motil Cytoskel 39:233–245
Wourms JP (1976) Annual fish oogenesis. I. Differentiation of the mature oocyte and formation of the primary envelope. Dev Biol 50:338–354
Yamagami K, Hamazaki TS, Yasumasu S, Masuda K, Iuchi I (1992) Molecular and cellular basis of formation, hardening, and breakdown of the egg envelope in fish. Int Rev Cytol 136:51–92
Yamagami K, Murata K, Hamazaki TS, Iuchi I (1994) Intrahepatic formation of SF substances, the precursors of egg envelope proteins, in response to estrogen administration in fish. In: Davey KG, Peter RE, Tobe SS (eds) Perspectives in comparative endocrinology. National Research Council of Canada, Ottawa, pp 316–324
Yamamoto K (1952a) Studies of the fertilization in the dog salmon (*Oncorhynchus keta*). (1). Morphology of normal fertilization. J Fac Sci Hokkaido Univ (Ser VI, Zool) 11:81–94
Yamamoto K (1952b) Studies of the fertilization in the egg of the flounder. II. The morphological structure of the micropyle and its behavior in response to sperm entry. Cytologia (Tokyo) 16:302–306
Yamamoto K (1953) Studies of the fertilization in the egg of the flounder. III. On the formation of the micropyle-ball. Annot Zool Jpn 26:133–137
Yamamoto K (1955) Studies on the formation of fish eggs. V. The chemical nature and the origin of the yolk vesicle in the oocytes of the herring, *Clupea pallasii*. Annot Zool Jpn 28:158–162
Yamamoto K (1956) Studies on the formation of fish eggs. III. Localization of polysaccharides in oocytes of *Liopsetta abscura*. J Fac Sci Hokkaido Univ Ser 6(12):391–399
Yamamoto T (1939) Changes of the cortical layer of the egg of *Oryzias latipes* at the time of fertilization. Proc Imp Acad Tokyo 15:267–271
Yamamoto T (1940) The change in volume of the fish egg at fertilization. Proc Imp Acad Tokyo 16:482–485
Yamamoto T (1944a) Physiological studies on fertilization and activation of fish eggs. Response of the cortical layer of the egg of *Oryzias latipes* to insemination and to artificial stimulation. Annot Zool Jpn 22:109–125
Yamamoto T (1944b) Physiological studies on fertilization and activation of fish eggs. II. The conduction of the "fertilization-wave" in the egg of *Oryzias latipes*. Annot Zool Jpn 22:126–136
Yamamoto T (1949) Physiological studies on fertilization and activation of fish eggs. IV. Fertilization and activation in narcotized eggs. Cytologia 15:1–7
Yamamoto T (1950) Photodynamic activation in the unfertilized eggs of fish (abstract, in Japanese). Zool Mag 59:19
Yamamoto T (1954) Physiological studies on fertilization and activation of fish eggs. V. The role of calcium ions in activation of *Oryzias* eggs. Exp Cell Res 6:56–68
Yamamoto T (1956) The physiology of fertilization in the medaka (*Oryzias latipes*). Exp Cell Res 10:387–393
Yamamoto T (1958) The physiology of fertilization in fish eggs. In: Dan K, Yamada T (eds) Developmental physiology (in Japanese). Baihukan, Tokyo, pp 73–135
Yamamoto T (1961) Physiology of fertilization in fish eggs. Int Rev Cytol 12:361–405
Yamamoto T (1962) Mechanism of breakdown of cortical alveoli during fertilization in the medaka, *Oryzias latipes*. Embryologia 7:228–251
Yamamoto TS (1957) Some experiments on chemical changes in the membrane of salmon eggs occurring at the time of activation. Jpn J Ichthyol 6:54–58
Yamamoto TS (1958) Biochemical property of the membrane of the herring egg, with special reference to the role of the micropyle in fertilization. J Fac Sci Hokkaido Univ Ser VI Zool 14:9–16
Yamamoto TS (1962) Activation of dog salmon egg by heavy metal ions. J Fac Sci Hokkaido Univ Ser VI Zool 15:148–156
Yamamoto TS (1967) Development of the dog salmon egg without breakdown of cortical alveoli. J Fac Sci Hokkaido Univ Ser VI Zool 16:186–196

Yamamoto TS, Kobayashi W (1992) Closure of the micropyle during embryonic development of some pelagic fish eggs. J Fish Biol 40:225–241
Yamashita M, Onozato H, Nakanoshi T, Nagahama Y (1990) Breakdown of the sperm nuclear envelope is a prerequisite for male pronucleus formation: direct evidence from gynogenetic crucian carp *Carassius auratuslangsdorfii*. Dev Biol 137:155–160
Yamashita M, Fukada S, Yoshikuni M, Bulet P, Hirai T, Yamaguchi A, Yasuda H, Ohba Y, Nagahama Y (1992a) M-phase-specific histone H1 kinase in fish oocytes: its purification, components and biochemical properties. Eur J Biochem 205:537–543
Yamashita M, Fukada S, Yoshikuni M, Bulet P, Hirai T, Yamaguchi A, Lou YH, Zhao Z, Nagahama Y (1992b) Purification and characterization of maturation-promoting factor in fish. Dev Biol 148:8–15
Yamashita M, Kajiura H, Tanaka T, Onoe S, Nagahama Y (1995) Molecular mechanisms of the activation of maturation-promoting factor during goldfish oocyte maturation. Dev Biol 168:62–75
Yanagimachi R (1957a) Some properties of sperm-activating factor in the micropyle area of the herring egg. Annot Zool Jpn 30:114–119
Yanagimachi R (1957b) Studies of fertilization in *Clupea pallasii*. II. Structure and activity of spermatozoa. Zool Mag (Tokyo) 66:222–225
Yanagimachi R (1957c) Studies on fertilization in *Clupea pallasi*. VI. Fertilization of the egg deprived of the membrane (in Japanese). Jpn J Ichthyol 6:41–47
Yanagimachi R (1972) Penetration of guinea-pig spermatozoa into hamster eggs in vitro. J Reprod Fertil 28:477–480
Yanagimachi R (1988) Mammalian fertilization. In: Knobil E, Neill J, Greenwald GS, Markert CL, Pfaff DW (eds) The physiology of reproduction. Raven Press, New York, pp 135–185
Yanagimachi R (1994) Mammalian fertilization. In: Knobil E, Neill J, Greenwald GS, Markert CL, Pfaff DW (eds) The physiology of reproduction, 2nd edn. Raven Press, New York, pp 189–317
Yanagimachi R, Cherr GN, Pillai MC, Baldwin JD (1992) Factors controlling sperm entry into the micropyles of salmonid and herring eggs. Dev Growth Differ 34:447–461
Yanagimachi R, Kanoh Y (1953) Manner of sperm entry in the herring egg, with special reference to the role of calcium ions in fertilization. J Fac Sci Hokkaido Univ Ser VI (11):487–494
Yaron Z (1971) Observations on the granulosa cells of *Acanthobrama terrae-sanctae* and *Tilapia nilotica* (Teleostei). Gen Comp Endocrinol 17:247–252
Yoshimoto Y, Iwamatsu T, Hirano K, Hiramoto Y (1986) The wave pattern of free calcium release upon fertilization in medaka and sand dollar eggs. Dev Growth Differ 28:583–596
Young G, Ueda H, Nagahama Y (1983) Estradiol-17β and 17α,20β-dihydroxy-4-pregnen-3-one production by isolated ovarian follicles of amago salmon (*Oncorhynchus rhodurus*) in response to mammalian pituitary and placental hormones and salmon gonadotropin. Gen Comp Endocrinol 52:329–335
Zhang Q, Arai K, Yamashita M (1998) Cytoplasmic mechanisms for triploid and haploid egg formation in the triploid loach *Misgurnus anguillicaudatus*. J Exp Zool 281:608–619
Zotin A (1958) The mechanism of hardening of the salmonid egg membrane after fertilization of spontaneous activation. J Embryol Exp Morphol 6:546–568

CHAPTER 4

Fertilization in Amphibians

YASUHIRO IWAO[1]

1 Introduction

Fertilization of a female gamete or oocyte by a male gamete or sperm is an indispensable step in sexual reproduction in animals. Since both sperm and oocytes are highly specialized to ensure development of offspring with a diploid genome, many molecules are involved in cell-cell adhesion, membrane fusion, and signaling cascades until the karyogamy between sperm and oocyte nuclei. This chapter focuses on cellular and molecular aspects of fertilization in amphibians, which are one of the best studied animals for fertilization in vertebrates. The mechanisms of polyspermy block to ensure diploid development are discussed in relation to extracellular interactions between the sperm and the oocytes as well as oocyte activation by the sperm.

The subclass Lissamphibia is composed of three orders: Urodela (salamanders and newts), Gymnophiona (caecilians, limbless amphibians), and Anura (frogs and toads) (Table 1). Gametogenesis is basically the same in these orders. Sperm produced in a testis are transported into a sperm duct. Although the sperm of most anurans acquire fertilizability in the testis, the testicular sperm in urodeles do not have sufficient fertilizability (McLaughlin and Humphries 1978). The sperm duct of urodeles is required not only for making spermatophores to transfer the sperm to the females, but also for maturation of sperm. The oocytes in an ovary are surrounded by thecal and inner follicle cells (Dumont 1972). The oocytes arrested at meiotic prophase have grown up more than 1000-fold in volume, and have stored numerous yolk granules and a large amount of cellular components, such as mitochondria and ribosomes, for embryonic development (Laskey 1979). Ovulation is controlled by the secretion of gonadotropins from a pituitary during the breeding season. Full-grown oocytes, which are capable of responding to the gonadotropins, are ovulated into the body cavity, the coelom (Fig. 1). The ovulated oocytes surrounded by coelomic envelopes (CE) are transported to the opening of the oviduct, the ostium, by ciliary movement of the wall of the coelom. After passing through the pars recta of the oviduct, the oocytes are wrapped in

[1] Department of Biological Science, Faculty of Science, Yamaguchi University, 753-8512 Yamaguchi, Japan

Juan J. Tarín and Antonio Cano (Eds.)
Fertilization in Protozoa and Metazoan Animals

Table 1. Classification of amphibians. (Based on Duellman and Trueb 1986)

Class Amphibia	
Subclass Lissamphibia	
Order Urodela (Caudata)	
Suborder Sireniodea	
Family Sirenidae	*Siren intermedia*
Suborder Cryptobranchoidea	
Family Hynobidae	*Hynobius dunni, Hynobius nebulosus, Hynobius retardatus*
Family Cryptobranchidae	*Andrias japonicus, Cryptobranchus alleganiensis*
Suborder Salamandroidea	
Family Proteidae	*Necturus maculosus*
Family Dicamptodontidae	*Dicamptodon ensatus*
Family Amphiumidae	*Amphiuma tridactylum*
Family Salamandridae	*Cynops pyrrhogaster, Notophthalmus viridescens, Pleurodeles waltl*
Family Ambystomatidae	*Ambystoma jeffersonianum, Ambystoma laterale, Ambystoma mexicanum, Ambystoma tigrinum, Ambystoma texanum*
Family Plethodontidae	*Eurycea tynerensis*
Order Gymnophiona (Apoda)	
Family Ichtyophiidae	*Ichthyophis glutinosus*
Order Anura	
Archaeobatrachians	
Family Leiopelmatidae	*Ascaphus truei*
Family Discoglossidae	*Discoglossus pictus*
Family Pipidae	*Xenopus laevis*
Neobatrachians	
Family Leptodactylidae	*Eleutherodactylus coqui, Leptodactylus chaquensis*
Family Bufonidae	*Bufo americanus, Bufo arenarum, Bufo japonicus, Bufo vulgaris, Mertensophryne micranotis, Nectophrynoides* sp.
Family Ranidae	*Rana cameranoi, Rana esculenta, Rana japonica, Rana nigromaculata, Rana pipiens, Rana rugosa, Rana temporaria*

several jelly layers during the passage through the pars convoluta, and then temporarily stay in the ovisac until oviposition. In species exhibiting external fertilization, the jellied oocytes are inseminated by the sperm just after oviposition into water. The females in the urodeles exhibiting internal fertilization receive a ball of sperm, the spermatophore, deposited by a male in front of the females, following the species-specific courtship (Duellman and Trueb 1986). Since the female stores the spermatophore in spermatheca near the cloaca, insemination with the sperm released from the spermatheca occurs at the cloaca just before oviposition (Sever and Brizzi 1998).

In urodeles, which comprise 9 families containing 62 genera and 352 species (Duellman and Trueb 1986), about 90% of the species exhibit internal fertilization (Sever and Brizzi 1998) and some species are ovoviviparous or vivipa-

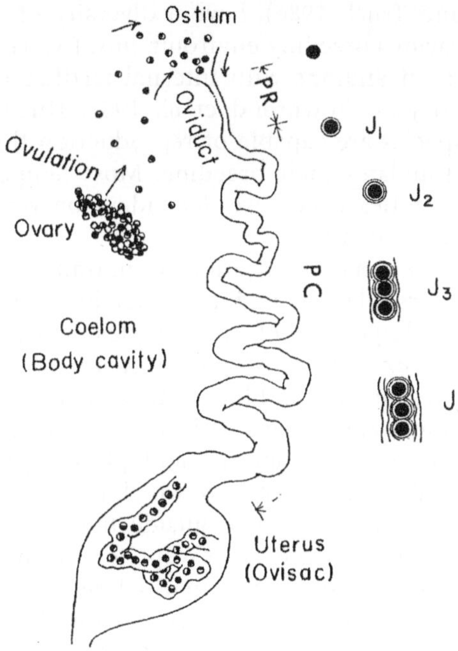

Fig. 1. Reproductive system of a female anuran, *Bufo japonicus*. *PR* Pars recta; *PC* pars convoluta; J_{1-4} inner to outer jelly layers. In this species, outer jelly layers form string of oocytes. (Katagiri 1987)

rous (Wake and Dickie 1998). External fertilization is observed in the families Cryptobranchoidae, Hynobiidae, and Sirenidae, and seems to be primitive in urodeles. Most urodeles exhibit annual seasonal reproduction, but biennial reproduction is characteristic of the terrestrial species of the family Plethodontidae (Duellman and Trueb 1986). Some urodeles attain reproductive maturity with larval external morphology and this is called neoteny. For example, *Amphiuma tridactylum*, *Necturus maculosus*, *Ambystoma mexicanum*, and *Eurycea tynerensis* are obligate neotenes in nature, but *Ambystoma tigrinum* and *Dicamptodon ensatus* are facultatively neotenic (Duellman and Trueb 1986). Some species of *Ambystoma* exhibit gynogenetic reproduction–development without participation of paternal genomes (Uzzell 1964). Little is known about fertilization in caecilians, which comprise 6 families containing 34 genera and 162 species (Duellman and Trueb 1986). The caecilians which live in moist soils of tropical forests have internal fertilization, and most species are ovoviviparous or viviparous (Wake and Dickie 1998). Females in the caecilians seem to have a biennial reproductive cycle (Wake 1980). The oocytes of *Ichthyophis glutinosus* (35 mm in diameter and 42 mm in length) are the largest amphibian oocytes known (Breckenridge and Jayasinghe 1979). Anurans, with 21 families containing 301 genera and 3438 species (Duellman and Trueb 1986), have a diversity of reproductive modes for adaptation to various breeding environments. External fertilization, however, is characteristic of anurans with internal fertilization in a few species having terrestrial oocytes (Townsend et al. 1981; Duellman and Trueb 1986). Most tropical species are capable of reproduction throughout the year, but usually rainfall stimulates their breeding. Most temperate species have cyclic reproductive activities which are dependent on rainfall and temperature (Duellman and Trueb 1986).

External fertilization in anurans and relatively easy artificial fertilization in urodeles having internal fertilization are useful in experiments. In addition, the large size of the oocytes which are tolerant of micromanipulations and extraction of cytoplasm provides us with useful systems to undertake physiological and biochemical analyses of fertilization mechanisms. Furthermore, amphibians utilize two distinct strategies to prevent polyploidy between monospermy and physiological polyspermy. Anurans, in general, exhibit monospermy in which only one sperm enters the oocyte and all other sperm are excluded outside the oocyte membrane or the extracellular investments. Urodeles, except for the monospermic Hynobiidae, exhibit physiological polyspermy in which several sperm enter the oocyte. Only one sperm nucleus is selected to fuse with the oocyte nucleus and the other extra sperm nuclei do not participate in later embryonic development. Thus, amphibian fertilization is one of the most useful systems to determine the role of polyspermy block mechanisms operating in the fertilization process. This chapter emphasizes the recent advances in the study of amphibian fertilization and polyspermy block (for previous work, see an intensive review by Elinson 1986).

2
Interactions of Sperm with Oocyte Investments and Oocyte Membrane

2.1
Interactions of Sperm with Jelly Layers

Amphibian oocytes are surrounded by several layers of investments: a vitelline envelope/coat (VE/VC) whose precursor, the coelomic envelope/coat (CE/CC), is synthesized and secreted by the oocytes during oogenesis (Yamaguchi et al. 1989; Cabada et al. 1996; Kubo et al. 1997) and three to six layers of jelly coat deposited on the VE during the passage through the oviduct (see Katagiri 1987) (Figs. 1–3A). Sperm must penetrate these investments before reaching the oocyte plasma membrane. The jelly layers are composed of highly glycosylated proteins (Katagiri 1973; Yurewicz et al. 1975; Jego et al. 1986; Carroll et al. 1991) with species-specific patterns of oligosaccharide chains (Strecker et al. 1995; Morelle et al. 1998). The jelly layers participate not only in protection of the

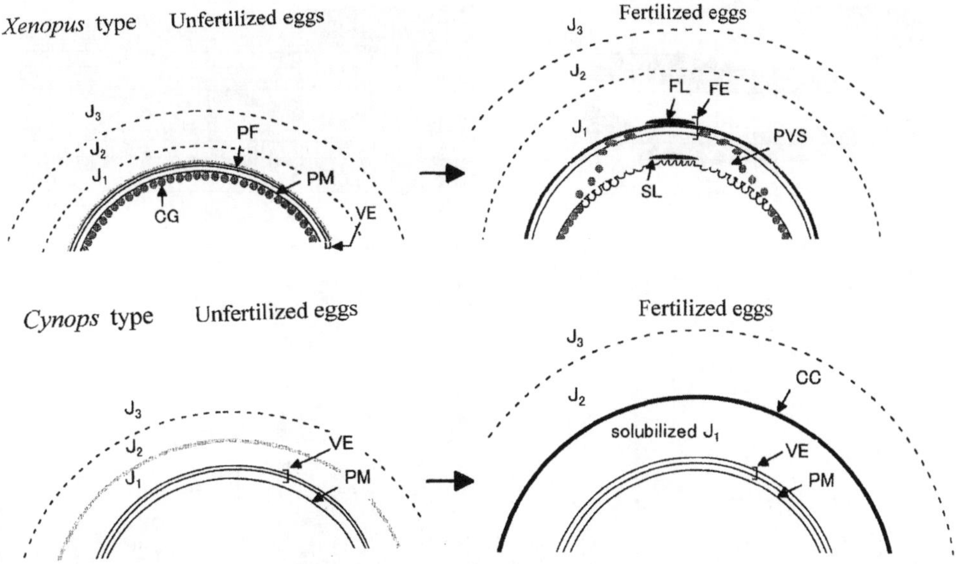

Fig. 2. Formation of fertilization envelope (*FE*) and perivitelline space in anuran oocytes (*Xenopus*-type) and formation of capsular chamber (*CC*) in urodele oocytes (*Cynops*-type). *CG* Cortical granules; *FL* fertilization layer; *J* jelly layers; *PF* pre-fertilization layer; *PM* oocyte plasma membrane; *PVS perivitelline space*; *SL* smooth layer; *VE* vitelline envelope

embryos against mechanical injury and bacterial infection during embryonic development, but also in the assurance of fertilization.

The jelly-free oocytes surrounded by the VE are not fertilizable, but can be fertilized in the presence of jelly components (Katagiri 1966a,b; Barbieri and Raisman 1969; Elinson 1971a; Stewart-Savage and Grey 1984; Heasman et al. 1991). An exception is the newt, *Cynops pyrrhogaster*, whose oocytes can be fertilized in the absence of jelly components (Matsuda and Onitake 1984b). In the frog, *Xenopus laevis*, the jelly layers are designated J_1, J_2, and J_3 from the innermost to the outermost (Fig. 2; Freeman 1968; Yurewicz et al. 1975), and each layer has a fibrillogranular ultrastructure (Bonnell et al. 1996). The jelly layers of anuran oocytes consist of two components (Bonnell et al. 1996; Arranz et al. 1997): fibrous high-molecular-weight glycoconjugates which are the major constituents of the jelly fiber network, and globular low-molecular-weight proteins which appear to function as diffusible sperm chemoattractant (al-Anzi and Chandler 1998). In the toad, *Bufo japonicus*, the jelly layers provide a microenvironment of divalent cations (Ishihara et al. 1984; Shimoda et al. 1994) which is indispensable for sperm–oocyte interactions. In addition, the physical architecture of the jelly layers appears to be important in orienting the sperm towards the surface of the VE (Omata 1993), because fertilization of jelly-less oocytes is facilitated in the presence of synthetic polymers, such as dextran, Ficoll, or PVP (Katagiri 1966b, 1973, 1974; Heasman et al.

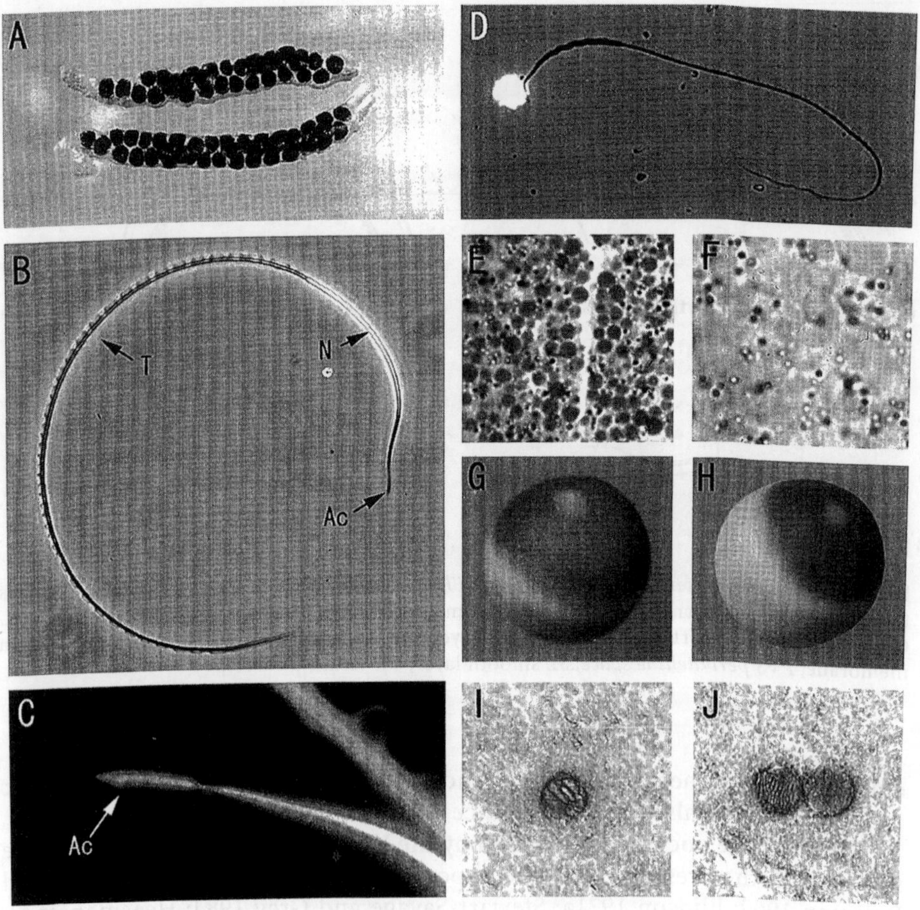

Fig. 3. A *H. nebulosus* oocytes with jelly coats before hydration. B A *C. pyrrhogaster* sperm (*Ac* acrosome; *N* nucleus; *T* tail). C Tip of *C. pyrrhogaster* sperm stained with acridine orange, showing acrosome (*Ac*). D A *C. pyrrhogaster* sperm on gelatin film, showing gelatin-lysis due to sperm acrosomal protease. E and F Cortex of *X. laevis* oocytes, showing cortical granules in the unfertilized oocyte (E) and their loss in the fertilized oocyte after exocytosis (F). G and H *Xenopus* oocytes before activation (G) and during cortical contraction (H), showing accumulation of cortical pigments on the animal pole. I and J DNA synthesis (incorporation of BrdU) in zygote nucleus (I) and accessory sperm nucleus (J) during first cell cycle in the polyspermic *C. pyrrhogaster* oocyte

1991). In the frog, *Discoglossus pictus*, the site of sperm entry into the oocytes is restricted to a dimple at the center of a large concavity of the animal hemisphere and a plug-like jelly on the dimple is necessary for sperm penetration (Campanella 1975; Talevi and Campanella 1988).

The motility of *X. laevis* sperm is suppressed in seminal plasma and initiated by a decrease in osmolality when released into hypotonic water (Inoda and Morisawa 1987), which is required for sperm penetration and passage through the jelly layers (Reinhart et al. 1998). Urodele sperm are quiescent in the spermatheca (Hardy and Dent 1986), but the sperm, deposited from the spermatheca on the oocyte jelly, can be activated by the hypotonic condition in the cloaca where the inseminated oocytes are deposited (Hardy and Dent 1986). A similar situation has been proposed in the frog, *Eleutherodactylus coqui*, which exhibits internal fertilization (Elinson 1987). In addition, sperm motility in the jelly layers is dependent upon some carbohydrate moieties recognized by wheatgerm agglutinin (Reinhart et al. 1998).

2.2 Sperm-Binding to the Vitelline Envelope

The VE of amphibian oocytes consists of networks of filaments (Grey et al. 1974, 1977; Yoshizaki 1984; Hedrick and Nishihara 1991), composed of several glycoproteins (gp) (Gerton and Hedrick 1986a; Takamune et al. 1987; Valz-Gianinet et al. 1991). Detailed ultrastructural observations demonstrate that the VE of *X. laevis* oocytes is composed of two layers: an inner, thin layer with lace-like filaments and an outer, thick layer with long, cable-like fibers (Larabell and Chandler 1988a,b, 1989a). The *X. laevis* VE consists of seven proteins ranging from 37 to 120kDa (Gerton and Hedrick 1986a,b; Bakos et al. 1990a). In *X. laevis*, more than 1000 sperm can bind to the surface of a single dejellied oocyte (Tian et al. 1997a). Not only the presence of two related gp69 and gp64 at insemination, but also the treatment of the oocytes with the antibodies against these glycoproteins, block the binding of sperm on the VE as well as fertilization (Tian et al. 1997a). The gp69/64 glycoprotein is a homologue of the mammalian sperm receptor ZP2 (34–37% identity) and the sperm–oocyte binding on the VE appears to require oligosaccharide chains of gp69/64 (Tian et al. 1999). The major gp41 is homologous to ZP3 glycoprotein in mammalian oocytes (35–39% identity) (Kubo et al. 1997; Yang and Hedrick 1997), but its role in *Xenopus* fertilization remains to be investigated. The VE of *B. japonicus* oocytes is composed of several glycoproteins ranging from 36 to 112kDa (Takamune et al. 1987; Lindsay et al. 1988; Omata and Katagiri 1996). It is estimated that 3000–5000 sperm can bind to a single dejellied oocyte of *B. japonicus* (Omata and Katagiri 1996). The carbohydrate moieties of the major 36- to 39-kDa glycoproteins appear to be involved in the Ca^{2+}-dependent binding of acrosome-unreacted sperm (Omata and Katagiri 1996).

Although coelomic oocytes, the ovulated oocytes into the coelom, cannot be fertilized without passing through the oviduct even in the condition in which the dejellied uterine oocytes are fertilizable by the supplement of jelly components (see Katagiri 1987), removal or partial digestion of CE surrounding the coelomic oocytes makes them fertilizable (Elinson 1973; Katagiri 1974; Stewart-Savage and Grey 1984). Passage through the oviduct renders the coelomic oocytes fertilizable (Cabada et al. 1978; Miceli et al. 1978; Katagiri et al. 1982; Takamune et al. 1986). The CE changes during the passage of the oocytes through the uppermost region of the oviduct, known as the pars recta. The pars recta consists of ciliated cells and non-ciliated cells containing electron-dense secretory granules (Yoshizaki and Katagiri 1981; Katagiri et al. 1982; Yoshizaki 1985). In the pars recta, the thick bundles of CE filaments are dispersed in parallel with the acquisition of sperm penetrability (Grey et al. 1977; Yoshizaki and Katagiri 1981; Katagiri et al. 1982; Larabell and Chandler 1989b; Hedrick and Nishihara 1991). This same alteration of the CE can be produced by treatment with an extract obtained from the pars recta (Cabada et al. 1978; Miceli et al. 1978; Yoshizaki and Katagiri 1981; Katagiri et al. 1982; Bakos et al. 1990b) or from its secretion granules (Katagiri 1986; Takamune et al. 1986, 1987). The passage through the pars recta or the treatment with its extract increases the sensitivity of the CE against the VE lysin of the sperm (Cabada et al. 1978; Miceli et al. 1978; Katagiri et al. 1982; Takamune et al. 1986; Yamasaki et al. 1988) to make the CE penetrable by sperm. In *Bufo arenarum*, the involvement of an oviducal, trypsin-like protease has been proposed in this transformation of the CE into the VE (CE/VE transition) (Miceli and Fernández 1982).

Macromolecular alteration in the CE also occurs during the passage through the oviduct. In *X. laevis*, the 43-kDa glycoprotein in the CE changes into the 41-kDa glycoprotein after passing through the pars recta (Gerton and Hedrick 1986a). In *B. japonicus*, the 40- to 52-kDa glycoproteins disappear, while a 39-kDa glycoprotein appears (Takamune et al. 1986, 1987). The CE/VE transition is caused by a tryptic protease in the pars recta granules (Takamune and Katagiri 1987; Bakos et al. 1990b; Hardy and Hedrick 1992), called oviductin in *X. laevis* (Hardy and Hedrick 1992; Kubo et al. 1999). Not only the treatment with the pars recta extract (Katagiri et al. 1999) but also limited hydrolysis of the CE with trypsin (Lindsay and Hedrick 1998) renders the oocytes fertilizable, mimicking the effect of the pars recta extract.

The number of sperm binding to the CE greatly increases after passage of the oocytes through the pars recta (Omata and Katagiri 1996; Tian et al. 1997b). In *B. japonicus*, the appearance of 39-kDa proteins in the VE correlates with the appearance of the sperm-binding capacity of the VE (Omata and Katagiri 1996). Although *X. laevis* sperm can bind to the VE but not to the CE, there is no detectable molecular weight (MW) change of gp69 and gp64 during the CE/VE transition (Tian et al. 1997b). The acquisition of sperm-binding capacity is probably due to the exposure of gp69/64 on the surface of the VE (Tian et al. 1997b). The hydrolysis of gp43 into gp41 by the oviductin (Hardy and Hedrick 1992) as well as the alteration of ultrastructure in the VE (Larabell and Chandler 1988a, 1989b) may be involved in this transformation. Interestingly, Cdc6, a protein that functions in initiation of DNA replication and cell cycle control, is involved in the acquisition of sperm binding during oocyte maturation (Tian et al. 1997c), suggesting a link between a regulator of the cell cycle and alterations in cell surface properties. In addition, glycoproteins with mannoside residues recognized by concanavalin A (Con A) may be involved in the interaction of sperm with the VE (Valz-Gianinet et al. 1991). A number of Con A-binding sites exist on the surface of the VE (del Pino and Cabada 1987) and fertilization of dejellied oocytes is inhibited by Con A (Cabada et al. 1978; Takamune et al. 1986). It has been recently reported that sperm of the urodele *C. pyrrhogaster* bind to the VE, mediated by heparin-like molecules on the sperm surface (Nakai et al. 1999).

2.3 Acrosome Reaction

The amphibian sperm have an acrosome on each sperm head (Fig. 3B,C), and undergo the acrosome reaction before reaching the oocyte plasma membrane. The ultrastructure of the acrosome has been investigated in anurans (Yoshizaki and Katagiri 1982; Yoshizaki 1987; Bernardini et al. 1988; Campanella et al. 1997) and in urodeles (Picheral 1977a). Although the loss of acrosomal caps in hypotonic solution is reported in *Leptodactylus chaquensis* (Raisman et al. 1980), the sperm appear to undergo the acrosome reaction under the influence of oocyte investments. In *B. japonicus*, the acrosome reaction occurs in the outer surface of the VE (Katagiri et al. 1982; Yoshizaki and Katagiri 1982; Omata 1993). Although the acrosome reaction can be induced by the treatment with the pars recta extract (Katagiri et al. 1982), the nature of an acrosome reaction-inducing substance on the VE remains to be determined. In addition to the pars recta secretions, the 39-kDa sperm-binding protein (Omata and Katagiri 1996) and the components of the jelly layers (Miceli et al. 1987; Omata 1993) appear to be important for the induction of the acrosome reaction. In *D. pictus*, the acrosome reaction is induced in the outermost jelly layer (Campanella et al. 1997). No ultrastructural observations during the acrosome reaction have been performed on *X. laevis*. The detection of the acrosome reaction by staining with a specific lectin (Takamune 1986) or with antibodies against the acrosomal matrix (Martinez and Cabada 1996) may be useful in further investigations. In the newt, *Pleurodeles waltl*, sperm undergo the acrosome reaction in the innermost jelly layers (Picheral 1977a,b).

The acrosome reaction causes the release of acrosomal proteases, observed in sperm of various species (Fig. 3D; Raisman and Barbieri 1969; Elinson 1971b; Raisman and Cabada 1977; Iwao and Katagiri 1982; Cabada et al. 1989; Iwao et al. 1994). In *B. japonicus*, a tryptic sperm protease which can digest the VE (VE lysin) has been purified and characterized (Yamasaki et al. 1988). Since the VE lysin is released from sperm by the treatment with the pars recta extract (Katagiri et al. 1982; Yamasaki et al. 1988), the acrosome-reacted sperm on the VE in response to the pars recta materials likely penetrate the VE in association with the digestion of the VE by the VE lysin. In the urodele *C. pyrrhogaster*, sperm have a unique, strong protease in the acrosomal region (Fig. 3D; Iwao et al. 1994; Mizote et al. 1999), but its activity as the VE lysin has not yet been determined.

2.4
Interactions Between Sperm- and Oocyte-Membranes

What kinds of molecules are involved in the interaction between sperm- and oocyte-membranes at fertilization? The fusion of the inner acrosomal membrane with the oocyte plasma membrane at the top of the sperm head has been confirmed in *D. pictus* (Campanella et al. 1997) and in *P. waltl* (Picheral 1977b). A protein containing a membrane-anchored metalloprotease domain, a disintegrin domain, and a cysteine-rich domain (xMDC16 protein) was identified in the cDNA library of *X. laevis* testis (Shilling et al. 1997). An antibody against xMDC16 recognizes a 56-kDa protein on the posterior surface, probably on the plasma membrane, of the head of oocyte jelly-activated sperm (Shilling et al. 1998). A predicted integrin-binding sequence (KTE) is localized in the disintegrin domain and the oligopeptides containing the KTE sequence inhibit fertilization (Shilling et al. 1997). The xMDC16 may bind to some integrins on the oocyte surface like mammalian sperm MDC protein, fertilin α/β (Almeida et al. 1995). A molecule with a virus fusion peptide-like sequence like fertilin in mammalian sperm (Blobel et al. 1992) has not yet been identified in amphibian sperm.

Although the receptor for xMDC16 remains to be determined, the putative oocyte receptors for sperm binding appear to be unequally distributed on oocyte plasma membranes between hemispheres, because the entry of anuran sperm is restricted to the animal hemisphere (Grey et al. 1982) or a vegetal hemisphere near the equator (Kline 1988), even in VE-free oocytes (Elinson 1973; Katagiri 1974). Since there is no significant difference in sperm binding to the VE between hemispheres (Tian et al. 1997a), the regional difference in sperm receptivity may be due to the difference in sperm adhesiveness on the oocyte plasma membrane between hemispheres (Stewart-Savage and Grey 1987) as well as to the higher sensitivity to the VE lysin of the animal hemisphere (Cabada et al. 1989). Urodele sperm can enter through the entire surface of their own oocytes as well as when cross-fertilizing anuran oocytes (Jaffe et al. 1983; Iwao 1985), but the number of sperm that enter around a vegetal pole is very low (Iwao et al. 1993). In *D. pictus*, fucosyl-containing glycoconjugates are distributed on the oocyte plasma membrane at the D1 region in the bottom of the animal dimple (Denis-Donini and Campanella 1977; Gualtieri and Andreuccetti 1996) where the fertilizing sperm enters (Campanella 1975; Talevi and Campanella 1988). Recently, it has been demonstrated that the 200- and 270-/260-kDa fucosyl glycoproteins which can bind to sperm plasma membrane are localized on the surface of the dimple region (Maturi et al. 1998). It will be interesting to determine whether those glycoproteins bind to the disintegrin sequences on the sperm.

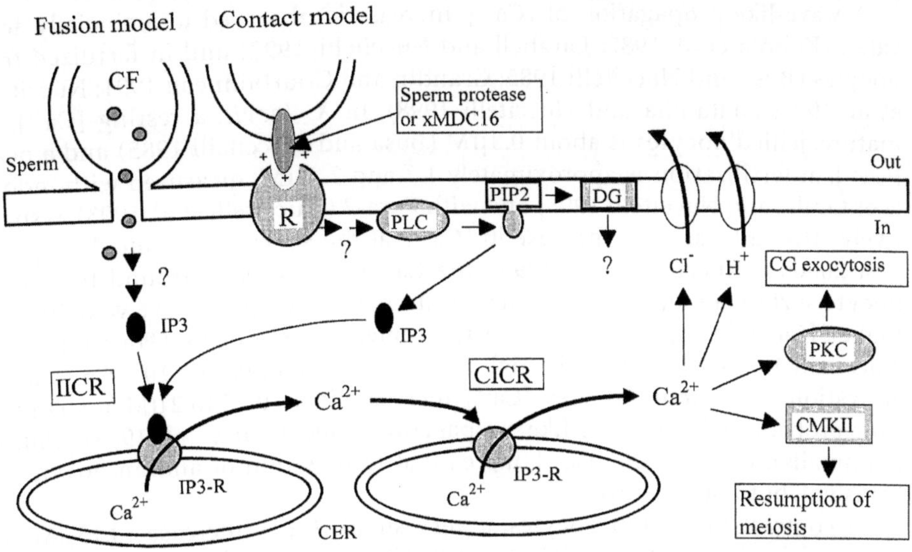

Fig. 4. Signaling pathways in activation of anuran oocytes. *CER* Cortical endoplasmic reticulum; *CF* cytoplasmic factor; *CG* cortical granule; *CICR* Ca^{2+}-induced Ca^{2+} release; *CMKII* calmodulin-dependent protein kinase II; *DG* diacylglycerol; *IICR* IP_3-induced Ca^{2+} release; *IP3* inositol 1,4,5-trisphosphate; *IP3-R* IP_3-receptor; *PIP2* phosphoinositol 4,5-bisphosphate; *PKC* protein kinase C; *PLC* phospholipase C; *R* receptor

3
Mechanism of Oocyte Activation

3.1
$[Ca^{2+}]_i$ Increase

Activation of anuran oocytes at fertilization consists of a series of morphological and biochemical changes in the oocyte: depolarization of the oocyte plasma membrane (fertilization potential), cortical granule exocytosis, spread of surface contraction waves, cortical contraction, and resumption of meiosis II (Fig. 4). The anuran sperm triggers a transient increase in intracellular free Ca^{2+} ($[Ca^{2+}]_i$), which is a primary signal for oocyte activation. The anuran oocytes can be artificially activated by increasing $[Ca^{2+}]_i$ in various ways: by treatment with calcium ionophore A23187 (Steinhardt et al. 1974; Iwao 1982; Charbonneau and Picheral 1983; Andreuccetti et al. 1984; Iwao and Masui 1995), pricking in Ca^{2+}-containing solutions (Wolf 1974b; Goldenberg and Elinson 1980; Iwao 1982; Andreuccetti et al. 1984; Talevi et al. 1985), and microinjection of Ca^{2+} into oocyte cytoplasm (Cross 1981). Activation reactions including the fertilization potential, cortical granule exocytosis, cortical contraction, and resumption of meiosis are all blocked by inhibition of intracellular Ca^{2+} activity with a Ca^{2+} buffer (Kline 1988).

A wave-like propagation of $[Ca^{2+}]_i$ increase is observed in artificially activated (Kubota et al. 1987; Larabell and Nuccitelli 1992) and in fertilized frog oocytes (Busa and Nuccitelli 1985; Grandin and Charbonneau 1991; Nuccitelli et al. 1993; Fontanilla and Nuccitelli 1998). In *X. laevis*, a resting $[Ca^{2+}]_i$ in mature, jellied oocytes is about 0.3 μM (Busa and Nuccitelli 1985) and a peak $[Ca^{2+}]_i$ at fertilization is approximately 1.2 and 2.2 μM, measured with indo-1 (Fontanilla and Nuccitelli 1998) and with Fura-2 (Nuccitelli et al. 1993), respectively. After a local, slow increase in $[Ca^{2+}]_i$ at the sperm entry site during the first 20–60 s (Nuccitelli et al. 1993), the Ca^{2+} wave spreads around the entire oocyte surface about 5 min after fertilization, with a higher wave velocity and higher $[Ca^{2+}]_i$ in the cortex than in the center of the oocyte (Fontanilla and Nuccitelli 1998). The relatively high $[Ca^{2+}]_i$ is maintained for 10–15 min after activation. In *D. pictus* oocytes, $[Ca^{2+}]_i$ increases from 0.25 to 2 μM at fertilization probably with a wave-like propagation (Nuccitelli et al. 1988). Unlike mammalian oocytes, the $[Ca^{2+}]_i$ increase at fertilization of anuran oocytes is transient, but not repetitive.

In contrast to anuran oocytes, a very small $[Ca^{2+}]_i$ increase (0.15 μM) is observed in that of the urodele *P. waltl* (Grandin and Charbonneau 1992a). There is a small $[Ca^{2+}]_i$ increase about 40 min after fertilization of the urodele *C. pyrrhogaster*, and the injection of 1,2-bis (2 aminophenoxy) ethane-N,N,N',N'-tetraacetic acid (BAPTA) into unfertilized oocytes inhibits their activation (Yamamoto et al. 1999). Activation of urodele oocytes can be induced by Ca ionophore (Charbonneau et al. 1983; Iwao and Masui 1995), while most urodele oocytes are resistant to activation by pricking (Fankhauser 1967; Iwao and Masui 1995) except for *P. waltl* oocytes (Aimar and Labrousse 1975). Although the $[Ca^{2+}]_i$ increase induced by sperm is indispensable for oocyte activation, a longer increase in $[Ca^{2+}]_i$ appears to be necessary for the activation of urodele oocytes than anuran oocytes.

The intracellular pH ($[pH]_i$) of *X. laevis* oocytes increases by about 0.3 at fertilization (Webb and Nuccitelli 1981; Grandin and Charbonneau 1989, 1991) with a wave-like propagation after the Ca^{2+} wave (Grandin and Charbonneau 1992b). In the urodele *P. waltl*, fertilized oocytes exhibit a $[pH]_i$ increase of about 0.24 concomitant with disappearance of MPF activity (Grandin et al. 1991). Since artificial increase of intracellular pH does not stimulate protein synthesis (Grandin and Charbonneau 1989), the exact role of $[pH]_i$ increase at fertilization remains unclear.

If the transient $[Ca^{2+}]_i$ increase at fertilization is the primary trigger for initiation of oocyte development in amphibians, how does the fertilizing sperm trigger the $[Ca^{2+}]_i$ increase in anuran oocytes? Inositol 1,4,5-trisphosphate (IP_3) appears to be the second messenger for intracellular release of Ca^{2+} ions from intracellular stores (Fig. 4). Injection of IP_3 into unfertilized oocytes causes $[Ca^{2+}]_i$ release from the endoplasmic reticulum (Busa et al. 1985; Han and Nuccitelli 1990; Larabell and Nuccitelli 1992), followed by the opening of Cl^- channels (Nuccitelli et al. 1988). The $[Ca^{2+}]_i$ increase is inhibited by injection of heparin, a non-specific inhibitor of IP_3 receptor (Nuccitelli et al. 1993), or by injection of an antibody against phosphoinositol 4,5-bisphosphate (PIP_2) (Larabell and Nuccitelli 1992). PIP_2 produces IP_3 and diacylglycerol (DG) when hydrolyzed by phospholipases. Multiple spots of elevated $[Ca^{2+}]_i$, which do not initiate the Ca^{2+} wave, are induced by sperm in the heparin-injected oocytes (Nuccitelli et al. 1993; Fontanilla and Nuccitelli 1998).

IP_3 receptors are distributed in the cortical region of both hemispheres (Kume et al. 1993), and cortical endoplasmic reticulum develops between cortical granules (Gardiner and Grey 1983; Andreuccetti et al. 1984; Campanella et al. 1988). The mass of IP_3 in fertilized oocytes increases about three- to five-fold within 5 min after fertilization (Stith et al. 1993, 1994; Snow et al. 1996). The Ca^{2+}-induced Ca^{2+} release (CICR) for the Ca^{2+} wave appears to be mediated by IP_3 receptors on the cortical endoplasmic reticulum (Nuccitelli et al. 1993), but not by a cADP ribose/ryanodine receptor system (Whitaker and Swann 1993).

On the other hand, the amount of DG increases after fertilization (Stith et al. 1997), and DG may be involved in another signaling pathway in oocyte cytoplasm at fertilization, involving protein kinase C (PKC). An activator of PKC can induce cortical granule exocytosis and cortical contraction (Bement and Capco 1989; Grandin and Charbonneau 1991), in addition to sperm chromatin decondensation and nuclear envelope disassembly in the absence of intracellular Ca^{2+} activity (Bement and Capco 1991). PKC appears to act downstream of the Ca^{2+} wave at oocyte activation (Bement and Capco 1990). However, inhibitors of PKC can trigger oocyte activation, and PKC activators induce neither the $[pH]_i$ increase nor the $[Ca^{2+}]_i$ increase (Grandin and Charbonneau 1991). Further investigation seems to be necessary to understand the role of PKC in activation of amphibian oocytes.

There are at least two signal transduction pathways in the oocytes, a G-protein-linked cascade and a tyrosine-kinase cascade, which can be stimulated by the fertilizing sperm. When serotonin 1c receptor which is a G-protein-linked receptor (Kline et al. 1988) is exogenously introduced into *X. laevis* oocytes, all activation responses can be induced by treatment with serotonin. When epidermal growth factor (EGF) receptor, a tyrosine-kinase-linked receptor (Yim et al. 1994), is introduced, the oocytes can be activated by EGF. The G-protein of *X. laevis* oocytes is in the pertussis toxin-insensitive class (Kline et al. 1991). Injection of guanosine triphosphate (GTP)-γ-S, which stimulates G-proteins, causes oocyte activation in *X. laevis* (Kline et al. 1991) and the amount of α q family G-proteins increases during oocyte maturation (Gallo et al. 1996). However, it remains to be determined whether an endogenous receptor for sperm is coupled with those pathways (Runft et al. 1999).

Recently, a protein-tyrosine kinase (Xyk) related to Src-family protein kinases has been purified from *X. laevis* oocytes (Sato et al. 1996). Interestingly, the kinase activity of Xyk is elevated several-fold upon fertilization, concomitant with its translocation from the oocyte cortex to the cytosolic fraction (Sato et al. 1999). Fertilization of *X. laevis* is blocked by inhibitors of protein tyrosine kinase (Glahn et al. 1998; Sato et al. 1998, 1999). Although the substrate for the Xyk in the oocyte has not yet been identified, the Xyk must play an important role in the cascade of oocyte activation before $[Ca^{2+}]_i$ increase, because the activation as well as the translocation of Xyk is induced by a fertilizing sperm, but not by Ca ionophore (Sato et al. 1999).

There are two distinct models to explain how the fertilizing sperm transmits a signal into oocyte cytoplasm: a contact model and a fusion model (Fig. 4). In the contact model, contact between the sperm and oocyte membranes forms a complex between a sperm agonist (ligand) and an oocyte receptor. The complex then transmits a signal causing $[Ca^{2+}]_i$ increase in the oocyte cytoplasm. The contact model is supported by the following evidence: (1) oocytes of *X. laevis* are activated by a purified, tryptic sperm protease of *C. pyrrhogaster* or by cathepsin B (Iwao et al. 1994, 1995; Mizote et al. 1999). The sperm protease may hydrolyze the oocyte receptor to allow the agonistic sperm peptide to react with the receptor. Alternatively, the complex between the sperm protease and the oocyte receptor may interact with lateral integrins, and then trigger the signaling; (2) the fertilization of denuded oocytes of *X. laevis* is blocked by the inhibitors of the sperm protease (Mizote et al. 1999); (3) the oocytes of *X. laevis* are activated by Arg-Gly-Asp (RGD)-containing peptides, RGDS (Iwao and Fujimura 1996), which are known to bind some integrins located on various cell membranes; (4) oligopeptides containing the predicted integrin-binding sequence, KTE, localized in the disintegrin domain of sperm xMDC16, cause oocyte activation in *X. laevis* (Shilling et al. 1998); (5) the RGDS peptide induces the activation as well as translocation of Xyk, as does the fertilizing sperm (Sato et al. 1999).

In contrast, there is little evidence in anurans for a fusion model of oocyte activation. This mode of transmission of a signal through a connection between sperm and oocyte after sperm–oocyte fusion has been proposed in mammalian oocyte activation (Parrington et al. 1996). In contrast to the aforementioned evidence for a contact model in anurans, some evidence suggests a fusion model of oocyte activation in urodeles. For instance, although *C. pyrrhogaster* oocytes exhibit a temporal, spike-like $[Ca^{2+}]_i$ increase during the early stage of oocyte activation (Yamamoto et al. 1999), a single local increase in $[Ca^{2+}]_i$ by pricking their plasma membrane does not activate the oocytes. Furthermore, *C. pyrrhogaster* oocytes are less sensitive to the sperm protease than *X. laevis* oocytes (Iwao et al. 1994). Although we have found that the injection of sperm extract into unfertilized *C. pyrrhogaster* oocytes induces oocyte activation accompanied by a long-lasting $[Ca^{2+}]_i$ increase (Yamamoto and Iwao 1998), further investigation is necessary to determine a role for a sperm cytosolic factor in amphibian oocyte activation.

3.2 Fertilization Potentials

The unfertilized oocytes of both anurans and urodeles have slightly negative membrane potentials (–10 to –30 mV) (see references in Table 2). At fertiliza-

Table 2. Fertilization potentials and block potentials in amphibians

Species	Peak of fertilization (activation) potentials (mV)	Voltage (mV) inhibiting fertilization[a]	References[b]
Anurans			
Bufo japonicus	+30	+10 to ~+20	1, 2, 4, 10, 13
Bufo americanus	+21	N.D.	3
Bufo vulgaris	ca. +30	N.D.	6
Rana pipiens	+17	+1 to ~+22	3
Rana temporaria	ca. +5	~+15	6
Rana cameranoi	+6	N.D.	17
Rana esculenta	ca. +30	N.D.	6
Rana japonica	+15	N.D.	18
Rana nigromaculata	+20	N.D.	18
Rana rugosa	+32	N.D.	2
Xenopus laevis	+3 to ~+19	0 to ~+20	5, 9, 16
Discoglossus pictus	+18	None	11, 12, 15
Urodeles			
Cynops pyrrhogaster	−23	None	8, 14
Pleurodeles waltl	−18 (no change)	None	6
Ambystoma mexicanum	−26 (no change)	N.D.	6
Notophthalmus viridescens	N.D.	None[c]	7
Andrias japonicus	N.D.	None[d]	18
Hynobius nebulosus	+47	+30 to ~+50	1

N.D. Not determined.

[a] Lowest voltage affecting fertilization and voltage for complete inhibition of fertilization.

[b] 1, Maéno (1959); 2, Ito (1972); 3, Cross and Elinson (1980); 4, Iwao et al. (1981); 5, Grey et al. (1982); 6, Charbonneau et al. (1983); 7, Jaffe et al. (1983); 8, Iwao (1985); 9, Webb and Nuccitelli (1985); 10, Iwao (1987); 11, Nuccitelli et al. (1988); 12, Talevi and Campanella (1988); 13, Iwao (1989); 14, Iwao and Jaffe (1989); 15, Talevi (1989); 16, Iwao et al. (1994); 17, Erdoğan et al. (1996); 18, Iwao (unpubl.).

[c] Voltage-independent fertilization with *X. laevis* eggs.

[d] Polyspermy in cross with *X. laevis* eggs.

tion, anuran oocytes exhibit a positive-going shift of membrane potential (depolarization) of the oocyte plasma membrane or a change in membrane potential, called the fertilization potential (Table 2). The fertilization potential in anuran oocytes is mediated mainly by opening of Cl^- channels (halide ion channels) (Ito 1972; Grey et al. 1982; Webb and Nuccitelli 1985; and references in Table 2), accompanied by opening of K^+ channels (Jaffe and Schlichter 1985; Jaffe et al. 1985; Erdoğan et al. 1996) and voltage-activated Na^+ channels (Jaffe and Schlichter 1985; Peres and Mancinelli 1985). The opening of Cl^- channels in the oocyte plasma membrane is caused by the increase in $[Ca^{2+}]_i$ (Cross 1981; Young et al. 1984) and the initiation of $[Ca^{2+}]_i$ increase corresponds to the onset of the fertilization potential (Nuccitelli et al. 1993). The opening of Cl^- channels (Jaffe et al. 1985; Kline and Nuccitelli 1985) as well as K^+ channels (Jaffe et al. 1985) propagates across the oocytes from the sperm entry site concomitant with the wave of $[Ca^{2+}]_i$ increase at fertilization. The membrane potential of fertilized oocytes returns to negative levels 10–15 min after activation. The contribution of ATP-activated Na^+ channels to the fertilization potential (Kupitz and Atlas 1993) has not yet been confirmed (Nuccitelli 1994). The distribution of Cl^- channels in *D. pictus* oocytes is restricted to the oocyte plasma membrane at the dimple where the sperm penetrates (Talevi et al. 1985; Nuccitelli et al. 1988).

By contrast, in urodeles, the membrane potential of *P. waltl* oocytes does not change or exhibits a small negative-going shift (hyperpolarization) at fertilization (Charbonneau et al. 1983). The oocytes of *C. pyrrhogaster* exhibit small transient negative-going potentials corresponding to each sperm entry (Iwao 1985), but the ion channels mediating those small potential changes remain unclear. A remarkable exception is the monospermic salamander, *Hynobius nebulosus* (Iwao 1989), whose oocytes exhibit a large positive-going fertilization potential beyond +40 mV, mediated by the opening of Cl^- channels. The oocytes of *H. nebulosus* are easily activated by pricking, electric shock, or Ca ionophore.

3.3 Cortical Reactions and Fertilization Envelope Formation

In anuran oocytes, the wave of $[Ca^{2+}]_i$ increase leads to exocytosis of cortical granules localized beneath the oocyte plasma membrane (Figs. 2, 3E,F). The wave of cortical granule exocytosis spreads from the site of sperm entry or needle pricking (see Elinson 1986). For example, it takes about 1 min for the wave to cross the equator and about 2.5 min for it to reach the opposite pole in *B. japonicus* oocytes (Iwao 1982).

Cortical granules contain lectins and proteases which cause transformation of the VE into the fertilization envelope (FE). In *X. laevis*, the contents of cortical granules are deposited into the perivitelline space after exocytosis; they diffuse through the VE and reach the J_1/VE interface (Fig. 2; Grey et al. 1974; Wolf 1974a). The cortical granule (CG) lectin precipitates with a ligand in the J_1 layer, in the presence of Ca^{2+} to form a fertilization layer (F layer) on the VE (Wyrick et al. 1974; Nishihara et al. 1986). This occurs within 4–8 min after fertilization (Larabell and Chandler 1990). The CG lectin is a mixture of large glycoproteins (MW 539 and 655 kDa; Nishihara et al. 1986) containing 38 and 48 kDa subunits (Chamow and Hedrick 1986). The CG lectin specifically recognizes a terminal α-galactose of the ligand in the J_1 layer (MW >250 kDa) (Wyrick et al. 1974; Quill and Hedrick 1996). Another source of the ligand for the CG lectin, however, has been proposed in *X. laevis* oocytes (Yoshizaki 1984; Yoshizaki and Katagiri 1984). These authors found that a pre-fertilization (PF) layer localizes on the outer surface of the VE, not in the J_1 jelly layer (Fig. 2). The PF layer is formed by a secretion from the cells at the caudal portion of the pars recta. Two types of CG lectin, which have heterogenous subunits (Yoshizaki 1989), precipitate with a ligand (105 kDa) in the PF layer to form the F layer (Yoshizaki 1994). In either case, the electron-dense fertilization envelope (FE) in *X. laevis* oocytes is composed of the F layer and the altered VE, described below (Grey et al. 1974; Yoshizaki and Katagiri 1984). In addition, a smooth (S) layer forms on the tips of the microvilli facing the perivitelline space as a consequence of cortical granule exocytosis (Fig. 2; Larabell and Chandler 1988a,b, 1990).

The VE of anuran oocytes is also altered morphologically and biochemically at fertilization (Grey et al. 1974; Wyrick et al. 1974; Miceli et al. 1987; Lindsay et al. 1988; Bakos et al. 1990a; Hedrick and Nishihara 1991). In *X. laevis*, the thick bands of the cable-like fibers in the outer layer of the VE are transformed into concentric fibrous sheets (Larabell and Chandler 1988a,b) within 2–3 min after activation (Larabell and Chandler 1989a). Two VE components in *X. laevis*, gp69 and gp64, are converted to gp66 and gp61 in the FE, respectively (Gerton and Hedrick 1986b; Tian et al. 1997a,b). In *B. japonicus*, gp65 and gp61 of the VE are hydrolyzed to gp62 and gp58, respectively (Lindsay et al. 1988). A chymotrypsin-like protease, ovochymase, might be involved in the molecular-weight changes of the VE of *X. laevis* oocytes (Lindsay and Hedrick 1995). The inactive form of ovochymase localized in the perivitelline space of unfertilized oocytes (Lindsay et al. 1992) appears to be activated at oocyte activation by another trypsin-like protease probably released at cortical granule exocytosis (Lindsay and Hedrick 1989).

In *B. japonicus*, the hardening and hydrolysis of the VE are dependent upon Ca^{2+} and protease(s) (Lindsay et al. 1988). Although these changes occur independent of cortical granule exocytosis, their mechanisms remain unknown. The FE hardens, so that resistance to heating, chemicals, and proteolytic enzymes increases in the FE of various anurans (Raisman and Barbieri 1969; Miceli et al. 1977; Iwao and Katagiri 1982; Yamasaki et al. 1988; Bakos et al. 1990a).

The FE elevates away from oocyte surface probably due to an influx of water into the perivitelline space, because the fertilization envelope is impermeable to macromolecules (Schmell et al. 1983). Most anuran oocytes rotate with respect to gravity in the fluid-filled perivitelline space (Fig. 2), whereas the oocytes of *D. pictus* form a capsular chamber at the jelly layer/VE interface, in which the oocytes with the VE rotate according to gravity (Campanella et al. 1992). The filamentous, lens-shaped jelly coat at the animal pole is dissolved by an enzyme(s) with peroxidase activity, released at vacuole exocytosis (Campanella et al. 1988, 1992; Pitari et al. 1993). The rotation of oocytes in the perivitelline space or the capsular chamber seems to be important for the axis formation of embryos which is sensitive to gravity (Houliston and Elinson 1992) as well as protection from ultraviolet irradiation in nature.

Since urodele oocytes lack cortical granules (Hope et al. 1963; Picheral 1977a,b; Iwao 1989), the formation of the fertilization envelope as a consequence of cortical granule exocytosis is not expected. However, a change in the VE or deposition of macromolecules into the perivitelline space, not mediated by cortical granules, may occur in urodele oocytes, because the perivitelline space appears in *C. pyrrhogaster* oocytes 3–4h after fertilization. In urodele oocytes, the rotation with respect to gravity occurs in the capsular chamber formed by hydration and dissolution of the innermost jelly layer (Salthe 1963; Picheral 1977a,b) independent of oocyte activation (Fig. 2). The boundary with the outer jelly layers forms an osmotic barrier that allows water to be retained in the capsular chamber. This barrier is likely made by a lectin–ligand interaction between secretions of the anterior and middle oviducal regions (Jego et al. 1986).

Cortical granule exocytosis is followed by the spreading of a surface contraction wave which consists of a combination of elongation of microvilli on the oocyte surface and smoothing of the oocyte surface (see Elinson 1986). The contraction of the pigmented cortex in the animal hemisphere towards the animal pole is a clearly visible sign of oocyte activation in anurans and is called the cortical contraction (Fig. 3G,H; Elinson 1975; Stewart-Savage and Grey 1982). The contraction of an actin–myosin network in the oocyte cortex appears to be involved in the Ca^{2+}-regulated cortical contraction (Christensen et al. 1984; Campanella et al. 1990). In urodeles, surface microvilli also undergo a cyclic change of elongation/regression from fertilization to the first few division cycles (Aimar 1991). However, there is no drastic change in the distribution of cortical pigments at oocyte activation.

A small fertilization body (cone) is formed on the surface at each site of sperm entry (Elinson and Manes 1978; Charbonneau and Picheral 1983; Talevi and Campanella 1988; Talevi 1989), likely involving a reorganization of cortical filamentous actin (Chow and Elinson 1993). All sperm components, including a tail, are incorporated into the oocyte through the fertilization cone. The accumulation of cortical pigment at each site of sperm entry is observed in both anurans (Elinson and Manes 1978; Paleèek et al. 1978; Stewart-Savage and Grey 1982) and urodeles (Iwao 1989; Iwao et al. 1997), probably due to the activity of transportation along the microtubules of sperm asters (Stewart-Savage and Grey 1982; Schroeder and Gard 1992; Elinson and Paleèek 1993; Iwao et al. 1997).

3.4 Resumption of Meiosis and Sperm Nuclear Decondensation

The $[Ca^{2+}]_i$ increase at fertilization causes the resumption of meiosis. The unfertilized amphibian oocytes are arrested at the second meiotic metaphase (MII) by cytostatic factor (CSF) (Masui and Markert 1971; Meyerhof and Masui 1979) which maintains maturation-promoting factor (MPF) at high levels in oocyte cytoplasm (see Masui 1992). Activity of not only MPF (Gautier 1987; Aimar 1991; Iwao et al. 1993; Sakamoto et al. 1998), but also CSF (Sawai and Higuchi 1989; Iwao and Masui 1995) is also found in urodele oocytes.

MPF, which is responsible for entry into M phase, consists of cdc2 kinase and cyclin B (Lohka et al. 1988; Gautier et al. 1990). A Mos protein, the gene product of c-Mos, is a key component of CSF (Sagata et al. 1989). Although CSF is sensitive to Ca^{2+} (Meyerhof and Masui 1977), destruction of cyclin B occurs earlier than that of Mos at activation in *X. laevis* oocytes (Watanabe et al. 1991) and *C. pyrrhogaster* oocytes (Yamamoto et al. 1999). It is proposed that the $[Ca^{2+}]_i$ increase stimulates calmodulin-dependent protein kinase II to cause cyclin B degradation (Lorca et al. 1993). Cyclin B appears to be destroyed by proteasomes after ubiquitination (Aizawa et al. 1996; Tokumoto et al. 1997), which is probably regulated by a large ubiquitin ligase complex, termed the anaphase-promoting complex (Peters et al. 1996). Thus, the activated oocyte completes meiosis, so that a swelling oocyte pronucleus is formed after the emission of the second polar body.

The nucleus of amphibian sperm is composed of highly condensed chromatin containing sperm nuclear basic proteins (SBPs) (Poccia 1986). After entry into oocyte cytoplasm, the sperm undergo nuclear envelope breakdown and nuclear decondensation to form a sperm pronucleus. The change into the pronucleus is dependent upon the presence of contents of the germinal vesicle in oocyte cytoplasm after germinal vesicle breakdown (Skoblina 1974; Katagiri and Moriya 1976). The mechanisms of nuclear decondensation and pronucleus formation have been investigated with a cell-free system using an oocyte extract of *X. laevis* (Lohka and Masui 1983) or *B. japonicus* (Iwao and Katagiri 1984; Ohsumi and Katagiri 1991). The SBPs in *B. japonicus* (Ohsumi and Katagiri 1991) as well as SP2–6 in *X. laevis* (Philpott et al. 1991) are eliminated from the sperm chromatin by nucleoplasmin which is abundant in oocyte cytoplasm. The dependency of pronucleus formation on the germinal vesicle contents corresponds with the localization of nucleoplasmin in the germinal vesicle of full-grown oocytes (Wedlich and Dreyer 1988). The sperm pronucleus is accompanied by a sperm aster, which originates from a centrosome containing a pair of sperm-derived centrioles (Elinson and Paleèek 1993; Iwao et al. 1997). However, mitochondria of the incorporated sperm are unlikely to participate in embryonic development (Dawid and Blackler 1972).

4 Mechanisms for Polyspermy Block

4.1 Role of Polyspermy Block

Since fertilization in most amphibians occurs externally, sperm and oocytes are released into water in ponds or in streams. Approach of the sperm towards the oocytes is greatly dependent upon environmental circumstances. Numerous sperm are usually released from a male to fertilize an oocyte. For example, at mating of the Japanese giant salamander, *Andrias japonicus*, the water in a stream becomes milky with released sperm. A large number of sperm at fertilization ensures successful collision between sperm and oocytes, but puts the oocytes in danger of polyspermy. Polyspermic anuran oocytes undergo multipolar cleavage and cease development (see Elinson 1986), because sperm asters exclude each other and each forms a bipolar spindle at the first cleavage (Render and Elinson 1986; Iwao et al. 1997). In monospermic species, one oocyte must incorporate with one sperm for development with diploid genomes. Even in physiologically polyspermic oocytes, one sperm nucleus must be selected to participate in embryonic development. To avoid the risk of polyspermy, amphibians have evolved elaborate mechanisms to block polyspermy. The mechanisms for polyspermy block are so efficient that polyspermy is completely prevented even when the oocytes are artificially inseminated with 10^3 sperm in 1 μl, which is approximately the same volume as the *X. laevis* oocyte. Internally fertilized urodele oocytes appear to be inseminated with a very limited number of sperm released from spermatheca, because extra sperm are seldom in the jelly layers or around the vitelline envelope. Indeed, the *C. pyrrhogaster* female, with a spermatophore, can oviposit more than 100 fertilized oocytes over a month without participation of another male.

The strategies for the polyspermy block vary between species. The oocytes of anurans and monospermic urodeles have typical extracellular blocks to polyspermy. Only one sperm fuses with the oocyte, and the rest of the sperm are prevented from entering. Anuran oocytes have three different ways of blocking polyspermy: a 'very slow block' in the jelly layers, a 'slow block' on the fertilization envelope, and a 'fast block' on the oocyte plasma membrane. Urodele species showing physiological polyspermy, however, differ from this pattern because they exhibit neither the 'slow block' on the fertilization envelope nor the 'fast block' on the plasma membrane. They exhibit the 'very slow block' to polyspermy in the jelly layers and another mechanism called the 'very slow, cytoplasmic block' which consists of selecting a single sperm nucleus to form a diploid individual and rejecting the other supernumerary sperm nuclei.

4.2
Reduction in Number of Sperm in Oocyte Investments

While the jelly layers are required for fertilization as mentioned earlier, a relatively small number of sperm reach the VE within several minutes after insemination (Reinhart et al. 1998). When an immature anuran oocyte that lacks polyspermy block mechanisms is inseminated, 1–20 sperm enter the oocyte at intervals of 2–10 s (Schlichter and Elinson 1981; Iwao 1987). The delay in arrival of the second sperm appears to provide enough time for the oocyte to switch on the polyspermy block system. Mature oocytes from *H. nebulosus*, a typical urodele species that exhibits monospermy in normal conditions, undergo polyspermy under heavy concentration of sperm (Iwao 1989). The limitation on sperm penetration into the jelly layers is probably due to jelly hydration and swelling, when released into hypotonic conditions such as a freshwater. Consequently, sperm cannot pass through the swollen jelly layers within 10–15 min after immersion. This kind of prevention of sperm entry is reported not only in anurans (Katagiri 1962; Elinson 1971a; Wolf and Hedrick 1971), but also in urodeles (McLaughlin and Humphries 1978; Matsuda and Onitake 1984a). In urodeles, the outer wall of the capsular chamber (CC) which forms after jelly hydration may be a barrier to sperm entry (Fig. 2). The block in the jelly layers is 'a very slow, but permanent block' to polyspermy in amphibians.

4.3 Fast, Electrical Block to Polyspermy

The fast, electrical block to polyspermy operates at the oocyte plasma membrane of anuran and monospermic urodele oocytes. Fertilization of denuded anuran oocytes, lacking the VE, is monospermic (Elinson 1973; Katagiri 1974), suggesting rapid changes in oocyte plasma membrane after the entry of the primary sperm. This fast block is temporary, because the fertilized oocytes become fertilizable again about 30 min later (Yamasaki and Katagiri 1991). The rise in oocyte membrane potential to a positive level within 1 s suggests that the positive-going fertilization potential operates as the fast, electrical block to polyspermy. The relationship between the positive fertilization potential and the fast block to polyspermy was demonstrated in anuran amphibians (Cross and Elinson 1980), as it had previously been shown in sea urchins (Jaffe 1976). When the membrane potential of unfertilized oocytes is kept negative by adding halide ions to the medium, a negative-going fertilization potential occurs and the oocytes undergo polyspermy (Cross and Elinson 1980; Grey et al. 1982). Polyspermy induced by halide ions occurs with natural mating in *X. laevis* (Grey et al. 1982). The order of effectiveness of halide ions ($I^- > Br^- > Cl^- > F^-$) causing the negative potential (Cross and Elinson 1980; Grey et al. 1982; Webb and Nuccitelli 1985) corresponds to the effectiveness in induction of polyspermy, originally reported by Bataillon (1919).

Further evidences for the electrical block are provided from experiments with voltage-clamp at fertilization (Cross and Elinson 1980; Charbonneau et al. 1983; Jaffe et al. 1983; Iwao 1989; Iwao and Jaffe 1989; Iwao et al. 1994). When the membrane potential of an unfertilized *X. laevis* oocyte is held at a positive level, fertilization (oocyte activation) is prevented, so that neither a change in the holding current nor a cortical reaction is observed. After the clamp is removed, the oocyte becomes fertilizable and undergoes activation, exhibiting a positive-going fertilization potential. When the membrane potential of unfertilized oocytes is held at a negative level, fertilization is not blocked, and a large amount of inward current (fertilization current) is needed to keep the membrane potential at the clamped level. Under these conditions, several sperm enter the oocyte. Fertilization is not affected under ±0 mV, but penetration of sperm as well as oocyte activation are completely prevented by clamping at +20 mV (Table 2). A membrane potential of less than ±0 mV in unfertilized oocytes allows sperm to enter the oocyte, but the positive-going fertilization potentials amounting to about +19 mV prevent the extra sperm from entering the activated oocytes. A similar correlation between the amount of positive-going fertilization potential and that of holding potentials to prevent fertilization is observed in various species (Table 2). Thus, the entry of sperm into oocytes of monospermic species is greatly dependent upon the oocyte membrane potential, and this situation is referred to as 'voltage-dependent fertilization'. There is, however, an interesting exception: the oocytes of *D. pictus* elicit a Cl-dependent, positive-going fertilization potential, but fertilization is independent of voltage, and polyspermy occurs (Talevi 1989).

In contrast to the voltage-dependent fertilization of anurans, the oocytes of physiologically polyspermic urodeles do not exhibit a positive-going fertilization potential (Table 2). Voltage clamping of the oocyte membrane to the positive level do not affect fertilization of *P. waltl* (Charbonneau et al. 1983) and *C. pyrrhogaster* (Iwao and Jaffe 1989). These are 'voltage-independent fertilizations'. The presence of voltage-independent fertilization corresponds to the negative-going fertilization potentials or no fertilization potential in these species.

In cross-fertilization between the oocytes of a voltage-dependent species (*X. laevis*) and the sperm of a voltage-independent species (*Notophthalmus viridescens* or *C. pyrrhogaster*), the oocytes exhibit a positive-going fertilization potential, but fertilization is polyspermic and voltage-independent (Jaffe et al. 1983; Iwao et al. 1994). The reverse combination, cross-fertilization of polyspermic urodele oocytes with monospermic anuran sperm, has not yet succeeded. However, cross-fertilization between a voltage-independent species (*C. pyrrhogaster* oocytes) and a voltage-dependent species (*H. nebulosus* sperm) is voltage-dependent (Iwao and Jaffe 1989). A converse combination (*H. nebulosus* oocytes and *C. pyrrhogaster* sperm) is voltage-independent (Iwao and Jaffe 1989). Furthermore, homologous fertilization of *B. japonicus* is prevented at +20 mV, and that of *H. nebulosus* is prevented at +40 mV (Table 2), whereas the cross-fertilization of *B, japonicus* oocytes by *H. nebulosus* sperm is prevented at +40 mV, but not at +20 mV (Iwao and Jaffe 1989). These findings strongly support the hypothesis that the component responsible for the voltage-dependency is a molecule on the sperm plasma membrane, but not on the oocyte membrane (Iwao and Jaffe 1989).

A molecule with a positive charge like a virus fusion peptide is expected to be on the sperm membrane (Jaffe et al. 1983; Iwao and Jaffe 1989). In this respect, activation of *X. laevis* oocytes by the peptide containing RGD sequence is independent of voltage (Iwao and Fujimura 1996), suggesting that a voltage-dependency is mediated by part of a putative sperm agonist different from RGD sequence. However, xMDC16 oligopeptides containing KTE sequence activate *X. laevis* oocytes in a voltage-dependent manner (Shilling et al. 1998). A basic amino acid residue in the sequence (lysine) with a positive charge may be responsible for the voltage-dependent activation.

4.4
Slow Polyspermy Block by the Fertilization Envelope

Many sperm are stopped at the surface of the FE when anuran oocytes are inseminated with a dense sperm suspension. This observation indicates that the FE is an effective barrier preventing extra sperm from penetrating into the perivitelline space. As described previously, both morphological and biochemical features of the VE are greatly altered after fertilization, so that sperm cannot penetrate into the FE (Grey et al. 1976). *X. laevis* sperm are unable to bind to the FE probably due to the alteration of two components of the VE (gp69/64 into gp66/61) which are responsible for sperm binding (Tian et al. 1997a,b). It has been recently reported that proteolytic cleavage of the gp69/64 protein by a cortical granule protease results in loss of sperm binding to the activated oocytes (Tian et al. 1999). In addition, the FE loses sensitivity against the VE lysin (Raisman and Barbieri 1969; Iwao and Katagiri 1982; Cabada et al. 1987; Yamasaki et al. 1988). In *B. japonicus*, the loss of the VE lysin sensitivity which occurs 3–5 min after activation (Yamasaki and Katagiri 1991) is probably mediated by the contents of cortical granules (Cabada et al. 1987; Yamasaki and Katagiri 1991) and corresponds with the morphological alteration in the VE 2–4 min after fertilization (Larabell and Chandler 1990). Thus, a slow block to polyspermy in anurans is accomplished by at least two changes of the FE: a decrease in sperm binding and resistance to the VE lysin. These changes occur 4–8 min after fertilization (Larabell and Chandler 1990) which coincides with the time of establishment of the slow block as determined in the absence of a fast block (Grey et al. 1982).

In addition, some further changes of the oocyte plasma membrane also appear to block the fertilization of *X. laevis* oocytes. Receptivity of denuded oocytes to sperm is lost during the first 10 min following fertilization, independent of the fast, electrical block (Stewart-Savage and Grey 1987). This loss may be due to the formation of the smooth (S) layer at the perivitelline space/VE interface 10–15 min after fertilization (Larabell and Chandler 1988a,b, 1990) and a reduction in surface adhesivity (Stewart-Savage and Grey 1987).

Urodele oocytes, which lack cortical granules, appear unlikely to have the slow block to polyspermy at the VE. Indeed, the CG lectin of *X. laevis* oocytes can precipitate with a jelly component of the frogs *B. japonicus* (Hedrick and Katagiri 1988) and *R. japonica*, but not with that of the urodeles *C. pyrrhogaster* and *H. nebulosus* (Yonemoto and Iwao, unpubl.).

4.5
Acquisition of Polyspermy Block During Oocyte Maturation in Anurans

The block to polyspermy in monospermic anuran oocytes is acquired during oocyte maturation. In amphibians, a full-grown oocyte with a large nucleus (germinal vesicle) is arrested at the first meiotic prophase (PI). Meiosis is reinitiated in response to a progesterone-like hormone released from follicle cells (see Masui and Clark 1979; Masui 1992). The oocyte undergoes germinal vesicle breakdown (GVBD) and proceeds to the first meiotic metaphase (MI). After the emission of the first polar body (PBI), the oocyte again arrests at the second meiotic metaphase (MII) until fertilization. Ovulation occurs at around MI, and the oocytes complete maturation during the passage through the oviduct. The oocytes in the ovisac, the lowest portion of the oviduct, reach MII, so they are designated as mature, uterine oocytes. Insemination of immature oocytes around MI–PBI causes polyspermy (Katagiri 1974; Elinson 1977; Schlichter and Elinson 1981; Iwao 1987), indicating that the polyspermy block is not completed until oocyte maturation.

The rise in positive-going fertilization (activation) potentials in immature oocytes is much slower than that in mature oocytes (Iwao et al. 1981; Schlichter and Elinson 1981; Iwao 1982, 1987; Talevi et al. 1985). Immature oocytes have more negative membrane potential before activation and the level of the fertilization potential is quite low in comparison to mature oocytes, sometimes below ±0 mV (Schlichter and Elinson 1981; Iwao 1987). Immature oocytes exhibit several spike potentials, whose number corresponds with the number of sperm entries (Schlichter and Elinson 1981; Iwao 1987). Conversely, only one spike potential followed by a long-lasting fertilization potential is observed in mature, monospermic oocytes of *B. japonicus* (Iwao 1987). Both the long-lasting, positive-going fertilization potential as well as the spike potentials are mediated by opening Cl^- channels (Iwao 1987).

The propagation of cortical granule exocytosis in immature oocytes is slower than in mature ones (Iwao 1982). Since there is no difference in response of cortical granules to Ca^{2+} ions, when they are isolated between MI and MII (Goldenberg and Elinson 1980), the slower propagation of cortical granule exocytosis in immature oocytes seems to be due to the slower propagation of the Ca^{2+} wave. The acquisition of the ability to propagate exocytosis of cortical granules as well as the large fertilization potential correlate with the ability to form a Ca^{2+} transient in response to artificial stimuli (Ido et al. 1997). The acquisition of ability to propagate $[Ca^{2+}]_i$ increase coincides with the formation of the cortical endoplasmic reticulum at MI (Campanella et al. 1984; Charbonneau and Grey 1984). The cortical endoplasmic reticulum is a major Ca^{2+} store with IP_3 receptors (Kume et al. 1993).

4.6 Cytoplasmic Block to Polyspermy in Physiologically Polyspermic Oocytes

The oocytes of most urodeles are naturally polyspermic. The number of sperm incorporated into the oocyte varies between species, but 2–20 sperm normally enter the oocyte (Smith 1912; Bataillon and Tchou Su 1930; Fankhauser 1932; Kaylor 1937; Fankhauser and Moore 1941a; Street 1948; Labrousse 1971;

Picheral 1977b; McLaughlin and Humphries 1978; Wakimoto 1979; Charbonneau et al. 1983; Iwao et al. 1985, 1993). The oocytes are tolerant of 5–20 sperm per oocyte, but the maximum number of sperm for normal development is different between species. There is no extracellular block to polyspermy in physiologically polyspermic oocytes, except for the limitation imposed by the swollen jelly coats.

The behavior of sperm and oocyte nuclei in polyspermic urodele oocytes has been described in various species (Fankhauser 1932; Fankhauser and Moore 1941a; Wakimoto 1979), and we have confirmed these observations in *C. pyrrhogaster* oocytes (Iwao et al. 1985, 1993, 1997; Iwao and Elinson 1990; Fig. 5). These oocytes normally incorporate 7–8 sperm over the entire oocyte surface. All sperm nuclei undergo nuclear decondensation to form sperm pronuclei with sperm asters developing from their centrosomes (Fig. 5A,C). After completion of the second meiotic division, the oocyte pronucleus beneath the plasma membrane moves towards the center of the animal hemisphere (Fig. 5B,D). Inhibition of the movement of all pronuclei by nocodazole suggests the involvement of microtubules in migration of the oocyte pronucleus (Iwao et a. 1997), but the exact molecular mechanism is unclear. Microtubules of the astral fibers, growing from sperm centrosomes, appear to be important for the movement of sperm pronuclei in anuran oocytes (Briedis and Elinson 1982) and in oocyte extract (Reinsch and Karsenti 1997). However, the mechanism for only one sperm nucleus to move towards the oocyte nucleus remains unknown.

In *C. pyrrhogaster*, the asters of extra, accessory sperm nuclei eventually degenerate before cleavage (Iwao et al. 1997). The sperm asters in the animal hemisphere, about 28 µm in diameter, are larger than those in the vegetal hemisphere (Iwao et al. 1997). However, the size of the sperm asters in *C. pyrrhogaster* is much smaller than those in *X. laevis* which occupy nearly the entire surface of the animal hemisphere, about 600 µm in diameter (Schroeder and Gard 1992; Elinson and Paleèek 1993; Iwao et al. 1997). Since there is no significant difference in the amount of tubulin between the animal and vegetal hemispheres and *C. pyrrhogaster* sperm can form larger asters in *X. laevis* oocytes (Iwao et al. 1997), the size of sperm asters appears to be determined by some factors for microtubule polymerization in oocyte cytoplasm, such as γ-tubulin (Félex et al. 1994; Gard 1994). The γ-tubulin is abundant at the periphery of the germinal vesicle located in the animal hemisphere of full-grown oocytes and the distribution of cortical γ-tubulin is polarized along the animal–vegetal axis (Gard 1994). The degradation of the sperm aster is also determined by the factors in the oocyte cytoplasm, because the asters of *C. pyrrhogaster* sperm do not degenerate in *X. laevis* oocytes (Iwao et al. 1997). The smaller asters in physiologically polyspermic oocytes appear to be required for the migration of the principal sperm nucleus towards the center of the animal hemisphere, because in polyspermic anuran oocytes, large sperm asters exclude each other and the oocyte nucleus cannot make

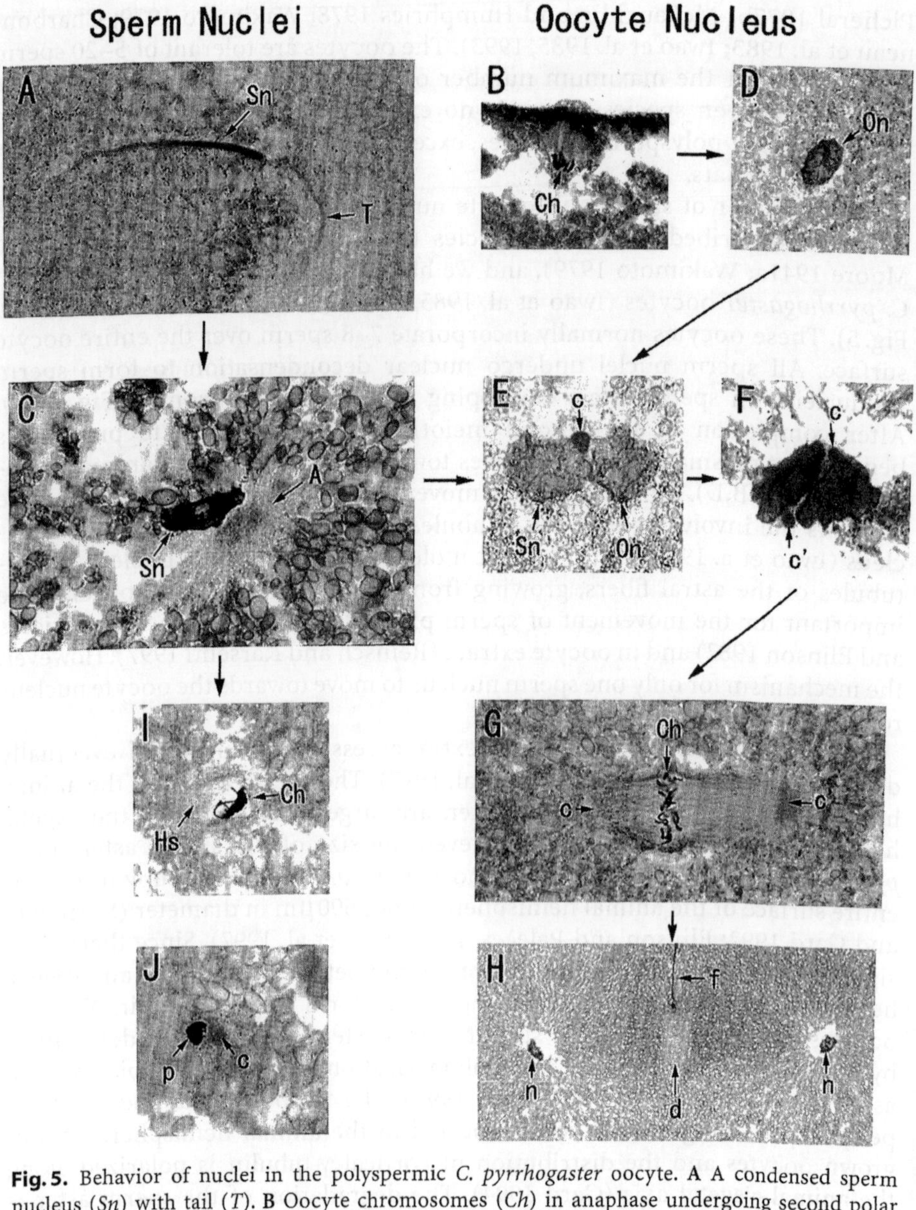

Fig. 5. Behavior of nuclei in the polyspermic *C. pyrrhogaster* oocyte. **A** A condensed sperm nucleus (*Sn*) with tail (*T*). **B** Oocyte chromosomes (*Ch*) in anaphase undergoing second polar body emission. **C** Decondensed sperm pronucleus with sperm aster (*A*). **D** Decondensed oocyte nucleus (*On*) without aster. **E** Contact between oocyte nucleus and principal sperm nucleus with centrosome (*c*). **F** Duplication of centrosome (*c, c'*) in prometaphase zygote nucleus. **G** Metaphase spindle of zygote chromosomes (*Ch*). **H** Formation of cleavage furrow (*f*) following diastema (*d*) between two cleavage nuclei (*n*). **I** and **J** Degenerating accessory sperm chromosome or chromatins with a half spindle (*Hs*) accompanied by a haploid set of chromosomes or with pycnotic chromatin (*p*)

contact with any sperm nuclei in the oocytes that received 4–5 sperm (Iwao et al. 1997).

The accessory sperm nuclei remain at the periphery of the oocyte, but all sperm pronuclei as well as the oocyte nucleus enter S phase (Fig. 3I,J; Labrousse 1971; Wakimoto 1979; Iwao et al. 1985, 1993). In the accessory sperm nuclei, the onset of DNA synthesis is delayed and its rate is slower than in the oocyte nucleus or in the primary sperm nucleus near the oocyte nucleus (Iwao et al. 1993). When the zygote nucleus enters pro-metaphase, its centrosome, derived from the primary sperm nucleus, divides into two and the centrosomes move to form a bipolar spindle for the first cleavage (Fig. 5E,F). In contrast, the centrosomes of the accessory sperm nuclei never divide (Fig. 5I,J), but it is unknown whether or not their centrioles are duplicated. When the zygote nucleus enters M phase (Fig. 5G), most of the accessory sperm nuclei are arrested at the pronucleus stage. Around first cleavage (Fig. 5H), some accessory sperm nucleus near the zygote nucleus forms a half spindle with a haploid set of chromosomes (Fig. 5I), but the rest of the accessory sperm nuclei undergo pycnosis (Fig. 5J). Eventually, all accessory sperm chromatins degenerate and their centrosomes are dispersed into the oocyte cytoplasm. In this way, the physiologically polyspermic oocytes complete the 'very slow, cytoplasmic block to polyspermy'.

Two distinct hypotheses were proposed more than 60 years ago to explain the mechanism of degeneration of accessory sperm nuclei in urodele oocytes. Fankhauser (1932) proposed two factors emanating from the oocyte- or the zygote nucleus, although he found later that the suppression of accessory sperm is a more complicated process (Fankhauser 1934a,b; Fankhauser and Moore 1941b). The first factor is an activator which enhances the development of one sperm nucleus so that it forms a larger sperm aster, assuring the contact with the oocyte nucleus. The second factor is an inhibitor which causes the degeneration of the accessory sperm nuclei. The second factor is based on Fankhauser's intensive study of the ligation of fertilized oocytes, in which some accessory sperm nuclei can be rescued. We confirmed his experiments with halves of the *C. pyrrhogaster* oocyte just after fertilization (Iwao et al. 1985). One half (a diploid merogon) contains the zygote nucleus and some accessory sperm nuclei, whereas the other half (an andromerogon) contains only accessory sperm nuclei. In the diploid merogon, all accessory sperm nuclei degenerate, and normal bipolar cleavage occurs on time. In the andromerogon, some sperm nuclei escape from degeneration to form bipolar spindles and the oocyte half undergoes multipolar cleavage with some delay (Sawai and Satoh 1984; Iwao et al. 1985). Furthermore, when both oocyte halves are connected with a thin cytoplasmic bridge to inhibit the cytoplasmic flow, the degeneration of accessory sperm in the half without the zygote nucleus is inhibited (Iwao et al. 1985). These results support the experiments of Fankhauser's inhibitory factor released from the zygote nucleus and transmitted through cytoplasm.

In contrast, Bataillon and Tchou Su (1930) suggested that active cytoplasm, indispensable for mitosis, is localized around the oocyte nucleus. The zygote nucleus can enter M phase under the influence of the active cytoplasm, but the accessory sperm degenerate because their metabolic changes are asynchronous relative to the zygotic cell cycles. This hypothesis is based on the centrifugation of the oocytes to concentrate oocyte cytoplasm toward the animal hemisphere. The centrifugated oocytes undergo multipolar cleavage and some sperm nuclei enter mitosis without degeneration. However, since the behavior of sperm nuclei in androgenetic oocyte fragments is heterogeneous (Fankhauser 1934a; Fankhauser and Moore 1941b), it is difficult to distinguish Fankhauser's and Bataillon's hypotheses from those experiments (see Elinson 1986).

The behavior of sperm nuclei has been carefully examined in terms of the nuclear cycle with *C. pyrrhogaster* oocytes (Iwao et al. 1985, 1993; Iwao and Elinson 1990). When the zygote nucleus enters metaphase with a bipolar spindle in the center of the animal hemisphere, the accessory sperm nucleus nearest the zygote nucleus enters M phase and forms a haploid set of chromosomes with a monopolar spindle (Fig. 6). Sperm nuclei in the animal side of the equator show an early M phase state: condensed chromatin and an indistinct nuclear membrane. Sperm nuclei in the vegetal hemisphere show an interphase state: decondensed chromatin enclosed by a distinct nuclear membrane. Thus, the accessory sperm do not enter M phase probably due to the lack of some cytoplasmic factors. This implies the involvement of MPF, the dominant regulatory factor for entering M phase in the embryonic cell cycle (see Masui 1992).

To determine the role of MPF in the suppression of accessory sperm, a partially purified MPF-rich fraction was injected into the animal hemisphere of polyspermic oocytes just before prometaphase (Iwao and Elinson 1990). The nuclear cycle in the zygote nucleus is accelerated by MPF injection, leading to precocious cleavage. The accessory sperm nuclei located in the animal hemisphere or at the equator can enter M phase to form bipolar spindles leading to multipolar cleavage, but the sperm nuclei in the vegetal hemisphere degenerate. MPF activity in the animal hemisphere is higher than that in the vegetal hemisphere at M phase (Iwao et al. 1993). More active cdc2 kinase, a catalytic subunit of MPF, is distributed in the animal hemisphere than in the vegetal hemisphere (Iwao et al. 1993). In addition, cyclin B, a regulatory subunit of MPF, is preferentially localized in the cortex of the animal hemisphere (Sakamoto et al. 1998). Interestingly, the initiation of MPF activity in the animal hemisphere and its spreading towards the vegetal hemisphere are also observed in monospermic *X. laevis* oocytes, but the amount of MPF activity (histone H1 kinase activity) is not significantly different between hemispheres (Pérez-Mongiovi et al. 1998). This MPF distribution corresponds to undegeneration of sperm nuclei in the polyspermic *Xenopus* oocytes. This evidence supports the notion that high MPF activity around the zygote nucleus ensures normal progression of the nuclear cycle, while the accessory sperm nuclei

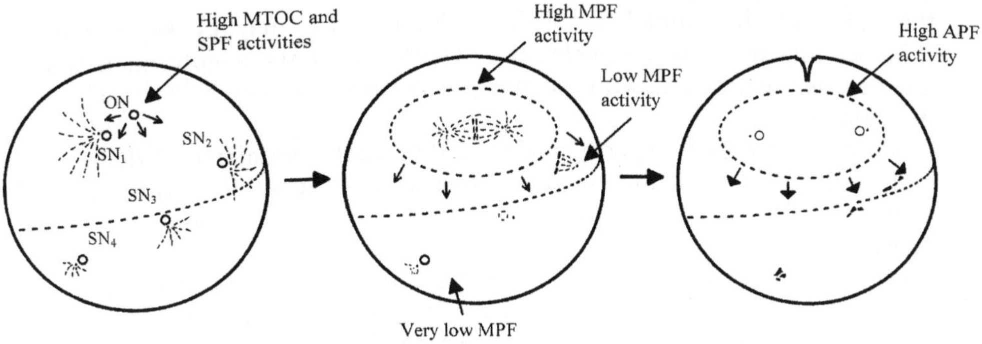

Fig. 6. MPF model for degeneration of accessory sperm nuclei in polyspermic *C. pyrrhogaster* oocytes. *ON* Oocyte nucleus; *SN* sperm nucleus; *MTOC* microtubule-organizing center; *SPF* S-phase-promoting factor; *MPF* M-phase-promoting factor; *APF* anaphase-promoting factor

in the vegetal hemisphere degenerate due to insufficient exposure of MPF activity (Fig. 6).

Why is the completion of S phase as well as the entry into M phase accelerated in the zygote nucleus? In *C. pyrrhogaster*, the progression of the cell cycle is delayed by inactivation of the oocyte nucleus with UV-irradiation (Iwao et al. 1993) as well as by the injection of an inhibitor of DNA polymerase, aphidicolin (Iwao et al. 1985), but not by the inactivation of sperm nuclei with UV-irradiation (Iwao et al. 1993). This situation is different from that of anuran oocytes in which the embryonic cell cycle progresses normally in the absence of DNA synthesis (Dasso and Newport 1990) or the oocyte nucleus (Debauvalle et al. 1988). The oocyte nucleus of urodeles appears to possess a feedback system for monitoring the state of DNA replication.

Since the enhancement of DNA synthesis occurs not only in the oocyte nucleus but also in the principal sperm nucleus (Fig. 3I,J; Wakimoto 1979; Iwao et al. 1993), the cytoplasm closely associated with them may have factors for promoting S phase. In this context, it is worth noting that the remains of nuclear materials derived from germinal vesicle breakdown during oocyte maturation are preferentially localized around the center of the animal hemisphere after oocyte maturation (Imoh and Miyazaki 1984). The germinal vesicle contains a large amount of DNA polymerase-α (Nagano et al. 1982) and is required for replacement of sperm basic proteins as described above. The cdc2/cdk2 kinases which promote S phase (Chevalier et al. 1995) are abundant in the animal hemisphere (Iwao et al. 1993). The appearance of MPF activity during oocyte maturation depends on the presence of germinal vesicles both in *X. laevis* (Iwashita et al. 1998) and in *C. pyrrhogaster* (Sakamoto et al. 1998). The germinal vesicles of *C. pyrrhogaster* oocytes contain abundant cdk7 kinase which might induce the activation of MPF by phosphorylating its catalytic subunit cdc2 kinase (Sakamoto et al. 1998). These evidences suggest that a large amount of S-phase-promoting factor is localized around the zygote nucleus in the center of the animal hemisphere (Fig. 6). This may be supported by evidence that the blastula nuclei, the descendant of the zygote nucleus, did not degenerate when introduced into the fertilized oocytes (Yamasaki and Iwao 1987).

Since the accessory sperm nuclei in the vegetal hemisphere degenerate even in the absence of the oocyte nucleus, their degeneration in the vegetal hemisphere is probably due to the lack of components required for nuclear division. However, another factor seems to be necessary to account for the degeneration of accessory sperm nuclei in the animal or the equatorial regions. We have proposed that the zygote nucleus releases a factor to inactivate MPF (Fig. 6). This factor is used to exit M phase, and it causes rapid degradation of MPF activity around the accessory sperm nuclei in the animal hemisphere or in the equator (Iwao et al. 1993). MPF is reportedly inactivated by degradation of cyclin B through ubiquitination and proteasome pathways, however, the nature of the MPF inactivator in *C. pyrrhogaster* needs to be identified to support our model.

5
Variations in Amphibian Fertilization

5.1
Polyspermy in Anurans

Most anurans have an elaborate dual block to polyspermy, the mechanisms of which are quite different from physiologically polyspermic urodeles. The three modern orders of amphibians (Lissamphibia) are considered to have a monophyletic origin, but they diverged from each other in the Paleozoic, 300–280 million years ago (my) (Duellman and Trueb 1986; Feller and Hedges 1998). Recent molecular analysis suggests a divergence of the anuran group from the urodele/caecilians group at the beginning of the Mesozoic (240 my) and between the urodele and caecilians groups later in the Mesozoic (190–160 my) (Feller and Hedges 1998). In more primitive amphibians, several species exhibit intriguing modes of fertilization and polyspermy block mechanisms. It is useful to know the fertilization of these species in order to understand how the ancestors of the current amphibians acquired those distinct mechanisms during evolution.

In anurans, the oldest groups are classified as archaeobatrachian frogs (Duellman and Trueb 1986; Hedges and Maxson 1993; Hillis et al. 1993). The family Discoglossidae is archaeobatrachian, and, as described earlier, the sperm entry site in the oocytes of *D. pictus* is restricted to the dimple at the animal hemisphere. In addition to the plug-like jelly on the dimple which is necessary for sperm penetration, the localization of Cl^- channels to produce a positive-going fertilization potential and the cortical granules are restricted to this area. This *Discoglossus*-type of fertilization is voltage-independent, and the oocytes appear to tolerate the incorporation of a small number of sperm (Talevi 1989). It is not clear whether the oocytes have the same mechanism for suppression of extra sperm as operate in physiologically polyspermic newt oocytes.

Another case of polyspermy in anurans is found in the Puerto Rican tree frog, *E. coqui* (family Leptodactylidae). Its fertilization occurs internally (Townsend et al. 1981), and froglets develop directly from a large, yolky oocyte without a free-living tadpole stage (Townsend and Stewart 1985). Sperm entry appears to be restricted to a small area in the animal pole and some oocytes are polyspermic, although the oocytes have cortical granules in both animal and vegetal hemispheres (Elinson 1987). A delay in the progression of nuclear cycles is observed in accessory sperm nuclei (Elinson 1987), which seems to be the similar situation in physiologically polyspermic newt oocytes. Since the diameter of oocytes of *E. coqui* is about 3.6 mm, their acquisition of polyspermy may be an evidence to support 'a large oocyte model' proposed by Elinson (1986), in which the restriction of oocyte cytoplasm around the animal pole makes the oocytes tolerant to polyspermy. However, the polyspermy and internal fertilization in *E. coqui* must have been acquired as a result of adaptation to terrestrial breeding and direct development with a different origin from that in urodeles. In this respect, it would be interesting to know the mode of internal fertilization in *Ascaphus truei*, some Nectophrynoides and *Mertensophryne micranotis* (Duellman and Trueb 1986).

5.2
Monospermy in Primitive Urodeles

In urodeles, the suborder Sirenidea as well as the suborder Cryptobranchoidea are classified as the earliest derived groups (Duellman and Trueb 1986; Larson 1991; Hedges and Maxson 1993). Cryptobranchoidea contains two families, Hynobiidae and Cryptobranchidae, which have external fertilization and exhibit interesting intermediate modes of fertilization. In the genus *Hynobius* (Hynobiidae), *H. nebulosus* oocytes exhibit monospermy (Iwao 1989). They have a fast, electrical block at the oocyte plasma membrane, but they lack the slow block mediated by cortical granule exocytosis (Iwao 1989). The polyspermy block in the oocyte cytoplasm, as observed in physiologically polyspermic urodele oocytes, does not operate in *H. nebulosus* oocytes. The *Hynobius*-type monospermy is characteristic of this family, with monospermic fertilization in *H. retardatus* (Makino 1934), *H. nebulosus* (Iwao 1989), and *H. dunni* (Iwao and Akiyama, unpublished). By contrast, in the Cryptobranchidae, fertilization of both the hellbender, *Cryptobranchus alleganiensis* (Smith 1912), and the Japanese giant salamander, *A. japonicus* (Kunitomo 1934; Iwao and Kuwahara, unpublished), is physiologically polyspermic with very slow cytoplasmic block. They probably lack the fast, electrical block, because of the polyspermic cross-fertilization between monospermic *X. laevis* oocytes with *A. japonicus* sperm (Iwao and Kuwahara, unpublished). Since the diameter of *A. japonicus* oocytes is over 5 mm, the increase in size of the oocytes may allow them to tolerate polyspermy, according to the large oocyte model (Elinson 1986).

In higher urodeles, for example the Ambystomatidae and the Salamandridae, fertilization is physiologically polyspermic and the oocytes have no extracellular block to polyspermy except for the block at the jelly layers, as in *Cynops*. Since more than 90% of species of urodeles exhibit internal fertilization in which the females inseminate the oocytes with the sperm stored in the spermatheca (Duellman and Trueb 1986) and most of them probably exhibit physiological polyspermy, the *Cynops*-type of polyspermy may have the advantage of efficient fertilization with very limited number of sperm released from the spermatheca. The assumption that the *Hynobius*-type of monospermy is the ancestral mode of urodele fertilization is consistent with the notion that external fertilization in Cryptobranchoidea appears to be an ancestral state of urodeles, but the external fertilization was unlikely obtained from evolutional loss of internal fertilization (Larson 1991). Since Sirenidea appear to diverge earlier than Cryptobranchoidea (Duellman and Trueb 1986; Hedges and Maxson 1993), it would be very interesting to examine their fertilization. All that is known, however, is that fertilization is external in *Siren intermedia* (Sever et al. 1996).

5.3 Sperm Nuclear Behavior in Gynogenetic Urodele Oocytes

Some natural unisexual groups exhibiting gynogenetic reproduction have evolved in urodeles (Uzzell 1964). In *Ambystoma* hybrids (*A. jeffersonianum*, *A. laterale*, *A. texanum*, and *A. tigrinum*) throughout eastern North America, offspring from triploid hybrid females often undergo gynogenetic development (Macgregor and Uzzell 1964; Bogart et al. 1989). The time of origin of the unisexual hybrid lineage is estimated to be about 4 my based on cytochrome B sequence divergence (Hedges et al. 1992). According to Sessions (1982), a diploid hybrid (JL) between a female of *A. jeffersonianum* (JJ) and a male of *A. laterale* (LL) produced diploid oocytes probably due to meiotic reduplication during oogenesis. The diploid oocytes (JL) are fertilized by sperm (J) of *A. jeffersonianum* or sperm (L) of *A. laterale*, which result in an allotriploid of JJL or JLL, respectively. After fertilization by sperm of diploid males, the resulting triploid oocytes usually develop gynogenetically, because the sperm nucleus is rejected when exposed to low temperatures characteristic of the breeding season (Bogart et al. 1989). The observation of nuclear behavior in the oocytes, however, demonstrates that the principal sperm nucleus is selectively eliminated after forming the zygote nucleus. Suppression of accessory sperm nuclei occurs as in other polyspermic urodeles (Elinson et al. 1992). The figures of degenerating principal sperm nuclei in the triploid *Ambystoma* oocytes (Elinson et al. 1992) are similar to those of *Cynops* sperm nuclei that have been irradiated with UV before insemination (Iwao et al. 1993). Since MPF must be abundant around the zygote nucleus, the failure of DNA synthesis in the principal sperm nucleus induced by low environmental temperature may cause its degeneration in the triploid *Ambystoma* oocytes.

Acknowledgements. I would like to express my sincere thanks to Dr. Richard P. Elinson and Dr. Chiaki Katagiri for their valuable comments on the manuscript. This work was supported by the Grant-in-Aid for Scientific Research (C) given by the Ministry of Education, Science, Sports and Culture to the author.

References

Aimar C (1991) Cyclic changes in amphibian egg microvilli occur during the division cycle: implication of MPF. Development 112:91–98

Aimar C, Labrousse J-P (1975) DNA synthesis and evolution, in presence of a somatic nucleus, of the female pronucleus after experimental activation of the egg of *Pleurodeles waltlii*. Dev Growth Differ 17:197–207

Aizawa H, Kawahara H, Tanaka K, Yokosawa H (1996) Activation of the proteasome during *Xenopus* egg activation implies a link between proteasome activation and intracellular calcium. Biochem Biophys Res Commun 218:224–228

al-Anzi B, Chandler DE (1998) A sperm chemoattractant is released from *Xenopus* egg jelly during spawning. Dev Biol 198:366–375

Almeida EA, Huovila A-PJ, Sutherland AE, Stephens LE, Calarco PG, Shaw LM, Mercurio AM, Sonnenberg A, Primakoff P, Myles DG, White JM (1995) Mouse egg integrin $\alpha6\beta1$ functions as a sperm receptor. Cell 81:1095–1104

Andreuccetti P, Denis-Donini S, Burrini AG, Campanella C (1984) Calcium ultrastructural localization in *Xenopus laevis* eggs following activation by pricking or by calcium ionophore A 23187. J Exp Zool 229:295–308

Arranz SE, Albertali IE, Cabada MO (1997) *Bufo arenarum* egg jelly coat: purification and characterization of two highly glycosylated proteins. Biochem J 323:307–312

Bakos MA, Kurosky A, Hedrick JL (1990a) Physicochemical characterization of progressive changes in the *Xenopus laevis* egg envelope following oviductal transport and fertilization. Biochemistry 29:609–616

Bakos MA, Kurosky A, Hedrick JL (1990b) Enzymatic and envelope-converting activities of pars recta oviductal fluid from *Xenopus laevis*. Dev Biol 138:169–176

Barbieri FD, Raisman JS (1969) Non-gametic factors involved in the fertilization of *Bufo arenarum* oocyte. Embryologia 10:363–372

Bataillon E (1919) Analyse de l'activation par la technique des oeufs nus et la polyspermie expérimentale chez les Batraciens. Ann Sci Nat Zool Ser 10 (3):1–38

Bataillon E, Tchou Su (1930) Etudes analytiques et expérimentales sur les rythmes cinétiques dans l'oeuf. Arch Biol 40:439–540

Bement WM, Capco DG (1989) Activators of protein kinase C trigger cortical granule exocytosis, cortical contraction, and cleavage furrow formation in *Xenopus laevis* oocytes and eggs. J Cell Biol 108:885–892

Bement WM, Capco DG (1990) Protein kinase C acts downstream of calcium at entry into the first mitotic interphase of *Xenopus laevis*. Cell Regul 1:315–326

Bement WM, Capco DG (1991) Parallel pathways of cell cycle control during *Xenopus* egg activation. Proc Natl Acad Sci USA 88:5172–5176

Bernardini G, Andrietti F, Cametini M, Casson MP (1988) *Xenopus* spermatozoon: correlation between shape and motility. Gamete Res 20:165–175

Blobel CP, Wolfsberg TG, Turck CW, Myles DG, Primakoff P, White JM (1992) A potential fusion peptide and integrin ligand domain in a protein active in sperm–egg fusion. Nature 356:248–252

Bogart JP, Elinson RP, Licht LE (1989) Temperature and sperm incorporation in polyploid salamanders. Science 246:1032–1034

Bonnell BS, Reinhart D, Chandler DE (1996) *Xenopus laevis* egg jelly coats consist of small diffusible proteins bound to a complex system of structurally stable networks composed of high-molecular-weight glycoconjugates. Dev Biol 174:32–42

Breckenridge WR, Jayasinghe S (1979) Observations on the eggs and larvae of *Ichthyophis glutinosus.* Ceylon J Sci (Biol Sci) 13:187–202

Briedis A, Elinson RP (1982) Suppression of male pronuclear movement in frog eggs by hydrostatic pressure and deuterium oxide yields androgenetic haploid. J Exp Zool 222:45–57

Busa WB, Nuccitelli R (1985) An elevated free cytosolic calcium wave follows fertilization in eggs of the frog, *Xenopus laevis.* J Cell Biol 100:1325–1329

Busa WB, Ferguson JE, Joseph SK, Williamson JR, Nuccitelli R (1985) Activation of frog (*Xenopus laevis*) eggs by inositol trisphosphate. I. Characterization of calcium release from intracellular stores. J Cell Biol 101:677–682

Cabada MO, MarianoMI, JS Raisman (1978) Effect of trypsin inhibitors and concanavalin A on the fertilization of *Bufo arenarum* coelomic oocytes. J Exp Zool 204:409–416

Cabada MO, Mariano MI, Gomez MI (1987) Cortical granules products and fertility prevention in *Bufo arenarum* oocytes. J Exp Zool 241:359–367

Cabada MO, Manes ME, Gomez MI (1989) Spermatolysins in *Bufo arenarum*: their activity on oocyte surface. J Exp Zool 249:229–234

Cabada MO, Sanchez Ruera AN, Genta HD, Sanchez SS, Barisone GA (1996) Vitelline envelope formation during oogenesis in *Bufo arenarum.* Biocell 20:77–86

Campanella C (1975) The site of spermatozoon entrance in the unfertilized egg of *Discoglossus pictus* (Anura): an electron microscopy study. Biol Reprod 12:439–447

Campanella C, Andreuccitti P, Taddei C, Talvi R (1984) The modifications of cortical endoplasmic reticulum during in vitro maturation of *Xenopus laevis* oocytes and its involvement in cortical granule exocytosis. J Exp Zool 229:283–293

Campanella C, Talevi R, Kline D, Nuccitelli R (1988) The cortical reaction in the egg of *Discoglossus pictus*: a study of the changes in the endoplasmic reticulum at activation. Dev Biol 130:108–119

Campanella C, Chaponnier C, Quaglia L, Gualtieri R, Gabbiani G (1990) Different cytoskeletal organization in two maturation stages of *Discoglossus pictus* (Anura) oocytes: thickness and stability of actin microfilaments and tropomyosin immunolocalization. Mol Reprod Dev 25:130–139

Campanella C, Amore F, Pitari G, Fusco C, Maurizi G, Dupré S (1992) Post-fertilization changes in *Discoglossus pictus* (Anura) eggs result in the formation of a capsular chamber where the egg rotates. Int J Dev Biol 36:413–422

Campanella C, Carotenuto R, Infante V, Maturi G, Atripaldi U (1997) Sperm–egg interaction in the painted frog (*Discoglossus pictus*): an ultrastructural study. Mol Reprod Dev 47:323–333

Carroll EJ Jr, Wei SH, Nagel GM, Ruibai R (1991) Structure and macromolecular composition of the egg and embryo jelly coats of the anuran *Lepidobatrachus laevis.* Dev Growth Differ 33:37–43

Chamow SM, Hedrick JL (1986) Subunit structure of a cortical granule lectin involved in the block to polyspermy in *Xenopus laevis* eggs. FEBS Lett 206:353–357

Charbonneau M, Grey RD (1984) The onset of activation responsiveness during maturation coincides with the formation of the cortical endoplasmic reticulum in oocytes of *Xenopus laevis.* Dev Biol 102:90–97

Charbonneau M, Picheral B (1983) Early events in anuran amphibian fertilization: an ultrastructural study of changes occurring in the course of monospermic fertilization and artificial activation. Dev Growth Differ 25:23–37

Charbonneau M, Moreau M, Picheral B, Vilain JP, Guerrier P (1983) Fertilization of amphibian eggs: a comparison of electrical responses between anurans and urodeles. Dev Biol 98:304–318

Chevalier S, Tassan JP, Cox R, Philippe M, Ford C (1995) Both cdc2 and cdk2 promote S phase initiation in *Xenopus* egg extracts. J Cell Sci 108:1831–1841

Chow RL, Elinson RP (1993) Local alteration of cortical actin in *Xenopus* eggs by the fertilizing sperm. Mol Reprod Dev 35:69–75

Christensen K, Sautere R, Merriam RW (1984) Role of soluble myosin in cortical contractions of *Xenopus* eggs. Nature 310:150–151

Cross NL (1981) Initiation of the activation potential by an increase in intracellular Ca^{2+} in eggs of the frog, *Rana pipiens*. Dev Biol 85:380–384

Cross NL, Elinson RP (1980) A fast block to polyspermy in frogs mediated by changes in the membrane potential. Dev Biol 75:187–198

Dasso M, Newport JW (1990) Completion of DNA replication is monitored by a feedback system that controls the initiation of mitosis in vitro: studies in *Xenopus*. Cell 61:811–823

Dawid IB, Blackler AW (1972) Maternal and cytoplasmic inheritance of mitochondrial DNA in *Xenopus*. Dev Biol 29:152–161

Debauvalle MC, Doree M, Bravo R, Karsenti E (1988) Role of nuclear material in the early cell cycle of *Xenopus* embryos. Cell 52:525–533

del Pino EJ, Cabada MO (1987) Lectin binding sites in the vitelline envelope of *Bufo arenarum* oocytes: role in fertilization. Gamete Res 17:333–342

Denis-Donini S, Campanella C (1977) Ultrastructural and lectin binding changes during the formation of the animal dimple in oocytes of *Discoglossus pictus* (Anura). Dev Biol 61:140–152

Duellman WE, Trueb L (1986) Biology of amphibian. McGraw-Hill, New York

Dumont JN (1972) Oogenesis in *Xenopus laevis* (Daudin). I. Stages of oocyte development in laboratory maintained animals. J Morphol 136:153–180

Elinson RP (1971a) Fertilization of partially jellied and jellyless oocytes of the frog, *Rana pipiens*. J Exp Zool 176:415–428

Elinson RP (1971b) Sperm lytic activity and its relation to fertilization in the frog *Rana pipiens*. J Exp Zool 177:207–218

Elinson RP (1973) Fertilization of frog body cavity eggs enhanced by treatment affecting the vitelline coat. J Exp Zool 183:291–302

Elinson RP (1975) Site of sperm entry and a cortical contraction associated with egg activation in the frog *Rana pipiens*. Dev Biol 47:257–268

Elinson RP (1977) Fertilization of immature frog eggs: cleavage and development following subsequent activation. J Embryol Exp Morphol 37:187–201

Elinson RP (1986) Fertilization of amphibians: the ancestry of the block to polyspermy. Int Rev Cytol 101:59–100

Elinson RP (1987) Fertilization and aqueous development of the Puerto Rican terrestrial-breeding frog, *Eleutherodactylus coqui*. J Morphol 193:217–224

Elinson RP, Manes ME (1978) Morphology of the site of sperm entry on the frog egg. Dev Biol 63:67–75

Elinson RP, Paleček J (1993) Independence of two microtubule systems in fertilized frog eggs: the sperm aster and the vegetal parallel array. Rouxs Arch Dev Biol 202:224–232

Elinson RP, Bogart JP, Licht LE, Lowcock LA (1992) Gynogenetic mechanisms in polyploid hybrid salamanders. J Exp Zool 264:93–99

Erdoğan S, Loğoğlu G, Özgünen T (1996) The ionic basis of membrane potential changes from before fertilization through the first cleavage in the egg of the frog *Rana cameranoi*. Gen Physiol Biophys 15:371–387

Fankhauser G (1932) Cytological studies on egg fragments of the salamander *Triton*. II. The history of the supernumerary sperm nuclei in normal fertilization and cleavage of fragments containing the egg nucleus. J Exp Zool 62:185–235

Fankhauser G (1934a) Cytological studies on egg fragments of the salamander *Triton*. III. The early development of the sperm nuclei in egg fragments without the egg nucleus. J Exp Zool 67:159–215

Fankhauser G (1934b) Cytological studies on egg fragments of the salamander *Triton*. V. Chromosome number and chromosome individuality in the cleavage mitoses of merogonic fragments. J Exp Zool 67:349–393

Fankhauser G (1967) I. System: procurement, maintenance, and use (urodeles). In: Wilt FH, Wessells NK (eds) Methods in developmental biology. TY Crowell, New York, pp 85–99

Fankhauser G, Moore C (1941a) Cytological and experimental studies of polyspermy in the newt, *Triturus viridescens*. I. Normal fertilization. J Morphol 68:347–385
Fankhauser G, Moore C (1941b) Cytological and experimental studies of polyspermy in the newt, *Triturus viridescens*. II. The behavior of the sperm nuclei in androgenetic eggs (in the absence of the egg nucleus). J Morphol 68:387–423
Félex M-A, Antony C, Wright M, Maro B (1994) Centrosome assembly in vitro: role of γ-tubulin recruitment in *Xenopus* sperm aster formation. J Cell Biol 124:19–31
Feller AE, Hedges SB (1998) Molecular evidence for the early history of living amphibians. Mol Phylogenet Evol 9:509–516
Fontanilla RA, Nuccitelli R (1998) Characterization of the sperm-induced calcium wave in *Xenopus* eggs using confocal microscopy. Biophys J 75:2079–2087
Gallo CJ, Jones TLZ, Aragay AM, Jaffe LA (1996) Increased expression of α family G-proteins during oocyte maturation and early development of *Xenopus laevis*. Dev Biol 177:300–308
Gard DL (1994) γ-tubulin is asymmetrically distributed in the cortex of *Xenopus* oocytes. Dev Biol 161:131–140
Gardiner MD, Grey RD (1983) Membrane junctions in *Xenopus* eggs: their distribution suggests a role in calcium regulation. J Cell Biol 96:1159–1163
Gautier J (1987) The role of the germinal vesicle for the appearance of maturation-promoting factor activity in the axolotl oocyte. Dev Biol 123:483–486
Gautier J, Minshull J, Lohka M, Glotzer M, Hunt T, Maller JL (1990) Cyclin is a component of maturation-promoting factor from *Xenopus*. Cell 60:487–494
Gerton GL, Hedrick JL (1986a) The coelomic envelope to vitelline envelope conversion in eggs of *Xenopus laevis*. J Cell Biochem 30:341–350
Gerton GL, Hedrick JL (1986b) The vitelline envelope to fertilization envelope conversion in eggs of *Xenopus laevis*. Dev Biol 116:1–7
Glahn D, Mark SD, Behr RK, Nuccitelli R (1998) Tyrosine kinase inhibitors block sperm-induced egg activation in *Xenopus laevis*. Dev Biol 205:171–180
Goldenberg M, Elinson RP (1980) Animal/vegetal difference in cortical granules exocytosis during activation of the frog egg. Dev Growth Differ 22:345–356
Grandin N, Charbonneau M (1989) Intracellular pH and the increase in protein synthesis accompanying activation of *Xenopus* eggs. Biol Cell 67:321–330
Grandin N, Charbonneau M (1991) Intracellular pH and intracellular free calcium responses to protein C kinase activators and inhibitors in *Xenopus* eggs. Development 112:461–470
Grandin N, Charbonneau M (1992a) Intracellular free Ca^{2+} changes during physiological polyspermy in amphibian eggs. Development 114:617–624
Grandin N, Charbonneau M (1992b) The increase in intracellular pH associated with *Xenopus* egg activation is a Ca^{2+}-dependent wave. J Cell Sci 101:55–67
Grandin N, Rolland J-P, Charbonneau M (1991) Changes in intracellular pH following egg activation and during the early cell cycle of the amphibian *Pleurodeles waltlii* coincide with changes in MPF activity. Biol Cell 72:259–267
Grandison AGC, Ashe S (1983) The distribution, behavioral ecology and breeding strategy of the pygmy toad, *Mertensophryne micronous* (Lov.). Bull Br Mus Nat Hist Zool 45:85–93
Grey RD, Wlof DP, Hedrick JL (1974) Formation and structure of the fertilization envelope in *Xenopus laevis*. Dev Biol 36:44–61
Grey RD, Working PK, Hedrick JL (1976) Evidence that the fertilization envelope blocks sperm entry in eggs of *Xenopus laevis*: interaction of sperm with isolated envelope. Dev Biol 54:52–60
Grey RD, Working PK, Hedrick JL (1977) Alteration of structure and penetrability of the vitelline envelope after passage of eggs from coelom to oviduct in *Xenopus laevis*. J Exp Zool 201:73–83
Grey RD, Bastiani MJ, Webb DJ, Schertel ER (1982) An electrical block is required to prevent polyspermy in eggs fertilized by natural mating of *Xenopus laevis*. Dev Biol 89:475–484
Gualtieri R, Andreuccitti P (1996) Localization of egg-surface carbohydrates and fertilization in *Discoglossus pictus* (Anura). Cell Tissue Res 283:173–181

Han J-K, Nuccitelli R (1990) Inositol 1,4,5-trisphosphate-induced calcium release in the organelle layers of the stratified, intact egg of *Xenopus laevis*. J Cell Biol 110:1103–1110
Hardy DM, Hedrick JL (1992) Oviductin. Purification and properties of the oviducal protease that processes the molecular weight 43000 glycoprotein of the *Xenopus laevis* egg envelope. Biochemistry 31:4466–4472
Hardy MP, Dent JN (1986) Regulation of motility in sperm of the red-spotted newt. J Exp Zool 240:385–396
Heasman J, Holwill S, Wylie CC (1991) Fertilization of cultured *Xenopus* oocytes and use in studies of maternally inherited molecules. Methods Cell Biol 36:213–230
Hedges SB, Maxson LR (1993) A molecular perspective on lissamphibian phylogeny. Herpetol Monogr 7:27–42
Hedges SB, Bogart JP, Maxson LR (1992) Ancestry of unisexual salamanders. Nature 356:708–710
Hedrick JL, Katagiri C (1988) *Bufo japonicus japonicus* and *Xenopus laevis* egg jellies contain structurally related antigens and cortical granule lectin ligands. J Exp Zool 245:78–85
Hedrick JL, Nishihara T (1991) Structure and function of the extracellular matrix of anuran eggs. J Electron Microsc Tech 17:319–335
Hillis DM, Ammerman LK, Dixon MT, de Sá RO (1993) Ribosomal DNA and the phylogeny of frogs. Herpetol Monogr 7:118–131
Hope J, Humphries AA, Bourne GM (1963) Intrastructural studies on developing oocytes of the salamander *Triturus viridescens*. J Ultrastruct Res 9:302–324
Houliston E, Elinson RP (1992) Microtubules and cytoplasmic reorganization in the frog egg. Curr Top Dev Biol 26:53–70
Ido M, Okamoto S, Iwao Y (1997) Acquisition of fertilizability during *Xenopus* oocyte maturation. Zool Sci 14 (Suppl):64
Imoh H, Miyazaki Y (1984) Distribution of the germinal vesicle material during progesterone-induced oocyte maturation in *Xenopus* and *Cynops*. Dev Growth Differ 26:157–165
Inoda T, Morisawa M (1987) Effect of osmolality on the initiation of sperm motility in *Xenopus laevis*. Comp Biochem Physiol 88A:539–542
Ishihara K, Hosono J, Kanatani H, Katagiri C (1984) Toad egg-jelly as a source of divalent cations essential for fertilization. Dev Biol 105:435–442
Ito S (1972) Effects of media of different ionic composition on the activation potential of anuran egg cells. Dev Growth Differ 14:217–227
Iwao Y (1982) Differential emergence of cortical granule breakdown and electrophysiological responses during maturation of toad oocytes. Dev Growth Differ 24:467–477
Iwao Y (1985) The membrane potential changes of amphibian eggs during species- and cross-fertilization. Dev Biol 111:26–34
Iwao Y (1987) The spike component of the fertilization potential in the toad, *Bufo japonicus*: changes during meiotic maturation and absence during cross-fertilization. Dev Biol 123:559–565
Iwao Y (1989) An electrically mediated block to polyspermy in the primitive urodele *Hynobius nebulosus* and phylogenetic comparison with other amphibians. Dev Biol 134:438–445
Iwao Y, Elinson RP (1990) Control of sperm nuclear behavior in physiologically polyspermic newt eggs: possible involvement of MPF. Dev Biol 142:301–312
Iwao Y, Fujimura T (1996) Activation of *Xenopus* eggs by RGD-containing peptides accompanied by intracellular Ca^{2+} release. Dev Biol 177:558–567
Iwao Y, Jaffe LA (1989) Evidence that the voltage-dependent component in the fertilization process is contributed by the sperm. Dev Biol 134:446–451
Iwao Y, Katagiri C (1982) Properties of vitelline coat lysin from toad sperm. J Exp Zool 219:87–95
Iwao Y, Katagiri C (1984) In vitro induction of sperm nucleus decondensation by cytosol from mature toad eggs. J Exp Zool 230:115–124
Iwao Y, Masui Y (1995) Activation of newt eggs in the absence of Ca^{2+} activity by treatment with cycloheximide or D_2O. Dev Growth Differ 37:641–651
Iwao Y, Ito S, Katagiri C (1981) Electrical properties of toad oocytes during maturation and activation. Dev Growth Differ 23:89–100

Iwao Y, Yamasaki H, Katagiri C (1985) Experiments pertaining to the suppression of accessory sperm in fertilized newt eggs. Dev Growth Differ 27:323–331

Iwao Y, Sakamoto N, Takahara K, Yamashita, M, Nagahama Y (1993) The egg nucleus regulates the behavior of sperm nuclei as well as cycling of MPF in physiologically polyspermic newt eggs. Dev Biol 160:15–27

Iwao Y, Miki A, Kobayashi M, Onitake K (1994) Activation of *Xenopus* eggs by an extract of *Cynops* sperm. Dev Growth Differ 36:469–479

Iwao Y, Kobayashi M, Miki A, Kubota HY, Yoshimoto Y (1995) Activation of *Xenopus* eggs by *Cynops* sperm extract is dependent upon both extra- and intra-cellular Ca activities. Zool Sci 12:573–581

Iwao Y, Yasumistu K, Narihira M, Jiang J, Nagahama Y (1997) Changes in microtubule structures during the first cell cycle of physiologically polyspermic newt eggs. Mol Reprod Dev 47:210–221

Iwashita J, Hayano Y, Sagata N (1998) Essential role of germinal vesicle material in the meiotic cell cycle of *Xenopus* oocytes. Proc Natl Acad Sci USA 95:4392–4397

Jaffe LA (1976) Fast block to polyspermy in sea urchin eggs is electrically mediated. Nature 261:68–71

Jaffe LA, Schlichter LC (1985) Fertilization-induced ionic conductances in eggs of the frog, *Rana pipiens*. J Physiol 358:299–319

Jaffe LA, Cross NL, Picheral B (1983) Studies of the voltage-dependent polyspermy block using cross-species fertilization of amphibians. Dev Biol 98:319–326

Jaffe LA, Kado RT, Muncy L (1985) Propagating potassium and chloride conductances during activation and fertilization of the egg of the frog, *Rana pipiens*. J Physiol 368:227–242

Jego P, Lerivray H, Chesnel A, Charbonneau M (1986) Urodele egg jelly and fertilization. In: Hedrick JL (ed) The molecular and cellular biology of fertilization. Plenum Press, New York, pp 205–233

Katagiri C (1962) On the fertilizability of the toad eggs, II. Change of the jelly envelopes in water. Jpn J Zool 13:365–373

Katagiri C (1966a) Fertilization of dejellied uterine toad eggs in various experimental conditions. Embryologia 9:159–169

Katagiri C (1966b) The fertilizing capability of frog sperm in the homologous and heterologous egg-jellies and polyvinylpyrrolidone (PVP). J Fac Sci Hokkaido Univ Ser VI 16:77–74

Katagiri C (1973) Chemical analysis of toad egg-jelly in relation to its "sperm-capacitating" activity. Dev Growth Differ 15:81–92

Katagiri C (1974) A high frequency of fertilization in premature and mature coelomic toad eggs after enzymic removal of vitelline membrane. J Embryol Exp Morphol 31:573–587

Katagiri C (1986) The role of oviducal secretions in mediating gamete fusion in the toad, *Bufo bufo japonicus*. Adv Exp Med Biol 207:151–166

Katagiri C (1987) Role of oviducal secretions in mediating gamete fusion in anuran amphibians. Zool Sci 4:1–14

Katagiri C, Moriya M (1976) Spermatozoon response to the toad oocytes matured after removal of germinal vesicle. Dev Biol 50:235–241

Katagiri Ch, Iwao Y, Yoshizaki N (1982) Participation of oviducal pars recta secretions in inducing the acrosome reaction and release of vitelline coat lysin in fertilizing toad sperm. Dev Biol 94:1–10

Katagiri C, Yoshizaki N, Kotani M, Kubo H (1999) Analyses of oviductal pars recta-induced fertilizability of coelomic eggs in *Xenopus laevis*. Dev Biol 210:269–276

Kaylor CT (1937) Experiments on androgenesis in the newt, *Triturus viridescens*. J Exp Zool 76:375–394

Kline D (1988) Calcium-dependent events at fertilization of the frog egg: injection of a calcium buffer blocks ion channel opening, exocytosis, and formation of pronuclei. Dev Biol 126: 346–361

Kline D, Nuccitelli R (1985) The wave of activation current in the *Xenopus* egg. Dev Biol 111:471–487

Kline D, Simocini L, Mandel G, Maue RA, Kado RT, Jaffe LA (1988) Fertilization events induced by neurotransmitters after injection of mRNA in *Xenopus* eggs. Science 241:464–467

Kline D, Kopf GS, Muncy LF, Jaffe LA (1991) Evidence for the involvement of a pertussis toxin-sensitive G-protein in egg activation of the frog, *Xenopus laevis*. Dev Biol 143:218–229

Kubo H, Kawano T, Tsubuki S, Kawashima S, Katagiri C, Suzuki A (1997) A major glycoprotein of *Xenopus* egg vitelline envelope, gp41, is a frog homolog of mammalian ZP3. Dev Growth Differ 39:405–417

Kubo H, Matsushita M, Kotani M, Kawasaki H, Saido TC, Kawashima S, Katagiri C, Suzuki A (1999) Molecular basis for oviductin-mediated processing from gp43 to gp41, the predominant glycoproteins of *Xenopus* egg envelopes. Dev Genet 25:123–129

Kubota HY, Yoshimoto Y, Yoneda M, Hiramoto Y (1987) Free calcium wave upon activation of *Xenopus* eggs. Dev Biol 119:129–136

Kume S, Muto A, Aruga J, Nakagawa T, Michikawa T, Furuichi T, Nakade S, Okano H, Mikoshiba K (1993) The *Xenopus* IP_3 receptor: structure, function, and localization in oocytes and eggs. Cell 73:555–570

Kunitomo K (1934) Fertilization in the egg of *Megalobatrachus japonicus* (in Japanese). Zool Mag 46:96

Kupitz Y, Atlas D (1993) A putative ATP-activated Na^+ channel involved in sperm-induced fertilization. Science 261:484–486

Labrousse JP (1971) Synthèse de l'adn dans l'úuf de l'amphibien urodèle *Pleurodeles waltlii* Michah. Ann Embryol Morphol 4:347–358

Larabell CA, Chandler DE (1988a) The extracellular matrix of *Xenopus laevis* eggs: a quick-freeze, deep etch analysis of its modification at fertilization. J Cell Biol 107:731–741

Larabell CA, Chandler DE (1988b) In vitro formation of the "S" layer, a unique component of the fertilization envelope in *Xenopus laevis* eggs. Dev Biol 130:356–364

Larabell CA, Chandler DE (1989a) Quick-freeze, deep-etch, rotary-shadow views of the extracellular matrix and cortical cytoskeleton of *Xenopus laevis* eggs. J Electron Microsc Tech 13:228–243

Larabell CA, Chandler DE (1989b) The coelomic envelope of *Xenopus laevis* oocytes egg: a quick-freeze, deep-etch analysis. Dev Biol 131:126–135

Larabell CA, Chandler DE (1990) Step wise transformation of the vitelline envelope of *Xenopus* eggs at activation: quick-freeze, deep-etch analysis. Dev Biol 139:263–268

Larabell C, Nuccitelli R (1992) Inositol lipid hydrolysis contributes to the Ca^{2+} wave in the activating egg of *Xenopus laevis*. Dev Biol 153:347–355

Larson A (1991) A molecular perspective on the evolutionary relationships of the salamander families. Evol Biol 25:211–277

Laskey RA (1979) Biochemical processes in early development. In: Bull AT, Lagnado JR, Thomsen JO, Tipton KF (eds) Companion to biochemistry, vol 2. Longman, London, pp 137–160

Lindsay LL, Hedrick JL (1989) Protease released from *Xenopus laevis* eggs at activation and their role in envelope conversion. Dev Biol 135:202–211

Lindsay LL, Hedrick JL (1995) Isolation and characterization of ovochymase, a chymotrypsin-like protease released during *Xenopus laevis* egg activation. Dev Biol 167:513–516

Lindsay LL, Hedrick JL (1998) Treatment of *Xenopus laevis* coelomic eggs with trypsin mimics pars recta oviductal transit by selectively hydrolyzing envelope glycoprotein gp43, increasing sperm binding to the envelope, and rendering eggs fertilizable. J Exp Zool 281:132–138

Lindsay LL, Yamasaki H, Hedrick JL, Katagiri C (1988) Egg envelope conversion following fertilization of *Bufo japonicus*. Dev Biol 130:37–44

Lindsay LL, Larabell CA, Hedrick JL (1992) Localization of a chymotrypsin-like protease to the perivitelline space of *Xenopus laevis* eggs. Dev Biol 154:433–436

Lohka MJ, Masui Y (1983) Formation in vitro of sperm pronuclei and mitotic chromosomes induced by amphibian ooplasmic components. Science 220:719–721

Lohka MJ, Hayes MK, Maller JL (1988) Purification of maturation-promoting factor, an intracellular regulator of early mitotic events. Proc Natl Acad Sci USA 85:3009–3013

Lorca T, Cruzalegul FH, Fesquet D, Cavadore J-C, Méry J, Means A, Dorée M (1993) Calmodulin-dependent protein kinase II mediates inactivation of MPF and CSF upon fertilization of *Xenopus* eggs. Nature 366:270–273

Macgregor HC, Uzzell TM (1964) Gynogenesis in salamanders related to *Ambystoma jeffersonianum*. Science 143:1043–1045

Maéno T (1959) Electrical characteristics and activation potential of *Bufo* eggs. J Gen Physiol 43:139–157

Makino S (1934) A cytological study on the maturation and fertilization of the egg of *Hynobius retardatus* (a urodele amphibian). J Fac Sci Hokkaido Imp Univ Ser 6 3:117–167

Martinez ML, Cabada MO (1996) Assessment of the acrosome reaction in *Bufo arenarum* spermatozoa by immunostaining: comparison with other methods. Zygote 4:181–190

Masui Y (1992) Towards understanding the control of the division cycle in animal cells. Biochem Cell Biol 70:920–945

Masui Y, Clark HJ (1979) Oocyte maturation. Int Rev Cytol 57:185–282

Masui Y, Markert CL (1971) Cytoplasmic control of nuclear behavior during meiotic maturation of frog oocyte. J Exp Zool 177:129–146

Matsuda M, Onitake K (1984a) Fertilization of the eggs of *Cynops pyrrhogaster* (Japanese newt) after immersion in water. Rouxs Arch Dev Biol 193:61–63

Matsuda M, Onitake K (1984b) Fertilization of newt coelomic eggs in the absence of jelly envelope material. Rouxs Arch Dev Biol 193:64–70

Maturi G, Infante V, Carotenuto R, Focarelli R, Caputo M, Campanella C (1998) Specific glycoconjugates are present at the oolemma of the fertilization site in the egg of *Discoglossus pictus* (anurans) and bind spermatozoa in an in vitro assay. Dev Biol 204:210–223

McLaughlin EW, Humphries AA Jr (1978) The jelly envelopes and fertilization of eggs of the newt, *Notophthalmus viridescens*. J Morphol 158:73–90

Meyer A, Dolven IS (1992) Molecules, fossils, and the origin of tetrapods. J Mol Evol 35:102–113

Meyerhof PG, Masui Y(1977) Ca and Mg control of cytoplasmic factors from *Rana pipiens* oocytes which cause metaphase and cleavage arrest. Dev Biol 80:489–494

Meyerhof PG, Masui Y (1979) Properties of a cytostatic factor from *Xenopus laevis* eggs. Dev Biol 72:182–187

Miceli DC, Fernández SN (1982) Properties of an oviducal protein involved in amphibian oocyte fertilization. J Exp Zool 221:357–364

Miceli DC, del Pino EJ, Barbieri FD, Mariano MI, Raisman JS (1977) The vitelline envelope-fertilization envelope transformation in the toad *Bufo arenarum*. Dev Biol 59:101–110

Miceli DC, Fernández SN, Raisman JS, Baribieri FD (1978) A trypsin-like oviducal proteinase involved in *Bufo arenarum* fertilization. J Embryol Exp Morphol 48:79–91

Miceli DC, Fernández SN, Mansilla ZC, Cabada MO (1987) New evidence of anuran oviducal pars recta involvement on gamete interaction. J Exp Zool 244:125–132

Mizote A, Okamoto S, Iwao Y (1999) Activation of *Xenopus* eggs by proteases: possible involvement of a sperm protease in fertilization. Dev Biol 208:79–92

Morelle W, Cabada MO, Strecker G (1998) Structural analysis of oligosaccharide-alditols released by reductive β-elimination from the jelly coats of the anuran *Bufo arenarum*. Eur J Biochem 252:253–260

Nagano H, Okano K, Ikegami S, Katagiri C (1982) Changes in intracellular localization of DNA polymerase-α during oocyte maturation of the toad, *Bufo japonicus*. Biochem Biophys Res Commun 106:683–690

Nakai S, Watanabe A, Onitake K (1999) Sperm surface heparin/heparan suldate is responsible for sperm binding to the uterine envelope in the newt, *Cynops pyrhogaster*. Dev Growth Differ 41:101–107

Nishihara T, Wyrick RE, Working PK, Chen Y, Hedrick JL (1986) Isolation and characterization of a lectin from the cortical granules of *Xenopus laevis* eggs. Biochemistry 25:6013–6020

Nuccitelli R (1994) Fertilization and ion channels. Science 263:988

Nuccitelli R, Kiline D, Busa WB, Talevi R, Campanella C (1988) A highly localized activation current yet widespread intracellular calcium increase in the egg of the frog, *Discoglossus pictus*. Dev Biol 130:120–132

Nuccitelli R, Yim DL, Smart T (1993) The sperm-induced Ca^{2+} wave following fertilization of *Xenopus* egg requires the production of Ins(1,4,5)P_3. Dev Biol 158:200–212

Ohsumi K, Katagiri C (1991) Characterization of the ooplasmic factor inducing decondensation of and protamine removal from toad sperm nuclei: involvement of nucleoplasmin. Dev Biol 148:295–305

Omata S (1993) Relative roles of jelly layers in successful fertilization of *Bufo japonicus*. J Exp Zool 265:329–335

Omata S, Katagiri C (1996) Involvement of carbohydrate moieties of the toad egg vitelline coat in binding with fertilizing sperm. Dev Growth Differ 38:663–672

Paleček J, Ubbels GA, Rzehak K (1978) Changes of the external and internal pigment pattern upon fertilization of the egg of *Xenopus laevis*. J Embryol Exp Morphol 45:203–214

Parrington J, Swann K, Shevchenko VI, Sesay AK, Lai FA (1996) Calcium oscillations in mammalian eggs triggered by a soluble sperm factor. Nature 379:364–368

Peres A, Mancinelli E (1985) Sodium conductance and the activation potential in *Xenopus laevis* eggs. Pflugers Arch 405:29–36

Pérez-Mongiovi D, Chang P, Houliston E (1998) A propagated wave of MPF activation accompanies surface contraction wave at first mitosis in *Xenopus*. J Cell Sci 111:385–393

Peters JM, King RW, Hoog C, Kirschner MW (1996) Identification of BIME as a subunit of the anaphase-promoting complex. Science 274:1199–1201

Philpott A, Leno GH, Laskey RA (1991) Sperm decondensation in *Xenopus* egg cytoplasm is mediated by nucleoplasmin. Cell 65:569–578

Picheral B (1977a) La fécondation chez le triton pleurodèle. I. La traversée des enveloppes de l'úuf par les spermatozodïes. J Ultrastruct Res 60:106–120

Picheral B (1977b) La fécondation chez le triton pleurodèle. II. La pénétration des spermatozodïes et la réaction locale de l'úuf. J Ultrastruct Res 60:181–202

Pitari G, Dupre S, Fusco C, Maurizi G, Campanella C (1993) Jelly plug dissolution in *Discoglossus pictus* eggs (Anura) involves peroxidase-like activity and oxidative opening of disulfide bonds. Zygote 1:61–69

Poccia D (1986) Remodeling of nucleoproteins during gamete genesis, fertilization, and early development. Int Rev Cytol 105:1–65

Quill TA, Hedrick JL (1996) The fertilization layers mediated block to polyspermy in *Xenopus laevis*: isolation of the cortical granule lectin ligand. Arch Biochem Biophys 333:326–332

Raisman JS, Barbieri FD (1969) Lytic effect of sperm suspensions on the vitelline membrane of *Bufo arenarum* oocytes. Acta Embryol Exp 1:17–26

Raisman JS, de Cunio RW, Cabada MO, del Pino EJ, Mariano MI (1980) Acrosome breakdown in *Leptodactylus chaquensis* (Amphibia Anura) spermatozoa. Dev Growth Differ 22:289–297

Reinhart D, Ridgway J, Chandler DE (1998) *Xenopus laevis* fertilisation: analysis of sperm motility in egg jelly using video light microscopy. Zygote 6:173–182

Reinsch S, Karsenti E (1997) Movement of nuclei along microtubules in *Xenopus* egg extracts. Curr Biol 7:211–214

Render J, Elinson RP (1986) Axis determination in polyspermic *Xenopus laevis* eggs. Dev Biol 115:425–433

Runft LL, Watras J, Jaffe LA (1999) Calcium release at fertilization of *Xenopus* eggs requires type I IP_3 receptors, but not SH_2 domain-mediated activation of PLCγ or G(q)-mediated activation of PLCβ. Dev Biol 214:399–411

Sagata N, Watanabe N, Vande Woude GF, Ikawa Y (1989) The c-mos protoncogene product is a cytostatic factor responsible for meiotic arrest in vertebrate eggs. Nature 342:512–518

Sakamoto I, Takahara K, Yamashita M, Iwao Y (1998) Changes in cyclin B during oocyte maturation and early embryonic cell cycle in the newt, *Cynops pyrrhogaster*: requirement of germinal vesicle for MPF. Dev Biol 195:60–69

Salthe SN (1963) The egg capsules in the amphibia. J Morphol 113:161–171
Sato K, Aoto M, Mori K, Akasofu S, Tokmakov AA, Sahara S, Fukami Y (1996) Purification and characterization of a Src-related p57 protein-tyrosine kinase from *Xenopus* oocytes. J Biol Chem 271:13250–13257
Sato K, Iwasaki T, Tamaki I, Aoto M, Tokmakov AA, Fukami Y (1998) Involvement of protein-tyrosine phosphorylation and dephosphorylation in sperm-induced *Xenopus* egg activation. FEBS Lett 424:113–118
Sato K, Iwao Y, Fujimura T, Tammaki I, Ogawa K, Iwasaki T, Tokmakov AA, Hatano O, Fukami Y (1999) Evidence for the involvement of a Src-related tyrosine kinase in the *Xenopus* egg activation. Dev Biol 209:308–320
Sawai T, Higuchi K (1989) Cytostatic effect of the cytoplasm of mature oocytes in the newt, *Cynops pyrrhogaster*. Zool Sci 6:709–714
Sawai T, Satoh T (1984) Furrow-inducing activity of extra sperm nuclei present in egg fragment without zygote nucleus in the newt eggs. Bull Yamagata Univ Nat Sci 11:95–99
Schlichter LC, Elinson RP (1981) Electrical responses of immature and mature *Rana pipiens* oocytes to sperm and other activating stimuli. Dev Biol 83:33–41
Schmell ED, Gulyas BJ, Hedrick JL (1983) Egg surface changes during fertilization and the molecular mechanism of the block to polyspermy. In: Hartmann JF (ed) Mechanism and control of animal fertilization. Academic Press, New York, pp 365–413
Schroeder MM, Gard DL (1992) Organization and regulation of cortical microtubules during the first cell cycle of *Xenopus* eggs. Development 114:699–709
Sessions SK (1982) Cytogenetics of diploid and triploid salamanders of the *Ambystoma jeffersonianum* complex. Chromosoma 84:599–621
Sever DM, Brizzi R (1998) Comparative biology of sperm storage in female salamanders. J Exp Zool 282:460–476
Sever DM, Rania LC, Krenz JD (1996) Reproduction of the salamander *Siren intermedia* Le Conte with special reference to oviducal anatomy and mode of fertilization. J Morphol 227:335–348
Shilling FM, Kratzschmar J, Cai H, Weskamp G, Gayko U, Leibow J, Myles DG, Nuccitelli R, Blobel CP (1997) Identification of metalloprotease/disintegrins in *Xenopus laevis* testis with potential role in fertilization. Dev Biol 186:155–164
Shilling FM, Magie CR, Nuccitelli R (1998) Voltage-dependent activation of frog eggs by a sperm surface disintegrin peptide. Dev Biol 202:113–124
Shimoda Y, Kitajima K, Inoue S, Inoue Y (1994) Isolation, structural determination, and calcium-binding properties of the major glycoprotein present in *Bufo japonicus* egg jelly. Eur J Biochem 223:223–231
Skoblina MN (1974) Behavior of sperm nuclei injected into intact oocytes undergoing the maturation and into oocytes matured after the germinal vesicle removal. Ontogenesis 5:334–340
Smith BG (1912) The embryology of *Cryptobranchus alleganiensis*, including comparisons with some other vertebrates. I. Introduction: the history of the egg before cleavage. J Morphol 23:61–157
Snow P, Yim DL, Leibow JD, Saini S, Nuccitelli R (1996) Fertilization stimulates an increase in inositol trisphosphate and inositol lipid levels in *Xenopus* eggs. Dev Biol 180:108–118
Steinhardt RA, Epel D, Carroll EJ Jr, Yanagimachi R (1974) Is calcium ionophore a universal activator for unfertilised eggs? Nature 252:41–43
Stewart-Savage J, Grey RD (1982) The temporal and spatial relationships between cortical contraction, sperm trail formation, and pronuclear migration in fertilized *Xenopus* eggs. Rouxs Arch Dev Biol 191:241–245
Stewart-Savage J, Grey RD (1984) Fertilization of investment-free *Xenopus* eggs. Exp Cell Res 154:639–642
Stewart-Savage J, Grey RD (1987) Loss of functional sperm entry into *Xenopus* eggs after activation is correlated with a reduction in surface adhesivity. Dev Biol 120:434–446

Stith BJ, Goalstone M, Silva S, Jaynes C (1993) Inositol 1,4,5-trisphosphate mass changes from fertilization through first cleavage in *Xenopus laevis*. Mol Biol Cell 4:435–443

Stith BJ, Espinoza R, Roberts D, Smart T (1994) Sperm increase inositol 1,4,5-trisphosphate mass in *Xenopus laevis* eggs preinjected with calcium buffers or heparin. Dev Biol 165:206–215

Stith BJ, Woronoff K, Espioza R, Smart T (1997) sn-1,2-diacylglycerol and choline increase after fertilization in *Xenopus laevis*. Mol Biol Cell 8:755–765

Strecker G, Wieruszeski J-M, Plancke Y, Boilly B (1995) Primary structure of 12 neutral oligosaccharide-alditols released from the jelly coats of the anuran *Xenopus laevis* by reductive β-elimination. Glycobiology 5:137–146

Streett JC (1948) Experiments on the organization of the unsegmented egg of *Triturus pyrrhogaster*. J Exp Zool 85:383–408

Takamune K (1986) Detection of acrosome-reacted toad sperm based on specific lectin binding to the inner acrosomal membrane. Gamete Res 18:215–233

Takamune K, Katagiri C (1987) The properties of the oviducal pars recta protease which mediates gamete interaction by affecting the vitelline coat of a toad egg. Dev Growth Differ 29:193–203

Takamune K, Yoshizaki N, Katagiri C (1986) Oviducal pars recta-induced degradation of vitelline coat proteins in relation to acquisition of fertilizability of toad eggs. Gamete Res 14:215–224

Takamune K, Lindsay LL, Hedrick JL, Katagiri C (1987) Comparative studies of *Bufo* and *Xenopus* vitelline coat molecular transformations induced by homologous and heterologous oviducal pars recta protease. J Exp Zool 244:145–150

Talevi (1989) Polyspermic eggs in the anuran *Discoglossus pictus* develop normally. Development 105:343–349

Talevi R, Campanella C (1988) Fertilization in *Discoglossus pictus* (Anura). I. Sperm–egg interactions in distinct regions of the dimple and occurrence of a late stage of sperm penetration. Dev Biol 130:524–535

Talevi R, Dale B, Campanella C (1985) Fertilization and activation potentials in *Discoglossus pictus* (Anura) eggs: a delayed response to activation by pricking. Dev Biol 111:316–323

Tian J, Gong H, Thomsen GH, Lennarz WJ (1997a) Gamete interactions in *Xenopus laevis*: identification of sperm binding glycoproteins in the egg vitelline envelope. J Cell Biol 136:1099–1108

Tian J, Gong H, Thomsen GH, Lennarz WJ (1997b) *Xenopus laevis* sperm–egg adhesion is regulated by modifications in the sperm receptor and the egg vitelline envelope. Dev Biol 187:143–153

Tian J, Thomsen GH, Gong H, Lennarz WJ (1997c) *Xenopus* cdc6 confers sperm binding competence to oocytes without inducing their maturation. Proc Natl Acad Sci USA 94:10729–10734

Tian J, Gong H, Lennarz WJ (1999) *Xenopus laevis* sperm receptor gp69/64 glycoprotein is a homolog of the mammalian sperm receptor ZP2. Proc Natl Acad Sci USA 96:829–834

Tokumoto T, Yamashita M, Tokumoto M, Katsu Y, Horiguchi R, Kajiura H, Nagahama Y (1997) Initiation of cyclin B degradation by the 26 S proteasome upon egg activation. J Cell Biol 138:1313–1322

Townsend DS, Stewart MM (1985) Direct development in *Eleuthrodactylus coqui*: a staging table. Copeia 1985:423–436

Townsend DS, Stewart MM, Pough FH, Brussard PF (1981) Internal fertilization in an oviparous frog. Science 212:469–471

Uzzell TM (1964) Relations of the diploid and triploid species of the *Ambystoma jeffersonianum* complex (Amphibia, Caudata). Copeia 1964:257–300

Valz-Gianinet JN, del Pino EJ, Cabada MO (1991) Glycoproteins from *Bufo arenarum* vitelline envelope with fertility-impairing effect on homologous spermatozoa. Dev Biol 146:416–422

Wake MH (1980) Reproduction, growth, and pupulation structure of the caecillian *Dermophis meicanunus*. Herpetologica 36:244–256

Wake MH, Dickie R (1998) Oviduct structure and function and reproductive modes in amphibians. J Exp Zool 282:477–506

Wakimoto BT (1979) DNA synthesis after polyspermic fertilization in the axolotl. J Embryol Exp Morphol 52:39–48
Watanabe N, Hunt T, Ikawa Y, Sagata N (1991) The meiotic release on fertilization of *Xenopus* eggs is independent of the inactivation of cytostatic factor (or Mos). Nature 352:247–248
Webb DJ, Nuccitelli R (1981) Direct measurement of intracellular pH changes in *Xenopus* eggs at fertilization and cleavage. J Cell Biol 91:562–567
Webb DJ, Nuccitelli R (1985) Fertilization potential and electrical properties of the *Xenopus laevis* egg. Dev Biol 107:395–406
Wedlich D, Dreyer C (1988) The distribution of nucleoplasmin in early development and organogenesis of *Xenopus laevis*. Cell Tissue Res 254:295–300
Whitaker M, Swann K (1993) Lighting the fuse at fertilization. Development 117:1–12
Wolf DP (1974a) The cortical granule reaction in living eggs of the toad, *Xenopus laevis*. Dev Biol 36:62–71
Wolf DP (1974b) The cortical response in *Xenopus laevis* ova. Dev Biol 40:102–115
Wolf DP, Hedrick JL (1971) A molecular approach to fertilization. II. Viability and artificial fertilization of *Xenopus laevis* gametes. Dev Biol 25:348–395
Wyrick RE, Nishihara T, Hedrick JL (1974) Agglutination of jelly coat and cortical granule components and the block to polyspermy in the amphibian *Xenopus laevis*. Proc Natl Acad Sci USA 71:2067–2071
Yamaguchi S, Hedrick JL, Katagiri C (1989) The synthesis and localization of envelope glycoproteins in oocytes of *Xenopus laevis* using immunocytochemical methods. Dev Growth Differ 31:85–94
Yamamoto S, Iwao Y (1998) A sperm factor which induces an increase of intracellular Ca^{2+} level in the newt eggs. Zool Sci 15 (Suppl):64
Yamamoto S, Yamashita M, Iwao Y (1999) Rise of intracellular Ca^{2+} level causes the decrease of cyclin B1 and Mos in the newt eggs at fertilization. Mol Reprod Dev 53:341–349
Yamasaki H, IwaoY (1987) Behavior of "accessory" somatic and sperm nuclei introduced into physiologically polyspermic newt eggs. J Exp Zool 243:331–338
Yamasaki H, Katagiri C (1991) Egg exudate-induced reduction of sperm lysin sensitivity in the vitelline coat after fertilization of *Bufo japonicus* and its participation in polyspermy block. J Exp Zool 258:404–413
Yamasaki H, Takamune K, Katagiri C (1988) Classification, inhibition, and specificity studies of the vitelline coat lysin from toad sperm. Gamete Res 20:287–300
Yang JC, Hedrick JL (1997) cDNA cloning and sequence analysis of the *Xenopus laevis* egg envelope glycoprotein gp43. Dev Growth Differ 39:457–467
Yim DL, Opresko LK, Wiley HS, Nuccitelli R (1994) Highly polarized EGF receptor tyrosine kinase activity initiates egg activation in *Xenopus*. Dev Biol 162:41–55
Yoshizaki N (1984) Immunoelectron microscopic demonstration of the pre-fertilization layer in *Xenopus* eggs. Dev Growth Differ 26:191–195
Yoshizaki N (1985) Fine structure of oviducal epithelium of *Xenopus laevis* in relation to its role in secreting egg envelopes. J Morphol 184:155–169
Yoshizaki N (1987) Isolation of spermatozoa, their ultrastructure, and their fertilizing capacity in two frogs, *Rana japonica* and *Xenopus laevis*. Zool Sci 4:193–196
Yoshizaki N (1989) Comparison of two lectins isolated from *Xenopus* cortical granules. Zool Sci 6:507–514
Yoshizaki N (1994) Identification and localization of a ligand molecule of *Xenopus* cortical granule lectins. Zool Sci 11:275–284
Yoshizaki N, Katagiri C (1981) Oviducal contribution to alteration of the vitelline coat in the frog, *Rana japonica*. Dev Growth Differ 23:495–506
Yoshizaki N, Katagiri C (1982) Acrosome reaction in sperm of the toad, *Bufo bufo japonicus*. Gamete Res 6:343–352
Yoshizaki N, Katagiri C (1984) Necessity of oviducal pars recta secretions for the formation of the fertilization layer in *Xenopus laevis*. Zool Sci 1:255–264

Young GPH, Young JD-E, Deshpande AK, Goldstein M, Koide SS, Cohn ZA (1984) A Ca^{2+}-activated channel from *Xenopus laevis* oocyte membranes reconstituted into planar bilayers. Proc Natl Acad Sci USA 81:5155–5159

Yurewicz EC, Oliphant G, Hedrick JL (1975) The macromolecular composition of *Xenopus laevis* egg jelly coat. Biochemistry 14:3101–3107

CHAPTER 5

Fertilization in Birds

GRAHAM J. WISHART[1] AND A. JANET HORROCKS[1]

1 Introduction to Avian Breeding Systems

The 9672 known species of birds (Sibley 1994) are widely distributed over every global habitat and, as might therefore be expected, display great diversity in their morphology, physiology and behaviour. This diversity is perhaps best manifested in avian breeding systems. In terms of mating behaviour, some species breed in small polygynous groups, others are relatively solitary, meeting only to mate in a 'lek', whilst more than 90% of avian species form breeding pairs, either in isolation or in colonies (Lack 1968). Notwithstanding, 'cuckoldry' is a normal feature of the mating behaviour of many of these apparently 'monogamous' pairs (Birkhead and Møller 1992a).

A unifying physiological feature of avian reproduction is the production of the calcified hard-shelled egg (we will use the term 'egg' only for this and refer to the follicular 'oocyte' and ovulated 'ovum'; see Fig. 1). Formation and laying the eggs sequentially, one at a time, keeps female body weight as low as possible to allow for flight, which, although not displayed by all birds, is at least a feature of their ancestors. In the oviduct, the envelopment of the ovum by the various layers of albumen, shell membrane, calcified shell and cuticle to form the egg takes at least 24h. This must be initiated quickly after ovulation, to minimize the interval between ovipositions in a clutch of eggs which are normally incubated concurrently. However, since the first layer of albumen blocks further sperm contact with the ovum, egg formation can only be expedited by minimising the time that the ovum is available for fertilization in the upper reaches of the oviduct. To ensure that spermatozoa are available to the ovum during this brief 'fertilization window', which may only last around 15min, hens store spermatozoa in specialized invaginations of the oviducal mucosa: the sperm storage tubules (Bakst et al. 1994). These have been found in all birds studied to date (Birkhead and Møller 1992b; Briskie and Montgomery 1993) and so, although not confined to the order Aves, oviducal sperm storage may also be considered as a characteristic feature of avian reproductive systems.

[1] Avian Reproduction Group, School of Science and Engineering, University of Abertay Dundee, Bell Street, Dundee DD1 1HG, Scotland, UK

Juan J. Tarín and Antonio Cano (Eds.)
Fertilization in Protozoa and Metazoan Animals

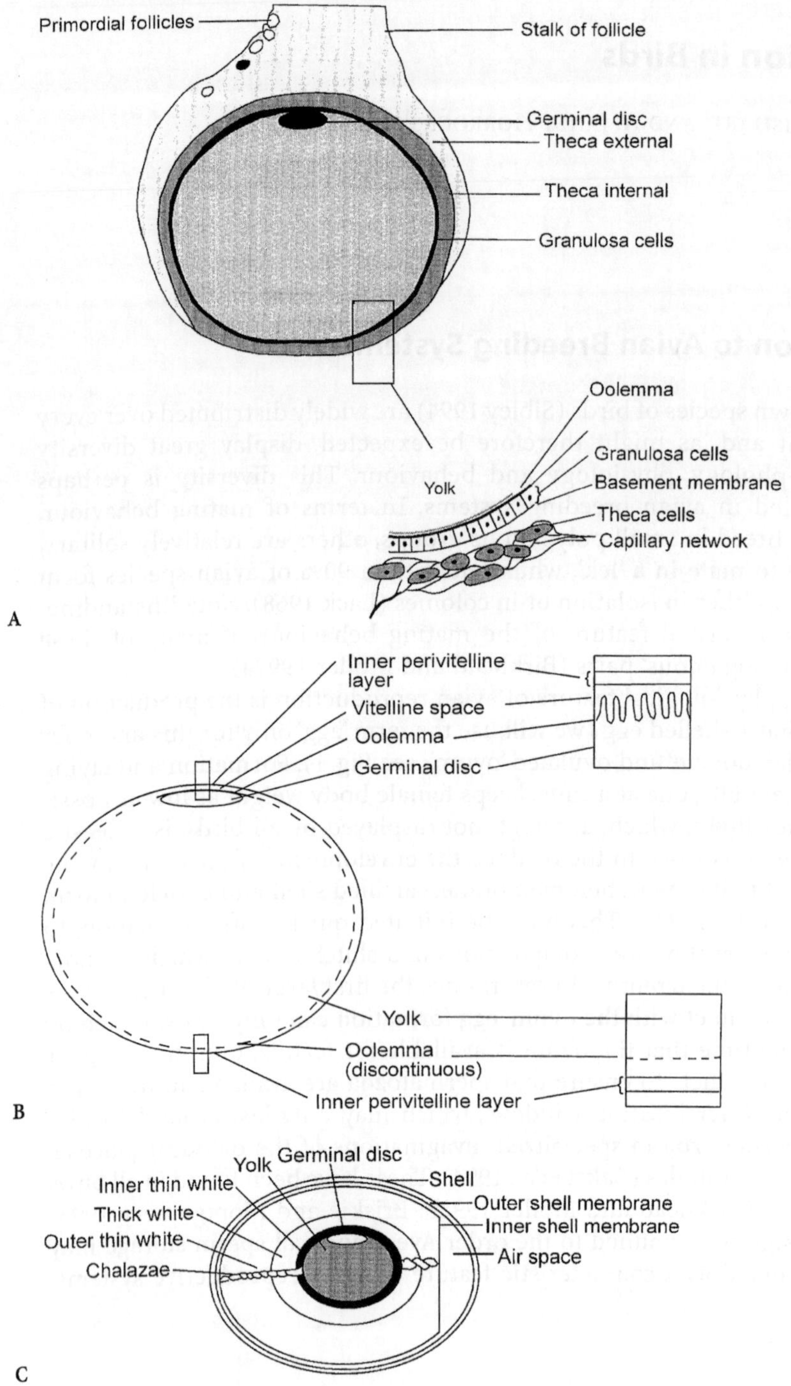

Fig. 1. An ovarian follicle (A), ovum (B) and calcified egg (C). No scale implied

The interval during which the female bird can continue to lay eggs after insemination is known as the fertile period (Lake 1975). This varies characteristically between species and is particularly lengthy in those birds in which mating is infrequent (Birkhead and Møller 1992b).

The brief interval during which spermatozoa and the ovum have the opportunity to interact in vivo and the finding that spermatozoa are capable of penetrating the ovum within a few minutes in vivo and in vitro have led to the idea that sperm capacitation is not necessary for avian fertilization (see Howarth 1984). The seemingly less complex nature of the events preceding avian sperm:egg interaction may be linked to the fact that avian sperm maturation in the male tract is not such a protracted, and apparently involved, process as that found in mammals (see Howarth 1995). So, it appears that the sequentially dependent events which form the overall strategy of avian reproductive systems are: flight; egg formation; brief sperm:ovum encounter; oviducal sperm storage; no sperm capacitation; and limited sperm maturation in the male tract. The physiology, cell and molecular biology of these will be covered in the following sections.

This review does not offer a comprehensive account of avian reproductive physiology but rather those aspects that impinge more directly on fertilization. There are therefore many omissions, the most significant of which is the fundamental subject of photoperiodic and consequent hormonal regulation of the reproductive systems. More comprehensive details of this and other aspects may be found in other volumes, for example, King and McLelland (1979, 1981), Hunton (1995) and Etches (1996). Most of the information that follows will refer to the domestic fowl, *Gallus gallus* ('chicken'), since, because of the agricultural importance of this animal and established husbandry practices, it is usually chosen as an experimental bird. Where no reference is made to species, it may be assumed the information pertains to chickens. Choosing a starting point for reviewing the cyclic reproductive process is always difficult (which comes first, the chicken or the egg?), but in this case we will start with the primordial gametes.

2 Gametes, Gametogenesis and Reproductive Systems

2.1 Primordial Germ Cells and Establishment of Gonads

It should be noted that during egg formation in the oviduct, embryonic development proceeds to the extent that the laid, fertilized egg contains an embryonic blastoderm of around 50,000 cells (Romanoff 1960). The somatic precursors of gametes, the primordial germ cells (PGCs), originate in the epiblast cell layer of these 'unincubated' embryos, from where they translocate to the hypoblast (Ginsburg and Eyal Giladi 1986). On incubation, following gastrulation, they are found, anterior to the embryo proper, in the outer region of the area pellucida, known as the germinal crescent (see Ginsburg 1997; Petitte et al. 1997). Here, around 200 cells can be identified histologically by their large size, eccentric nucleus and high glycogen content (Romanoff 1960). Following approximately 40 h of incubation, the PGCs enter the blood vessels, which have formed in the germinal crescent, and can be found in the circulation of embryos at 45–55 h of incubation. Towards the end of this period, PGCs leave the circulation and are incorporated, apparently following chemoattraction, into the developing gonadal anlage (Kuwana 1993). Sexual differentiation of the gonad occurs following the arrival of the PGCs. This, and proliferation of PGCs, takes place during 8–11 days of incubation in female embryos and a few days later in males. The somatic tissue of the gonad, rather than the PGCs, influences the development of ova or spermatogonial stem cells, since testis of the homogametic (ZZ) male can support development of artificially introduced female (heterogametic – WZ) W-containing spermatozoa (Kagami et al. 1995). During the past 10 years, there has been increased interest in transfer of PGCs between embryos, since the cells in vitro can be used as target cells for transgenesis (Petitte et al. 1997) and as a means of 'gene banking' and reconstruction of embryos from cryopreserved avian somatic cells (Naito et al. 1994).

2.2 Male Reproduction System and Sperm Production

2.2.1 Male Tract Anatomy

The main anatomical differences between the male reproductive system of birds and that of mammals are: (1) that the testes are situated within the body cavity, so that spermatogenesis takes place at the avian body temperature of 40–41 °C; (2) that the epididymis is much shorter, so that the period of post-testicular maturation is reduced; and (3) that there are no equivalent accessory glands, so that semen passes directly from the deferent ducts into the female tract. Additionally, birds do not have a penis and although some species have an intromittent organ, this does not deliver the semen (King 1981).

The paired testes are situated mid-dorsally within the abdominal cavity. Each testis consists of an interconnecting network of seminiferous tubules, with the spaces between the tubules filled with Leydig cells (Lake 1981). The internal surfaces of the seminiferous tubules are lined with Sertoli cells and spermatogonia, with spermatocytes and spermatids at various stages of development, and are arranged consecutively between, and in intimate contact with, the Sertoli cells. Formed spermatozoa are released from the Sertoli cells into the luminal fluid of the seminiferous tubules which drains into the tubules of the rete testis. From the rete testis the spermatozoa enter the epididymis – consisting of the efferent ducts, connecting ducts and the epididymal ducts

(Howarth 1995). In the epididymis, the sperm suspension is concentrated by fluid absorption and the contents of the epididymis then pass into the deferent duct which connects to papillae in the cloacal phallus, from which the semen is ejaculated. Some lymph-like fluid from the folds of the phallic region can be mixed with manually collected fluid, but it remains unclear whether this has any function (Froman 1995).

2.2.2 Maturation of Sperm Function

Unlike mammalian spermatozoa, avian spermatozoa from the testis can be motile in vitro and demonstrate fertilizing ability in vivo. Chicken sperm motility – or the capacity to become motile – is rather limited in testicular spermatozoa, increases in epididymal spermatozoa and reaches a maximum in those from the deferent duct (Munro 1938; Ashizawa and Sano 1990). Testicular spermatozoa from pigeon and drake, but not song sparrow, are also motile (Bedford 1979). Chicken testicular spermatozoa can also fertilize ova following intravaginal artificial insemination (AI) but at a very low rate, which is greater in epididymal spermatozoa and further increased in those from the deferent duct (Munro 1938). These findings led Munro to speculate that the maturation of sperm-fertilizing ability was simply a function of their motility. Howarth (1983) confirmed Munro's experimental findings and also showed that following surgical insemination of testicular spermatozoa into the magnum, 85% of all eggs laid were fertilized, whereas no fertile eggs were produced following intravaginal insemination. Deferent duct spermatozoa isolated from the turkey were also shown to be highly motile and fertile (Bakst and Cecil 1981).

There are two possible explanations for Howarth's (1983) results: firstly, that the greater effectiveness of testicular spermatozoa inseminated into the magnum may simply result from their increased numbers in the upper oviduct, which is an expected result of intramagnal insemination (Zavaleta and Ogasawara 1987); or, secondly, a sperm maturation factor, conferred in the epididymis, may be necessary for sperm migration through the lower oviduct. This may involve motility activation (Ashizawa et al. 1994a) and perhaps a surface characteristic, since proteins secreted by the epididymis/deferent duct were shown to be associated with deferent and epididymal, but not testicular, spermatozoa (Esponda and Bedford 1985). Whichever, it is clear that avian spermatozoa develop their fertilizing ability in the testis, and that sperm maturation within the excurrent ducts is more limited than in mammals. The shorter length of the avian epididymis and rapid migration of spermatozoa through the epididymis and deferent ducts (see Froman 1995; Howarth 1995) support this hypothesis.

2.3 Female Reproductive System

2.3.1 The Ovary and Ovulation

In females, the right ovary regresses to a rudimentary vestige after hatching. An inventory of the ovarian follicles of the mature hen lists: >1000 1-mm-diameter 'small white'; about 20 5-mm-diameter 'large white'; about 10 'small yellow'; and the 5–7 hierarchical follicles in the rapid growth phase. The F1 follicle is the largest of these and next to ovulate; the others, in order of decreasing size, are F2, F3 . . . , etc. The deposition of yellow yolk, which occurs in the latter stages, increases the size of the oocytes dramatically to around 36 mm in the chicken (Gilbert 1979; Etches 1996).

The follicular unit of the ovary consists of the ovum surrounded by granulosa cells and thecal cells (Fig. 1A). In earlier follicles, the thecal cells are primarily steroidogenic, secreting androgens and oestrogens from precursors synthesized by the granulosa cells. During the rapid growth phase, the monolayer of granulosa cells secretes progesterone, although significant output of progesterone is a feature of the F1 follicle only. The rise in plasma progesterone a few hours before ovulation stimulates the hypothalamic/pituitary axis to increase gonadotrophin production, which induces ovulation by weakening the connective tissue around the stigma of the follicle (Gilbert 1979; Etches 1996). The granulosa cells of the rapid growth phase follicles are also responsible for synthesizing proteins of the chicken (Waclawek et al. 1998) and quail (Kuroki and Mori 1995) inner perivitelline layer (IPVL) in which is a fibrous, glycoprotein layer that surrounds the ovum (Bellairs et al. 1963). By the time of ovulation the granulosa cells have become dislodged from the IPVL, which therefore represents the outer surface of the ovulated ovum (Fig. 1B).

In hierarchical follicles, the region of the female germplasm, arrested at prophase of the first meiotic division, is seen as the opaque 3-mm-diameter disc of the germinal disc (GD). During 6 h before ovulation, the germinal vesicle, in the centre of the GD, breaks down as meiosis resumes and the first meiotic division is completed. Around 1 h before ovulation, the first polar body is extruded, so that the sex of any offspring is determined by this time (Etches 1996).

2.3.2 The Oviduct and Egg Formation

As with the ovary, only the left oviduct is functional. The oviduct is composed of five major regions (see Fig. 2) and has seven layers of connective tissue, smooth muscle and mucosa (Gilbert 1979; Etches 1996). At the anterior

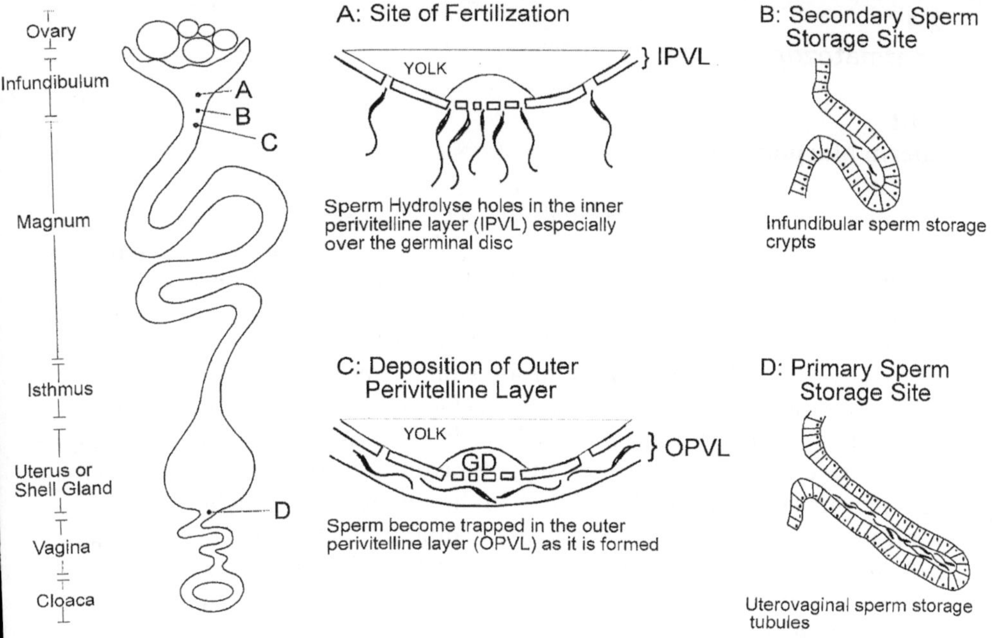

Fig. 2. Avian oviduct, highlighting sperm storage sites and nature of sperm:egg interaction. Not to scale

portion, the expanded, thinned funnel of the infundibulum is particularly active at the time of ovulation, surrounding the dominant ovarian follicle and engulfing the ovulated egg. The 'secondary' sperm storage sites are invaginations in the mucosal epithelium of the lower infundibulum. The magnum is the longest region of the oviduct and secretes the albumen layers, including the outer perivitelline layer (OPVL). The isthmus is a short region in which the shell membrane is deposited. The circumference of the oviduct then expands to form the uterus or shell gland. Fluid secreted from the shell gland is absorbed by the egg which swells to its final size. Calcification of the egg ensues, followed by the appropriate, characteristic pigmentation and finally the laying down of the outer waxy cuticle (Fig. 1C). The processing of the egg in the chicken shell gland takes around 20 h. Oviposition is initiated by relaxation of the smooth muscles of the uterovaginal junction and vaginal contraction (Gilbert 1979; Etches 1996). The mucosa of the anterior end of the vagina houses the primary sperm storage tubules (SSTs), whilst the vaginal mucosa posterior to the site of the SSTs appears to have a function in regulating sperm transport (Bakst et al. 1994).

2.4 Spermatozoa

2.4.1 Sperm Morphology

The morphology of avian spermatozoa is often described as 'sauropsid', since the shape of spermatozoa from the more primitive non-passerine avian orders is similar to those of reptiles (Fig. 3B). The non-passerine orders include all domesticated and related species, parrots, pigeons, raptors and gulls. Birds of the order Passeriformes ('perching and songbirds') have spermatozoa characterized by a helical configuration in some or all 'head and midpiece' organelles and have other characteristic structures (Fig. 3A). The considerable variations in features of sperm morphology have been considered, with some, albeit limited (McFarlane 1963, 1971), success, as indicators of avian phylogenetic relationships (Asa and Phillips 1987). In the context of this review, we will concentrate on non-passerine spermatozoa: although passerines represent about 57% of living avian species, we have comparatively little knowledge of their gamete physiology, and know virtually nothing of the mechanisms of their sperm:egg interaction.

Spermatozoa of non-passerine species are also described in the literature as fusiform (spindle-shaped) or vermiform (worm-shaped). Their general morphology is typified by that of the domestic fowl spermatozoon (Lake et al. 1968) and is illustrated in Fig. 3B,C. The nucleus is cylindrical, straight or comma-shaped, around 12 μm long and 0.6 μm in diameter. Anterior to this is the conical acrosome, approximately 2 μm in length. The perforatorium has projections that penetrate both the acrosome and the nucleus and, presumably, strengthens the links between these two structures. The outer surface of the acrosome is bounded by the outer acrosomal membrane and the surface apposed to the perforatorium by the inner acrosomal membrane (Bakst and Howarth 1975; Thurston and Hess 1987). The nucleus is surrounded by a double nuclear envelope and both the acrosome and nucleus are contained within a continuous plasma membrane (Lake et al. 1968). The anterior portion of the midpiece contains, at its core, the proximal and distal centrioles, from which the 9 + 2 microtubular pattern of the axoneme ensues. The midpiece is around 4 μm in length and contains 25–30 mitochondria arranged longitudinally in a helical pattern around the centriolar a nd axonemal core (Bakst and Howarth 1975; Thurston and Hess 1987). Spermatozoa of Columbiformes (McFarlane 1971) and quail (Woolley 1995) have, exceptionally and for unknown reasons, more than 1000 mitochondria in an extended midpiece. The flagellum of spermatozoa from most species may be expected to be between 70 and 90 μm in length, except for that of quail and Columbiform spermatozoa which at least double this length. The fibrous sheath that surrounds the axoneme of mammalian sperm flagellae is absent.

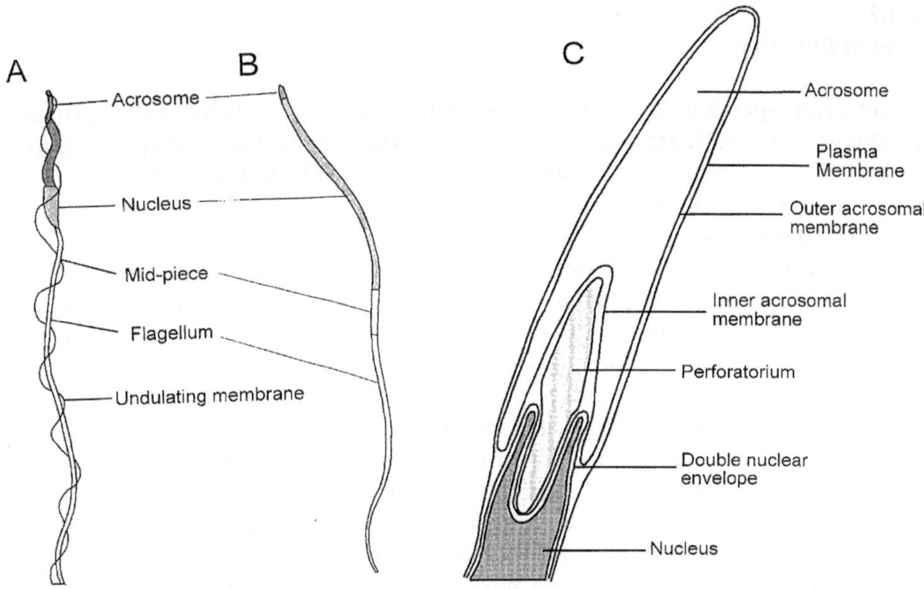

Fig. 3. Outline of morphological features of spermatozoa of passerine (**A**) and non-passerine (**B**) birds, with expanded view of domestic fowl sperm acrosomal region (**C**). For an appreciation of (approximate) scale, the non-passerine (domestic fowl) acrosome is 2 µm and nucleus is 12 µm in length, whilst acrosomal and nuclear lengths of the non-passerine example (canary) are 25 and 8 µm, respectively. Total lengths of flagellae in these examples are 150 and 80 µm for examples **A** and **B**, respectively. (Example **A** redrawn from Humphreys 1972)

Spermatozoa of passerine birds have a shorter nucleus, in the shape of a helical cylinder and a long tapering acrosome, which can be several times as long as the nucleus (Fig. 3A). The left-handed helical membrane extends over the whole length of the spermatozoon in most passerines and is continuous with the helix of the nucleus and the mitochondrion of the midpiece. This may extend along most of the length of the flagellum (McFarlane 1971; Asa and Phillips 1987).

The number of spermatozoa that can be collected from any one bird in a single ejaculate varies among species, from 5 billion in the turkey (Lake and Stewart 1978) to 3 million in some falcons (Gee 1995). The efficiency of manual semen collection undoubtedly depends on the method used, not to mention the individual bird and the human(s) involved in the process. Spermatozoa of poultry species can vary in their 'quality', in terms of their motility and integrity (Wishart 1995), and some species, such as cranes (Phillips et al. 1987) and falcons (Hartley 1999), produce pleiomorphic forms with variable nuclear shapes and sizes.

2.4.2
Sperm Motility

Intact avian spermatozoa of both passerine and non-passerine species show a clockwise (viewed from ahead) spinning motion about their progression axis as well as more complex wave motions (Vernon and Woolley 1999). The curvilinear velocity of chicken and turkey spermatozoa has been estimated to be around 50 µm/s (Holm and Wishart 1998).

Most work on the regulation of avian sperm motility regulation has been carried out on chicken spermatozoa. In simple buffered salt solutions at the avian body temperature of 40/41 °C, they are immotile, but regain motility when the temperature is lowered below 35 °C (see Ashizawa et al. 1994b). This temperature dependency is also found in drake spermatozoa, but not in any other species studied (Wishart and Wilson 1999), and has no known physiological significance. Motility of spermatozoa in vitro at 40 °C is stimulated by calcium, caffeine and phosphatase inhibitors (Ashizawa et al. 1994b). Calcium appears to be the primary positive regulator of chicken sperm motility and is the major motility-stimulating factor in fluids of both the male and female reproductive tracts (Ashizawa and Wishart 1992). The regulatory 'cascade' triggered by calcium has been shown to involve calmodulin, myosin light-chain kinase and protein phosphatase 1 and 2 (Ashizawa et al. 1996). Alkaline pH of both the external and internal environment also stimulates the motility of chicken spermatozoa (Ashizawa et al. 1994b). The motility of spermatozoa of other avian species shows considerable variation in its response to calcium, but inhibition of sperm motility in an environment of pH 7.0 or below and stimulation at pH 8.0 or above is found in spermatozoa from several other species (Holm and Wishart 1998; Holm 1999).

2.4.3
The Acrosome and Acrosomal Enzymes

The basic structure of the acrosome of a typical non-passerine spermatozoon can be seen in Fig. 3C. and is described above in Section 2.4.1. Our interpretation of electromicrographs from different sources (Bakst and Howarth 1975; Thurston and Hess 1987; L. Holm, unpubl. results) is that the inner and outer acrosomal membranes are continuous. The length of the acrosome in non-passerine birds is generally around 2 µm, whilst in passerine species (Fig. 3A) it is more variable and can be several times the length of the nucleus (McFarlane 1971). This may not have any particular relationship to enzymatic activity, since the quail acrosome is clearly longer than that of the domestic fowl (Woolley 1995) but appears to have less trypsin-like activity (Langford and Howarth 1974). In comparison to that of mammals, the proteolytic activity of the chicken sperm acrosome was particularly strong (Yanagimachi and Teichman 1972; Brown and Hartree 1976).

The hydrolytic enzymes assumed to originate from the acrosome of chicken spermatozoa include N-acetyl-β-D-glucosaminidase (McIndoe and Lake 1974) and the protease acrosin (Ho and Meizel 1970, 1976a). Acrosin is a sperm-specific protease with trypsin-like substrate specificity and inhibitor-binding characteristics, which preferentially hydrolyses peptide bonds on the carbonyl side of arginine. The enzyme has been isolated from both chicken (Ho and Meizel 1970, 1976a; Froman 1990) and turkey (Richardson et al. 1988) spermatozoa. The chicken enzyme showed a molecular weight of between 26 and 30 kDa (by gel filtration), depending on the concentration of buffer salts, and, when fractionated by DEAE Sephadex chromatography, was found to contain several isozymes of molecular weights 49.5, 47.4, 43.3, 25.6 and 27.7 kDa (determined by SDS PAGE). The turkey enzyme contained two isozymes that, on further analysis by SDS PAGE, revealed three polypeptides of MW 15.9, 13.9 and 11.7 kDa (Richardson et al. 1988). Analysis of the N-terminal sequence of chicken acrosin (L. Burgess, I. Dunn, J. Horrocks, unpubl. results) showed considerable similarity to mammalian, particularly rabbit, acrosin, thus suggesting (by analogy with the mammalian acrosin) that the observed multiple forms of acrosin were processing variants of the enzyme derived from a parent zymogen. Disruption of spermatozoa at pH 2.7 produced an inactive form of acrosin that could be activated by incubation at pH 8.0 (Brown and Hartree 1976), demonstrating that avian acrosin, like rabbit acrosin, is processed from an inactive zymogen. The role of acrosin in mammalian fertilization is currently the subject of some debate. Acrosin inhibitors will prevent sperm penetration of rabbit zona pellucida (Stambaugh et al. 1969) or the chicken inner perivitelline layer in vitro (Howarth and Digby 1973; Robertson 1999). Pre-incubation of chicken spermatozoa with the irreversible trypsin inhibitor tosyl-l-lysine chloromethyl ketone (TLCK) will also inhibit fertilization (Zanfeld et al. 1970; Palmer and Howarth 1973). However, mice carrying a mutation in the acrosin gene are fertile (Baba et al. 1994; Adham et al. 1997), leading to the suggestion that sperm penetration of the zona pellucida is unlikely to be a lytic event in mammals (Bedford 1998). In chickens, sperm depolymerize the inner perivitelline layer creating holes of around 9 μm diameter, which is significantly greater than the diameter of a single spermatozoon (0.5 μm). This observation, coupled with the resistance of the inner perivitelline layer to hydrolysis by rabbit acrosin (Ho and Meizel 1976b), would suggest that, in spite of physiochemical similarities, the mammalian and avian acrosins perform different functions during fertilization.

2.5 The Ovum

2.5.1 Morphology of Ovum at Ovulation

At ovulation, the ovum consists of the yolk mass, surrounded by the plasma membrane of the ovum – the oolemma – and is bounded on its outer surface by the IPVL (Fig. 1B). By far the greater volume of the ovum is taken up with the mass of yellow yolk. At the animal pole of the ovum, the 3-mm-diameter germinal disc containing the metaphase II meiotic nucleus, first polar body, mitochondria and glycogen granules is surrounded by white yolk spheres (Bakst and Howarth 1977a). The oolemma at the germinal disc is arranged into single and multiply-branched microvilli, which project into the perivitelline space (Bakst and Howarth 1977a) between the oolemma and IPVL. This region is also pitted at the periblastic region by several concavities of around 60–100 µm diameter and 20 µm depth, which have less expansive microvilli (Bakst and Howarth 1977a; Bakst 1978). Away from the germinal disc region, there are fewer microvilli, the perivitelline space does not exist and the oolemma is discontinuous, so that yolk spheres abut directly onto, and can pass a short distance into, the IPVL (Bakst and Howarth 1977a).

2.5.2 The Inner Perivitelline Layer

Surrounding ovum at ovulation is the acellular glycoprotein layer known as the IPVL . It supports the megalecithal ovum mechanically and presents a barrier to sperm penetration of the ovum and, as such, is considered equivalent to the mammalian zona pellucida. Structurally, the IPVL was shown to be about 2.7 µm deep, although thinner over the germinal disc, and to consist of an open lattice of fibres aligned parallel to the surface of the yolk (Bellairs et al. 1963). Later studies showed that the network was in fact filled with granular material (Bain and Hall 1969; Bakst and Howarth 1977a). Analysis of the whole perivitelline layer by SDS PAGE (Kido et al. 1975, 1976) revealed that it contained three major glycoproteins termed GP1, GP2 and GP3. Of these, GP1 and GP2 were physicochemically distinct, but GP2 and GP3 both had the same molecular weight when analysed by reducing SDS PAGE. GP1 was very insoluble and had a molecular weight of 27 kDa when assessed by ultracentrifugation (Kido et al. 1977), 32 kDa by non-reducing SDS PAGE or 43 kDa by reducing SDS PAGE (Back et al. 1982). GP2 had a molecular weight of 260 kDa by non-reducing SDS PAGE or 78 kDa by reducing SDS PAGE (Robertson et al. 1997a) and contained a high percentage of hydrophobic amino acids (Kido et al. 1977). Separation of the two layers (Kido and Doi 1988) showed that GP1 and GP2 were associated solely with the IPVL. GP1 cDNA has been sequenced and termed gp42 (Takeuchi et al. 1999) or chZP3 (Waclawek et al. 1998). In order to standardize nomenclature and avoid confusion with other glycoproteins or zonae, we will use IPVL1 to denote GP1, gp42 or chZPC and IPVL2 for GP2. The equivalents of IPVL1 and IPVL2 were identified in quail (Mori and Masuda 1993) with molecular weights of 33 and 175 kDa (determined by non-reducing SDS PAGE).

The amino acid sequence of IPVL1 (deduced from cDNA) and N-terminal sequence data showed that a 20 amino acid signal peptide was cleaved from a 435 amino acid transcript to produce a mature polypeptide of 415 amino acids (molecular weight 44,718 Da) (Takeuchi et al. 1999). However, Waclawek et al. (1998) identified a different N terminus for the IPVL1 polypeptide 26 amino acids downstream. The disparity in these results can be rationalized if the source of polypeptide is considered: Takeuchi et al. (1999) used follicles as a source of IPVL whilst Waclawek et al. (1998) used laid eggs. It has been noted by several authors in both chicken (Steele et al. 1994; Robertson et al. 1997a; Waclawek et al. 1998; Takeuchi et al. 1999) and quail (Mori and Masuda 1993) that IPVL1 from laid eggs has a slightly greater electrophoretic mobility than IPVL1 from follicular or freshly ovulated eggs. There is no evidence that the change in mobility is glycosylation-dependent (Waclawek et al. 1998), thus suggesting that N-terminal processing (possibly associated with deposition of OPVL) may be responsible for the apparent change in molecular weight. Both Waclawek et al. (1998) and Takeuchi et al. (1999) found that the IPVL1 polypeptide was smaller than predicted from the amino acid sequence and suggested that C-terminal processing of IPVL1 may also occur.

3
Oviducal Sperm Transport and Storage

During mating, the female presents her vaginal orifice via the cloaca and the male deposits semen at this site. For birds with an intromittent organ – and for artificial insemination – semen is deposited into the distal vagina. The majority of these spermatozoa are ejected from the vagina within 30 min (Howarth 1971b). Transport through the vagina appears to require that the spermatozoa are motile (Wishart 1994) and that they display appropriate surface characteristics, since migration is inhibited in spermatozoa with experimentally altered surface glycoproteins (Steele and Wishart 1996a,b), or from another species with different surface antigenicity (Steele and Wishart 1992a). Binding of vaginal immunoglobulin to the flagellar surface of the majority of spermatozoa occurs in the vagina and is associated with cell permeabilization, whereas spermatozoa found beyond the vagina are free of immunoglobulin and intact (Steele and Wishart 1992b).

At the uterovaginal junction (UVJ), spermatozoa enter the UVJ SSTs (see Fig. 2). These are epithelial invaginations of the UVJ mucosa, typically around 200 μm in length and 50 μm in diameter, although this varies among species (Birkhead and Møller 1992b; Briskie and Montgomery 1993). They consist of columnar epithelium cells surrounding a lumen of about 10 μm diameter. Up to 100 spermatozoa can enter each tubular lumen and remain, motionless or oscillating slowly, with their heads positioned towards the blind end of the tubule. The UVJ-SSTs represent the primary storage site of spermatozoa in the oviduct and so it is presumed that spermatozoa from the SSTs must be released to fertilize the eggs. Although entry of spermatozoa into the SSTs appears to be a random event, the mechanisms that maintain spermatozoa within the SSTs and the way in which they can exit from the SSTs remain unknown. It does appear that they are released continuously rather than episodically in synchrony with some event of the oviducal cycle, but further consideration of these mechanisms is beyond the scope of this review and may be found elsewhere (Zavaleta and Ogasawara 1987; Bakst et al. 1994).

Oviducal transport of spermatozoa from the UVJ-SSTs upwards through the shell gland and magnum is independent of sperm motility and appears to be due to antiperistaltic oviducal movements, although the presence of an egg in the magnum prevents passage of spermatozoa towards the infundibulum. At the infundibulum, there is a secondary storage site for spermatozoa. These infundibular crypts (Fig. 2) are shorter that the UVJ-SSTs, possess an epithelium consisting of secretory cells and, under most circumstances, contain very few spermatozoa (Bakst et al. 1994).

Quantitative approaches to sperm transport and storage in the oviduct have shown that the percentage of the inseminated dose of spermatozoa that enter the UVJ-SSTs is 1–2% in turkeys (Brillard and Bakst 1990), 1% (Brillard 1993) or 0.04% in chicken (McDaniel et al. 1997), 0.04% in the zebra finch (Birkhead et al. 1993) and around 1% in the Bengalese finch (Birkhead 1992). However, the actual numbers of stored spermatozoa differ with species, ranging from around 1 million in chickens and turkeys to 6000 in zebra finch. The number of spermatozoa that subsequently reach the infundibulum in chicken, turkey and zebra finch has been estimated to be around 1–2% of the spermatozoa within the SSTs (Birkhead and Møller 1992a; Wishart 1995).

Studies on the numbers of spermatozoa in different regions of the oviduct involve sacrifice of the bird and prevent any temporal studies of how this may change throughout the fertile period. However, the egg itself provides a record of the numbers of spermatozoa within the infundibulum, but further discussion of this first requires some consideration of the site and overall mechanism of sperm:egg interaction.

4 Sperm : Egg Interaction

4.1 Fertilization in Vivo and in Vitro

4.1.1 Site of Fertilization in Vivo

Tauber (1875) proposed that the infundibulum was the site of fertilization, based on observations of the presence of spermatozoa in that region. Studies of sperm penetration into the ovum showed that eggs taken from the body cavity (Olsen 1942) or infundibulum (Olsen and Nehrer 1948) of chickens and the infundibulum of turkeys (Olsen and Fraps 1944) had been fertilized within around 15 min of ovulation. Ova recovered from the lower infundibulum and proximal magnum have been invested with the OPVL (Bain and Hall 1969; Bakst and Howarth 1977b) which occludes the IPVL and inhibits fertilization of ova in vitro (Howarth and Digby 1973). Thus fertilization takes place in the body cavity or infundibulum (Fig. 2).

4.1.2 Capacitation

Spermatozoa inseminated into the vagina (Bobr et al. 1964) or infundibulum (Olsen and Nehrer 1948) following oviposition were able, in the first case, to traverse the oviduct and, in both cases, to fertilize an ovulated egg within around 15 min. Additionally, fertilization of ova by ejaculated spermatozoa in vitro was effected within around 15 min in both chickens (Howarth 1971a; Fujihara et al. 1973) and turkeys (Howarth and Palmer 1972). These findings have been taken as evidence that avian spermatozoa do not require a period of capacitation, since they are competent to fertilize the egg without a prolonged sojourn in the oviduct or in a 'capacitation medium' in vitro. Indeed, ejaculated spermatozoa are able to acrosome react and penetrate the isolated IPVL in vitro within 2.5 min (Robertson et al. 1997a).

Whilst these findings do not, in themselves, denote any functional changes in the spermatozoa associated with capacitation, there are other indications against avian sperm capacitation. Firstly, if capacitation of mammalian spermatozoa is viewed as a reversal of plasma membrane-associated changes which take place during sperm maturation in the excurrent male ducts, then these are likely to be of minor importance in birds, in which sperm have a rapid migration through the excurrent ducts and show fewer maturational changes (see Howarth 1995). Secondly, if the spermatozoa are to survive within the oviduct for protracted periods, then it would seem inappropriate for oviducal secretions to initiate the membrane destabilization of capacitation processes – although the SST environment could inhibit capacitation. Lastly, if the avian egg is only available for fertilization during 15 min or an even shorter period (see Sect. 4.2.5), then it is essential that the spermatozoa are activated quickly.

4.2 Sperm: Perivitelline Layer Interaction

4.2.1 Sperm: Inner Perivitelline Layer Interaction

Avian spermatozoa have to penetrate the IPVL to gain access to the oolemma. Early observations of the fibrous mesh of the IPVL led to the suggestion that the spermatozoa simply passed between these fibres, since transmission electron micrographs showed no apparent substance between the fibres of the IPVL (Bellairs et al. 1963; Fuji 1976). However, other studies (Bain and Hall 1969; Bakst and Howarth 1977a) found that the spaces between the fibres were filled with granular material, especially on its outer surface (Bakst and Howarth 1977a) so that simple penetration was unlikely. Okamura and Nishiyama (1978a) showed that the fabric of the IPVL was dissolved under the tip of the sperm head, creating a hole (tunnel) of 9 μm diameter. It seems likely that the holes were produced by sperm-associated hydrolytic enzymes such as acrosin, since no IPVL holes were observed in the presence of serine protease inhibitors such as soybean trypsin inhibitor (Robertson 1999).

The 'holes' in the IPVL are diagnostic of a successful sperm:egg interaction, since their formation requires that sperm bind to the IPVL and subsequently acrosome react, releasing the contents of their acrosomal vesicle. Sperm will form holes in vitro when incubated with isolated IPVL free of underlying yolk material (Koyanagi et al. 1988; Bramwell and Howarth 1992a; Steele et al. 1994; Robertson et al. 1997a). Formation of IPVL holes is initiated within 2.5 min of coincubation with spermatozoa (Robertson et al. 1997a) and their size increases with time (Wishart and Fairweather 1999). However, since there is no outer perivitelline layer to arrest hydrolysis in vitro, the holes enlarge, merge (Robertson et al. 1997a) and eventually fragment the IPVL. Spermatozoa encounter only the outer surface of the IPVL in vivo, whilst in a simple coincubation assay in vitro, spermatozoa can access both sides. If the IPVL is arranged so that spermatozoa have access to one side only, then they show a three-fold preference for the 'outer' side, to which they have access in vivo (Robertson 1999). Although this outer side contains the granular material (Bakst and Howarth 1977a), this does not establish that the granular material contains the sperm-binding protein, since antibody against IPVL1 has affinity for the large fibres of the IPVL (Waclawek et al. 1998) and will inhibit sperm:egg interaction (K. Connolly, G. Wishart, J. Horrocks and W. Schneider, unpubl. results).

In vitro, the IPVL from F1 to F5 follicular oocytes, freshly ovulated ova and laid eggs shows a similar response to sperm hydrolysis (Steele et al. 1994), so that the responsive elements have matured several days before ovulation and are unaffected by the cleavage of IPVL1 which occurs between then and the time of oviposition (Waclawek et al. 1998). The finding that solubilized IPVL from chicken, duck and turkey eggs could inhibit interaction of either chicken, drake or turkey spermatozoa and IPVL in vitro was taken as evidence of competitive inhibition across these species (Bramwell and Howarth 1992b). Although this interpretation may be invalid (see Sect. 4.2.3), the general principle of limited species specificity in avian sperm:IPVL interaction is upheld: the IPVL from several poultry (Steele et al. 1994) and other avian (Robertson 1999) species can be hydrolysed by chicken spermatozoa in vitro, indicating that the species specificity of mammalian zona:sperm interaction (Yangimachi 1994) is not found in Aves. Notwithstanding the behavioural barriers to mating and species-specificity of vaginal migration of spermatozoa (see Sect. 3), this may account for the large number of avian hybrids which have been reported (Gray 1958).

4.2.2 The Avian Acrosome Reaction

As mentioned above, it was originally considered that avian spermatozoa might penetrate the IPVL simply by passing between its constituent fibres (Bellairs et al. 1963; Fujii 1976). This idea was also given credence by the finding that some spermatozoa which were in contact with the IPVL did not appear to have undergone an acrosome reaction (AR), since they only showed slight ruffling or 'fenestration' of the outer acrosomal membrane and the plasma membrane overlying the acrosome (Bakst and Howarth 1977b). However, with the description of sperm-induced hydrolysis of 'tunnels' in the IPVL (Okamura and Nishiyama 1978a), and the knowledge that sperm:IPVL interaction could be reduced by trypsin inhibitors (Palmer and Howarth 1973), the principle that an avian AR was an essential feature of IPVL penetration became established (see Howarth 1984). Okamura and Nishiyama (1978a) proposed a mechanism for the chicken AR which involved initial fusion of the outer acrosomal membrane with the plasma membrane overlying the acrosome, followed by subsequent vesiculation of these membranes, exposure of the acrosome and release of the acrosomal contents. The proposed mechanism was completed when the acrosomal matrix was detached, so that the perforatorium, surrounded by the inner acrosomal membrane, was presented at the anterior tip of the spermatozoon. The loss of the acrosome was used to demonstrate that the IPVL stimulates the AR during coincubation of spermatozoa with IPVL sheets from a freshly ovulated egg (Koyanagi et al. 1988). However, maximum AR induction was only achieved following 30–60 min incubation, whereas normal sperm:IPVL interaction in vivo takes less than 10 min (Wishart and Fairweather 1999) and sperm penetration of the IPVL in vitro takes less than 5 min (Robertson et al. 1997a). Coincubation with IPVL fragments for 5 min exposes peanut agglutinin lectin-binding sites on the sperm acrosomal matrix (Robertson and Wishart 1996), but the majority of these acrosomes are still attached to the spermatozoa (Robertson 1999). Additionally, the process of IPVL hydrolysis by spermatozoa can be viewed in vitro, during which time the spermatozoa appear to link to the IPVL via their acrosomes, which remain attached. Furthermore, spermatozoa can be shown to pass through the IPVL in vitro with their acrosomes still attached (Kerr 1997). If it is accepted that the acrosome reaction need not involve loss of the acrosomal tip, then this has further implications in the interpretation of the evidence of the mechanisms involving sperm penetration of the ovum (see Sect. 4.3.1).

4.2.3 Inner Perivitelline Layer Molecules Mediating Interaction with Spermatozoa

The IPVL mediates the initial interaction between the ovum and spermatozoa and must therefore contain the crucial factors controlling this interaction.

However, specific avian IPVL sperm-binding factors have yet to be identified. Pretreatment of chicken spermatozoa with acid-solubilized IPVL reduces the sperm-mediated hydrolysis of the IPVL in vitro (Howarth 1990) and if the carbohydrate moiety is removed from the solubilized IPVL by trifluoromethanesulphanic acid treatment, then the deglycosylated proteins are unable to prevent sperm:egg interaction (Howarth 1992). These results have been proposed to demonstrate that IPVL glycopeptides play an essential role in mediating sperm:IPVL interaction. However, the assumption that preincubation with IPVL extracts will prevent sperm interaction with fresh IPVL by competitively blocking IPVL-binding sites on the sperm surface is not necessarily valid. Preincubation could induce an AR which would both render the sperm unable to bind to fresh IPVL in the normal way and severely limit the longevity of the spermatozoa (Robertson 1999).

A more direct approach, employing glycosidase treatment of intact IPVL, demonstrated that N-linked but not O-linked glycans were important for initiating sperm hydrolysis of the IPVL. Additionally, preincubation of IPVL with wheatgerm agglutinin (but not concanavalin A) inhibited any sperm activity towards the IPVL. These studies were taken to demonstrate that the IPVL sperm-binding protein contained an N-linked glycan with core N-acetylglucosamine moieties (Robertson et al. 1997b, 1998).

Sperm binding is, however, only the initial stage of sperm:egg interaction and penetration of the IPVL also requires release of acrosomal enzymes. Initial investigations into AR-inducing factors were hampered by problems in assessing the acrosomal status of spermatozoa (see Sect. 4.2.2). Wishart and Robertson (1996) showed that fluorescein-labelled peanut agglutinin differentially bound to the acrosomal matrix of spermatozoa which had undergone the AR and provided a routine assay for identifying these spermatozoa. Exposure to the IPVL, either dispersed or solubilized by protease treatment, triggered the acrosome reaction within 2.5 min at 40 °C in a solution containing buffered NaCl and 5 mM calcium (Robertson 1999). Furthermore, incubation with IPVL N-linked oligosaccharides alone induced the AR (Robertson et al. 1998). In this respect, the avian sperm:IPVL system is significantly different from that of mammals: in the mouse, carbohydrates are implicated in sperm binding to zonae and induction of the AR, but carbohydrates isolated from the zona are alone incapable of inducing the AR (Wassarman 1999). If N-linked oligosaccharides are indeed responsible for induction of the avian AR, the question remains to which IPVL polypeptide are the oligosaccharides attached? A prime candidate is IPVL1 since it has extensive homology in the central region with other sperm-activating egg envelope glycoproteins of the ZPC family which contain a single, conserved, N-glycosylation site (Waclawek et al. 1998; Takeuchi et al. 1999).

4.2.4
Preferential Attraction of Spermatozoa to the Germinal Disc

When freshly ovulated ova were incubated with spermatozoa in vitro, the finding that the ovum ruptured at the animal pole was proposed to be evidence of a preferential hydrolysis of the IPVL by spermatozoa at this locus (Howarth and Digby 1973). However, other evidence suggested the possibility that the IPVL over the GD at the animal pole might be more susceptible to acrosin activity, rather than attracting more spermatozoa (Ho and Meizel 1975). A later electron microscopic study of the IPVL from an ovum incubated with spermatozoa in vitro reported, albeit qualitatively, that in fact more spermatozoa were hydrolysing the IPVL overlying the animal pole (Bakst and Howarth 1977b). Quantitatively, the points of hydrolysis in the IPVL over the GD at the animal pole following fertilization in vivo were shown to be around 14–20 times more frequent than in the IPVL from other regions of the chicken ovum (Bramwell and Howarth 1992c; Wishart 1997). A similar pattern was found in turkey (Wishart 1997) and, to a lesser extent, in quail (Birkhead and Fletcher 1994a) and several other species (Birkhead et al. 1994).

The reason for the preferential hydrolysis of the IPVL at the animal pole was considered to be chemotaxis (Howarth and Digby 1973) or a preponderance of IPVL sperm 'receptors' (Bramwell and Howarth 1992b). The latter seems unlikely since subsequent work showed that the density of IPVL holes produced in IPVL isolated from the animal pole of an ovum and incubated with spermatozoa in vitro was the same as for IPVL from other regions of the egg (Steele et al. 1994). Although spermatozoa can attack and hydrolyse the IPVL from both sides, the IPVL holes are three times more frequent when the outer surface is presented to the spermatozoa (Robertson 1999). Thus, the disparity in the density of holes formed by spermatozoa in the IPVL in situ and in isolation appears to be due to the fact that sperm:IPVL interaction away from the germinal disc is inhibited in intact ova compared to isolated IPVL, rather than stimulated at the animal pole of the intact ovum (Wishart and Robertson 1996). Although the IPVL from over the GD appears thinner and has smaller and more numerous fibres than that from other regions during its formation (Perry et al. 1978), other electron microscopic studies could not detect any difference between the two regions (Jensen 1969). Furthermore, there do not appear to be any qualitative or apparent quantitative differences in the peptides of the IPVL from either region (Steele et al. 1994).

Thus, some factors associated with the underlying ovum influence sperm:IPVL interaction. At the periblastic region of the germinal disc, lacunae or rounded concavities of 60–100 μm diameter and 20 μm depth have been identified (Bakst 1978), which look like they might contain a fluid. This region specifically underlies the IPVL where most holes are found and the idea that potential activators of the sperm AR might be available only at this location was considered by Bakst and Howarth (1977b). However, further work has shown that only the IPVL proteins and 5 mmol/l calcium are required to induce the chicken sperm AR (Robertson et al. 1997a). Furthermore, sufficient calcium is found in the fluid proposed to be that surrounding the ovum at fertilization (Ashizawa and Wishart 1992), so it seems unnecessary to look for localized soluble sperm-activating factors. Away from the germinal disc, the plasma membrane of the ovum is discontinuous and yolk material comes into contact with, and even penetrates, the IPVL (Bakst and Howarth 1977a). Studies in vitro have shown that removal of the IPVL from the ovum and washing it free of yolk releases the inhibition of sperm hydrolysis of IPVL-holes, although the mechanism and whether the yolk inhibits sperm binding, the AR or the hydrolytic activity of acrosin remain to be ascertained (Robertson 1999).

Since the female genetic material is found within the GD, and since spermatozoa attempting to penetrate the IPVL at other locations will be unlikely to have a chance of fertilizing the egg, the reason why spermatozoa should form IPVL holes away from the GD is unclear. It is clearly a waste of the sperm resource: although non-GD-IPVL holes are less dense, their number will exceed those at the 5-mm^2 area of the germinal disc by a factor of at least 100. One idea was that these non-GD-IPVL holes are formed after the time of fertilization, when spermatozoa trapped in the OPVL close to the IPVL disintegrate and release their acrosomal contents to hydrolyse the IPVL (Birkhead et al. 1994). Evidence for this is that in laid eggs there are more (OPVL) spermatozoa associated with these non-GD-IPVL holes (Steele et al. 1994; Wishart 1997). However, non-GD-IPVL holes are produced at a similar density to in vivo fertilized eggs, during 5-min incubation of intact ova with spermatozoa in vitro, so these may indeed be formed at the time of fertilization (Wishart and Robertson 1996; Robertson 1999).

4.2.5 The Outer Perivitelline Layer and Cessation of Sperm: Inner Perivitelline Layer Interaction

There does not appear to be a mechanism equivalent to the mammalian cortical reaction for arresting sperm hydrolysis of the IPVL. Instead, this function is not controlled by the ovum, but by the secretions of the oviduct, in the form of the OPVL. The OPVL is the first secretion of the oviduct involved in egg formation and is deposited around the IPVL (Fig. 2). It consists of several sublayers, each composed of proteinaceous fibrils, and varies in thickness from 3.0 to 8.5 μm (Bellairs et al. 1963; Jensen 1969).

The proteins of the outer layer have been identified as lysozyme (60% of dry weight), an insoluble ovomucin complex (Back et al. 1982) and two outer membrane-specific basic proteins – the 17.5-kDa VMO1 (Back et al. 1982) and the 6-kDa VMOII (Kido et al. 1992). Washing the whole perivitelline layer in high concentrations of salt removed all the OPVL components with the exception of ovomucin and the layer appeared capable of reforming when the salt was removed by dialysis (Back et al. 1982).

The deposition of the OPVL around the IPVL occurs in the distal infundibulum and the proximal magnum at around 30 min after ovulation (Bain and Hall 1969). This layer prevents further penetration of the IPVL by spermatozoa in vitro (Howarth and Digby 1973) and appears to have the same function in vivo (Bakst and Howarth 1977a). The spermatozoa which are found trapped in the fibres of the OPVL (Bobr et al. 1964) do not appear to have acrosome-reacted (Bakst and Howarth 1977a) and are therefore distinct from those spermatozoa that penetrate the IPVL. In laid chicken eggs there are around ten times more spermatozoa trapped in the OPVL than there are holes in an equivalent area of the IPVL (Wishart 1997), although in quail and a range of wild avian species their numbers appeared to be similar (Birkhead and Fletcher 1994a; Birkhead et al. 1994). Whilst this may reflect a greater efficiency in sperm:IPVL interaction in these other species, it should also be noted that OPVL-sperm can degenerate and become unrecognizable following egg storage, as has been demonstrated in turkeys (Wishart 1997). In chicken and turkey eggs, the OPVL-spermatozoa appear to be randomly distributed throughout the OPVL, including that from over the germinal disc (Wishart 1997), although in 27 species of wild birds the density of OPVL-sperm was, on average, 2.5 times greater over the germinal disc (Birkhead et al. 1994). A greater preponderance of OPVL-sperm over the germinal disc might reflect a greater density of spermatozoa around the IPVL in this region when the OPVL fibres are deposited, whilst an even distribution suggests that spermatozoa are randomly distributed around the ovum at the time of OPVL deposition. There may be species differences in these distributions.

Clearly spermatozoa do not hydrolyse the OPVL, and the OPVL presents a mechanical barrier preventing further access to the IPVL and also contains the trypsin inhibitors, ovomucin and ovoinhibitor (Howarth 1984). Preparations of OPVL can inhibit the hydrolysis of IPVL by spermatozoa in vitro and also inhibit the IPVL-induced AR (Wishart and Fairweather 1999).

Thus, the OPVL prevents spermatozoa reaching the IPVL, inhibits the sperm AR and also inhibits acrosin activity. The area of the IPVL holes produced by spermatozoa in vitro increases logarithmically (Wishart and Fairweather 1999) until the holes merge and the IPVL disintegrates (Robertson et al. 1997a) and, in the intact ovum, the result of this disintegration is extrusion of the yolk at the germinal disc (Howarth and Digby 1973). In vivo, the OPVL supports the integrity of the IPVL by limiting the size of the IPVL holes and also adds strength to the yolk investments.

4.3 Sperm : Ovum Interaction and Syngamy

4.3.1 Sperm Engulfment by Oolemma

Spermatozoa which have penetrated the IPVL at the germinal disc enter the perivitelline space, from which they can come into contact with the oolemma (Okamura and Nishiyama 1978a). Two possible mechanisms by which sperm interact and penetrate the oolemma have been proposed, based on the status of the AR in spermatozoa found in the perivitelline space (Okamura and Nishiyama 1978b). In the first, with the spermatozoa approaching perpendicular to the surface of the ovum, the inner acrosomal membrane overlying the perforatorium of acrosome-less spermatozoa makes contact with the oolemma. With pressure from forward motion of the spermatozoa, the oolemma fuses with the inner acrosomal membrane. The second mechanism, which involves engulfment by phagocytosis of spermatozoa, lying parallel to the surface of the ovum, has been proposed to be abnormal. This is mainly because it was assumed that such spermatozoa, which were observed to have intact plasma membranes and an acrosome in situ (Okamura and Nishiyama 1978b), were not likely to have undergone the AR and were therefore either an artefact of the in vitro system used to study the process (Okamura and Nishiyama 1978b) or of the excessive numbers of infundibular spermatozoa following intramagnal inseminations (Koyanagi and Nishiyama 1980). However, the scanning electron microscopic evidence of Howarth (1984) shows, very convincingly, the microvilli of the oolemma wrapping around the postacrosomal region of a spermatozoon in a similar way to that of mammalian sperm entry into the ovum. If it is accepted that avian spermatozoa can acrosome react sufficiently to penetrate the IPVL with their acrosome in situ and plasma membrane over the nucleus intact (see Sect. 4.2.2.), then the phagocytosis of spermatozoa by the oolemma may be the normal mechanism of sperm entry into the ovum. Up to 60 spermatozoa can enter the germinal disc (see Sect. 4.3.2) and no mechanisms limiting polyspermy at the plasma membrane level have been identified.

4.3.2
Pronuclear Formation and Syngamy

During 0.5 to 1.0 h after ovulation, spermatozoa penetrate and move deeper into the ooplasm, as the rod-shaped nucleus transforms through elipsoidal and pear-shapes to become spherical as the chromatin decondenses (Fofanova 1965; Perry 1987). The perforatorium swells, fragments and becomes separated from the nucleus. Vesicles then appear around the nucleus and fuse to form the nuclear envelope of the male pronucleus. At the same time as nuclear decondensation, the mitochondria swell, become detached from the midpiece and disintegrate, while the axoneme also detaches from the nucleus, but appears to remain relatively intact (Okamura and Nishiyama 1978b). During most of this period, the female pronucleus, at the centre of the germinal disc, remains at the metaphase stage of the second meiotic division and does not complete meiosis and extrude the second polar body until sperm transformation has reached an advanced stage, at 2 to 3 h after ovulation (Harper 1904; Perry 1987).

The maximum numbers of spermatozoa found within the germinal disc have been reported as 24 (Patterson 1910), 58 (Fofanova 1965) and 62 (Nakanishi et al. 1990) for the domestic fowl, and 25 for the pigeon (Harper 1904). It is technically difficult to identify all of the spermatozoa and so these and other reports may underestimate the total (Perry 1987). Greater numbers of IPVL holes have been found over the germinal disc (Wishart 1997), but we still do not know if each of these represents a spermatozoon which will penetrate the ovum: it is possible that not all spermatozoa that hydrolyse the IPVL will penetrate it and then the ovum. The IPVL holes over the germinal disc are (Elliott 1997), and the spermatozoa within the germinal disc around 1 h after ovulation appear to be (Fofanova 1965; Perry 1987; Waddington et al. 1998), randomly distributed throughout the area of the germinal disc as viewed from above.

At around 3 h after ovulation, two pronuclei, presumably the male and female pronuclei, can be observed paired in the centre of the germinal disc, whilst other sperm pronuclei seem to have moved towards the periphery of the germinal disc (Harper 1904; Fofanova 1965; Perry 1987; Waddington et al. 1998). The mechanism by which movement of the male pronuclei occurs has received little attention (Harper 1904) and remains obscure. By 4 h after ovulation, the zygote nucleus has formed and is in mitosis. The supernumerary pronuclei can also undergo at least one or two rounds of mitosis at the peripheral regions of the germinal disc (Harper 1904; Fofanova 1965).

4.3.3
Supernumerary Spermatozoa and Embryonic Development

The fate of the supernumerary spermatozoa is that they will fragment and disintegrate (Fofanova 1965), which would suggest that they do not have a function related to embryonic development. Indeed, it has been noted that embryonic development proceeds normally to oviposition in eggs that contain only one IPVL hole in the germinal disc region (Wishart 1997) – which presumable represents a single penetrating spermatozoon. The finding that in chickens and turkeys the probability of fertilization is 100% when 6 or more spermatozoa penetrate and about 50:50 when 3 spermatozoa penetrate the IPVL over the germinal disc (Bramwell et al. 1995; Wishart 1997) is consistent with a stochastic model in which the purpose of polyspermy is simply to increase the chance that a male pronucleus will migrate to the centre of the germinal disc and fertilize. However, it has been hypothesized that the supernumerary spermatozoa may contribute to normal or abnormal development of the embryo (Fofanova 1965). An increased frequency of early embryonic death has been associated with intramagnal inseminations and the resultant increase in the numbers of spermatozoa that interact with the ovum (Van Krey et al. 1966). Conversely, the increase in early embryonic death that occurs in eggs laid by hens towards the end of their fertile period (Lodge et al. 1971) is more likely to be related to low numbers of spermatozoa. Whilst excessive numbers of supernumerary spermatozoa might 'interfere' with normal embryonic development, low numbers of penetrating spermatozoa might result in some sort of egg 'activation' without embryonic development or it is possible that in the absence of fertilization, the subsequent divisions of penetrating spermatozoa may be less restrained. In this respect, most early abnormal 'embryonic' development was shown to be haploid and thought to be derived from the supernumerary spermatozoa (Fechheimer 1981).

5 Quantitative Aspects of Sperm: Egg Interaction, Fertility and the Fertile Period

The laid egg contains evidence of the numbers of spermatozoa which interact with the ovum in the infundibulum and, as such, is a useful 'reporter' of the numbers of oviducal spermatozoa. This information has enabled our understanding of how the fertile period is regulated by numbers of oviducal spermatozoa.

Sperm interaction with the egg may be quantified as the numbers of points of hydrolysis made by spermatozoa in the IPVL at the time of fertilization (Bramwell et al. 1995) and by the non-penetrating spermatozoa which become trapped in the forming OPVL (Wishart 1987). Both these parameters have been shown to correlate with the numbers of spermatozoa in the UVJ-SSTs in chickens (Brillard 1993; McDaniel et al. 1997) and turkey (Brillard and Bakst 1990). In subsequent eggs laid by the same female, these parameters have a species-specific rate of decline, which therefore represents the rate of loss of spermatozoa from the SSTs. From sperm:IPVL data, the daily rate of decline has been shown to be, for example, around 11% in turkeys, 30% in chickens (Wishart and Staines 1999); 40% in quail (Wishart 1988; Birkhead and Fletcher 1994b) and 40% in zebra finch (Birkhead et al. 1993). Clearly the length of the fertile period will depend on the numbers of spermatozoa stored in the SSTs as well as their rate of loss – and also the numbers of spermatozoa required to interact with the egg to ensure fertilization. In chickens, the rate of sperm loss from the SSTs (as reflected by sperm-in-eggs) is relatively constant, so that the fertile period is determined by the numbers of spermatozoa that enter the storage tubules (Wishart 1987). Among species, differences in body and ovum size and the numbers of SSTs must also be taken into account. The average body mass of a species was correlated with the size of the ovum, the number of spermatozoa associated with eggs (Birkhead et al. 1994) and, to some extent, the number of sperm storage tubules (Birkhead and Møller 1992b). Lastly, in chickens, quail and turkeys, the relationship between the density of spermatozoa interacting with the IPVL and the probability of fertilization is similar (Wishart 1988, 1997). Thus, the main determinant of the fertile period in any avian species is most likely to be the rate of egress of spermatozoa from the SSTs, as reflected in the decline in numbers of spermatozoa associated with sequentially laid eggs.

Acknowledgements. We are very grateful to Drs Murray Bakst, Laura Robertson and the Editors for their helpful comments on the manuscript.

References

Adham IM, Mayernia K, Engel W (1997) Spermatozoa lacking acrosin protein show delayed fertilization. Mol Reprod Dev 46:370–376

Asa CS, Phillips DM (1987) Ultrastructure of avian spermatozoa: a short review. In: Mohri H (ed) New horizons in sperm cell research. Japan Scientific Society Press, Tokyo/Gordon and Breach Scientific Publishers, New York, pp 365–373

Ashizawa K, Sano R (1990) Effects of temperature on the immobilisation and the initiation of motility of spermatozoa in the male reproductive tract of the domestic fowl. Comp Biochem Physiol 96A:297–301

Ashizawa K, Wishart GJ (1992) Factors from fluid of the ovarian pocket that stimulate sperm motility in domestic hens. J Reprod Fertil 95:855–860

Ashizawa K, Wishart GJ, Nakao H, Okino Y, Tsuzuki Y (1994a) Inhibition of temperature-dependent immobilisation of fowl spermatozoa by an increased intracellular pH. J Reprod Fertil 101:593–598

Ashizawa K, Wishart GJ, Tsuzuki Y (1994b) Sperm motility in the female reproductive tract. In: Proc 9th Eur Poult Congr, vol II, Glasgow, pp 173–176

Ashizawa K, Hashimoto K, Tsuzuki Y (1996) Regulation of fowl sperm flagellar motility by protein phosphatase type 1 and its relationship with dephosphorylation of axonemal and/or accessory cytoskeletal proteins. Biochem Biophys Res Commun 235:108–112

Baba T, Azuma S, Kashiwabara S, Toyoda Y (1994) Sperm from mice carrying a targeted mutation of the acrosin gene can penetrate the oocyte zona pellucida and effect fertilisation. J Biol Chem 269:31845–31849

Back JF, Bain JM, Vadehra DV, Burley RW (1982) Proteins of the outer layer of the vitelline membrane of hen's eggs. Biochim Biophys Acta 705:12–19

Bain JM, Hall JM (1969) Observations on the development and structure of the vitelline membrane of the hen's egg: an electron microscope study. Aust J Biol Sci 22:653–665

Bakst MR (1978) Scanning electron microscopy of the vitelline membrane of the hen ovum. J Reprod Fertil 52:361–364

Bakst MR, Cecil HC (1981) Changes in the characteristics of turkey ejaculated semen and ductus deferens semen with repeated ejaculations. Reprod Nutr Dev 21:1095–1103

Bakst MR, Howarth B Jr (1975) The head, neck and midpiece of cock spermatozoa examined with the transmission electron microscope. Biol Reprod 12:632–640

Bakst MR, Howarth B Jr (1977a) The fine structure of the hen's ovum at ovulation. Biol Reprod 17:361–369

Bakst MR, Howarth B Jr (1977b) Hydrolysis of the hen's perivitelline layer by cock sperm. Biol Reprod 17:370–379

Bakst MR, Brillard J-P, Wishart GJ (1994) Sperm selection, transport and storage in the avian oviduct. Poult Sci Rev 5:143–177

Bedford JM (1979) Evolution of the sperm maturation and sperm storage functions of the epididymis. In: Fawcett DW, Bedford JM (eds) The spermatozoon: maturation, motility, surface properties and comparative aspects. Urban and Schwarzenberg, New York, pp 7–21

Bedford JM (1998) Mammalian fertilization misread? Sperm penetration of the eutherian zona pellucida is unlikely to be a lytic event. Biol Reprod 59:1275–1287

Bellairs R, Harkness M, Harkness RD (1963) The vitelline membrane of the hen's egg: a chemical and electron microscope study. J Ultrastruct Res 8:339–359

Birkhead TR (1992) Sperm storage and the fertile period in the Bengalese finch. Auk 109:620–625

Birkhead TR, Fletcher F (1994a) Numbers of spermatozoa attached to and penetrating perivitelline layers of Japanese quail eggs. Auk 136:101–104

Birkhead TR, Fletcher F (1994b) Storage and release of sperm from the sperm storage tubules in Japanese quail *Coturnix japonica.* Ibis 136:101–105
Birkhead TR, Møller AP (1992a) Sperm competition in birds, evolutionary causes and consequences. Academic Press, London
Birkhead,TR, Møller AP (1992b) Numbers and size of sperm storage tubules and the duration of sperm storage in birds. Biol J Linn Soc 45:363–372
Birkhead TR, Pellatt EJ, Fletcher F (1993) Selection and utilization of spermatozoa in the reproductive tract of the female zebra finch *Taeniopygia guttata.* J Reprod Fertil 99:593–600
Birkhead TR, Sheldon BC, Fletcher F (1994) A comparative study of sperm:egg interaction in birds. J Reprod Fertil 101:353–361
Bobr WL, Ogasawara FX, Lorenz FW (1964) Distribution of spermatozoa in the oviduct and fertility in domestic birds. II. Transport of spermatozoa in the fowl oviduct. J Reprod Fertil 8:49–58
Bramwell RK, Howarth B (1992a) Cross reactivity of sperm-binding protein from chicken, turkey and duck oocytes. Poult Sci 7:1927–1932
Bramwell RK, Howarth B (1992b) Preferential attachment of cock spermatozoa to the perivitelline layer directly over the germinal disc of the hen's ovum. Biol Reprod 47:1113–1117
Bramwell RK, Howarth B (1992c) Quantitative determination of spermatozoa penetration of the perivitelline layer overlying the germinal disc. Poult Sci 71:140
Bramwell RK, Marks HL, Howarth B (1995) Quantitative determination of spermatozoa penetration of the perivitelline layer of the hen's ovum as assessed on oviposited eggs. Poult Sci 74:1875–1883
Brillard J-P (1993) Sperm storage and transport following natural mating and artificial insemination. Poult Sci 77:923–928
Brillard J-P, Bakst MR (1990) Quantification of spermatozoa in the sperm-storage tubules of turkey hens and its relation to sperm numbers in the perivitelline layer of eggs. Biol Reprod 43:271–275
Briskie JV, Montgomery R (1993) Patterns of sperm storage in relation to sperm competition in passerine birds. Condor 95:424–454
Brown CR, Hartree EF (1976) Comparision of neutral proteinase activities in cock and ram spermatozoa and observations on a proacrosin in cock spermatozoa. J Reprod Fertil 46:155–164
Elliott M (1997) An investigation into the spatial density and distribution of hydrolysed holes in chicken eggs. BSc Honours Mathematical Sciences Thesis, University of Abertay Dundee, Dundee
Esponda P, Bedford JM (1985) Surface of the rooster spermatozoon changes in passing through the Wolffian duct. J Exp Zool 234:441–449
Etches RJ (1996) Reproduction in poultry. CAB International, Oxon
Fechheimer NS (1981) Origins of heteroploidy in chicken embryos. Poult Sci 60:1365–1371
Fofanova KV (1965) Morphological data on polyspermy in chickens. Fed Proc Trans (Suppl) 24:T239–T247
Froman DP (1990) Chicken acrosin: extraction and purification. Poult Sci 69:812–817
Froman DP (1995) Biology of semen production and ejaculation. In: Bakst MR, Wishart GJ (eds) Proc 1st Int Symp AI Poult, Poult Sci Assoc, Savoy, Illinois, pp 21–38
Fujihara N, Tanaka K, Nishiyama H (1973) In vitro fertilization of fowl ova. Jpn J Zootech Sci 11:564–568
Fujii S (1976) Scanning electron microscopical observation on the penetration mechanism of fowl spermatozoa into the ovum in the process of fertilisation. J Fac Fish Anim Husb Hiroshima Univ 15:85–92
Gee GF (1995) Artificial insemination and cryopreservation of semen from nondomestic birds. In: Bakst MR, Wishart GJ (eds) Proc 1st Int Symp AI Poult, Poult Sci Assoc, Savoy, Illinois, pp 262–279
Gilbert AB (1979) Female genital organs. In: King AS, McLelland J (eds) Form and function in birds, vol 1. Academic Press, New York, pp 237–360

Ginsburg M (1997) Primordial germ cell development in avians. Poult Sci 76:91–91

Ginsburg M, Eyal Giladi H (1986) Temporal and spatial aspects of the gradual migration of primordial germ cells from the epiblast into the germinal crescent in the avian embryo. J Embryol Exp Morphol 95:53–71

Gray AP (1958) Bird hybrids: a check list with bibliography. Tech Bull 13. Commonwealth Agricultural Bureau, Buckinghamshire

Han JY, Seo DS, Shoffner RN (1994) Primordial germ cells in Aves. Asian Aust J Anim Sci 7:459–466

Harper EH (1904) The fertilization and early development of the pigeon's egg. Am J Anat 3:349–386

Hartley PS (1999) The cryopreservation of semen from non-domestic avian species. MPhil Thesis, University of Abertay Dundee, Dundee

Ho J, Meizel S (1970) Electrophoretic detection of multiple forms of trypsin-like activity in spermatozoa of the domestic fowl. J Reprod Fertil 23:177–179

Ho J, Meizel S (1975) Multiple molecular forms of avian acrosin: differences in their kinetic properties. FEBS Lett 56:115–119

Ho J, Meizel S (1976a) Biochemical characterisation of avian spermatozoan acrosin and comparison of its properties to those of bovine trypsin and mammalian acrosin. Comp Biochem Physiol 54B:213–218

Ho J, Meizel S (1976b) Hydrolysis of the hen egg vitelline membrane by cock sperm acrosin and other enzymes. J Exp Zool 194:429–438

Holm L (1999) A comparative study of avian oviducal sperm storage with special reference to factors which regulate sperm motility. PhD Thesis, Swedish University of Agricultural Sciences, Uppsala

Holm L, Wishart GJ (1998) The effect of pH on the motility of spermatozoa from chicken, turkey and quail. Anim Reprod Sci 54:45–54

Howarth B Jr (1971a) An examination for sperm capacitation in the fowl. Biol Reprod 3:338–341

Howarth B Jr (1971b) Transport of spermatozoa in the reproductive tract of the turkey hen. Poult Sci 50:84–89

Howarth B Jr (1983) Fertilizing ability of cock spermatozoa from the testis, epididymis and vas deferens following intramagnal insemination. Biol Reprod 28:586–590

Howarth B Jr (1984) Maturation of spermatozoa and mechanism of fertilisation. In: Cunningham FJ, Lake PE, Hewitt D (eds) Reproductive biology of poultry. Longham Group, Harlow, pp 161–174

Howarth B (1990) Research note: sperm–egg interaction: perivitelline layer possesses receptor activity for spermatozoa. Poult Sci 69:1012–1015

Howarth B (1992) Carbohydrate involvement in sperm:egg interaction in the chicken. J Receptor Res 12:255–265

Howarth B (1995) Physiology of reproduction: the male. In: Hunton P (ed) Poultry production. World animal science subseries C, vol 9. Elsevier, Amsterdam, pp 243–270

Howarth B Jr, Digby ST (1973) Evidence for the penetration of the vitelline membrane of the hen's ovum by a trypsin-like acrosomal enzyme. J Reprod Fertil 33:123–125

Howarth B Jr, Palmer MB (1972) An examination of the need for sperm capacitation in the turkey. J Reprod Fertil 16:445–455

Humpheys PN (1972) Brief observations on the semen and spermatozoa of certain passerine and non-passerine birds. J Reprod Fertil 29:327–336

Hunton P (1995) Poultry production. World animal science subseries C, vol 9. Elsevier, Amsterdam

Jensen C (1969) Ultrastructural changes in the avian vitelline membrane during embryonic development. J Embryol Exp Morphol 21:467–484

Kagami H, Clark ME, Verrinder Gibbins AM, Etches RJ (1995) Sexual differentiation of chimeric chickens containing ZZ and ZW cells in the germline. Mol Reprod Dev 42:379–387

Kerr R (1997) Sperm viability during sperm:egg interaction in vitro. BSc Honours Biotechnology Thesis, University of Abertay Dundee, Dundee

Kido S, Doi Y (1988) Separation and properties of the inner and outer layers of the vitelline membrane of hen's eggs. Poult Sci 67:476–486

Kido S, Janado M, Nunora H (1975) Macromolecular components of the vitelline membrane of hen's egg I. Membrane structure and deterioration with age. J Biochem (Tokyo) 78:261–268

Kido S, Janado M, Nunora H (1976) Macromolecular components of the vitelline membrane of hen's egg II. Physiochemical properties of glycoprotein 1. J Biochem (Tokyo) 79:1351–1356

Kido S, Janado M, Nunora H (1977) Macromolecular components of the vitelline membrane of hen's egg III. Physiochemical properties of glycoprotein 2. J Biochem (Tokyo) 81:1543–1548

Kido S, Morimoto A, Kim F, Doi Y (1992) Isolation of a novel protein from the outer layer of the vitelline membrane. J Biochem (Tokyo) 286:17–22

King AS (1981) Phallus. In: King AS, McLelland J (eds) Form and function in birds, vol 2. Academic Press, London, pp 107–147

King AS, McLelland J (1979) Form and function in birds, vol 1. Academic Press, New York

King AS, McLelland J (1981) Form and function in birds, vol 2. Academic Press, New York

Koyanagi F, Nishiyama H (1980) Phagocytosis of spermatozoa by the ovum of the domestic fowl, *Gallus gallus,* at the time of fertilization. Cell Tissue Res 206:55–63

Koyanagi F, Masuda S, Nishiyama H (1988) Acrosome reaction of cock spermatozoa incubated with the perivitelline layer of the hen's ovum. Poult Sci 67:1770–1774

Kuroki M, Mori M (1995) Origin of 33 kDa protein of vitelline membrane of the quail egg: immunological studies. Dev Growth Differ 37:545–550

Kuwana T (1993) Migration of avian primordial germ cells towards the gonadal anlage. Dev Growth Differ 35:237–243

Lack D (1968) Ecological adaptations for breeding in birds. Chapman and Hall, London

Lake PE (1975) Gamete production and the fertile period with particular reference to domesticated birds. Symp Zool Soc Lond 35:225–244

Lake PE (1981) Male genital organs. In: King AS, McLelland J (eds) Form and function in birds, vol 2. Academic Press, London, pp 1–61

Lake PE, Stewart JM (1978) Artificial insemination in poultry. Ministry of Agriculture, Fisheries and Food, Bull 213. HM Stationery Office, London

Lake PE, Smith W, Young D (1968) The ultrastructure of the ejaculated fowl spermatozoon. Q J Exp Physiol 53:356–366

Langford BB, Howarth B Jr (1974) A trypsin-like enzyme in acrosomal extracts of chicken, turkey and quail spermatozoa. Poult Sci 53:834–837

Lodge JR, Fechheimer NS, Jaap RG (1971) Relationship of in vivo sperm storage interval to fertility and embryonic survival in the chicken. Biol Reprod 5:156–159

McDaniel CD, Bramwell RK, Howarth BJ (1997) Development of a new fluorescence technique for quantifying the total number of spermatozoa stored in the uterovaginal junction of hens. J Reprod Fertil 109:173–179

McFarlane RW (1963) The taxonomic significance of avian sperm. Proc 13th Int Ornithol Cong, Ithaca, New York, 1962, pp 91–102

McFarlane RW (1971) The ultrastructure and phylogenetic significance of avian spermatozoa. PhD Thesis, University of Florida, Gainesville

McIndoe W, Lake PE (1974) The distribution of some hydrolytic enzymes in the semen of the domestic fowl, *Gallus domesticus.* J Reprod Fertil 40:359–365

Mori M, Masuda N (1993) Proteins of the vitelline membrane of Quail (*Coturnix coturnix japonica*) eggs. Poult Sci 72:1566–1572

Munro SS (1938) Functional changes in fowl sperm during their passage through the excurrent ducts of the male. J Exp Zool 79:71–92

Naito M, Tajima A, Tagami T, Yasuda Y, Kuwana T (1994) Preservation of chick primordial germ cells in liquid nitrogen and subsequent production of viable offspring. J Reprod Fertil 102:321–325

Nakanishi A, Utsumi K, Iritani A (1990) Early nuclear events of in vitro fertilization in the domestic fowl (*Gallus domesticus*). Mol Reprod Dev 26:217–221

Okamura F, Nishiyama H (1978a) The passage of spermatozoa through the vitelline membrane in the domestic fowl, *Gallus gallus*. Cell Tissue Res 188:497–508
Okamura F, Nishiyama H (1978b) Penetration of spermatozoa into the ovum and transformation of the sperm nucleus into the male pronucleus in the domestic fowl, *Gallus gallus*. Cell Tissue Res 190:89–98
Olsen MW (1942) Maturation, fertilization and early cleavage in the hen's egg. J Morphol 70:513–533
Olsen MW, Fraps RM (1944) Maturation, fertilization and early cleavage of the egg of the domestic turkey. J Morphol 74:297–309
Olsen MW, Nehrer BH (1948) The site of fertilization in the domestic fowl. J Exp Zool 109:355–366
Palmer MB, Howarth B Jr (1973) The requirement of a trypsin-like acrosomal enzyme for fertilization in the domestic fowl. J Reprod Fertil 35:7–12
Patterson JT (1910) Early development of the hen's egg. J Morphol 21:101–134
Perry MM (1987) Nuclear events from fertilization to the early cleavage stages in the domestic fowl (*Gallus domesticus*). J Anat 150:99–109
Perry MM, Gilbert AB, Evans AJ (1978) The structure of the germinal disc region of the hen's ovarian follicle during the rapid growth phase. J Anat 127:379–392
Petitte JN, Karagenÿ L, Ginsburg M (1997) The origin of the avian germ line and transgenesis in birds. Poult Sci 76:1084–1092
Phillips DM, Asa CS, Stover J (1987) Ultrastructure of spermatozoa of the white-naped crane. J Submicrosc Cytol 19:489–494
Richardson ME, Bodine AB, Froman DP, Thurston RJ (1988) Turkey acrosin. 1. Isolation, purification and partial characterisation. Biol Reprod 38:645–651
Robertson L (1999) Sperm:egg interaction in birds: assays and mechanisms. PhD Thesis, University of Abertay Dundee, Dundee
Robertson L, Wishart GJ (1996) Detection of the acrosome reaction of chicken spermatozoa. J Reprod Fertil Abstr Ser 17:115
Robertson L, Brown HL, Staines HJ, Wishart GJ (1997a) Characterization and application of an avian in vitro spermatozoa-egg interaction assay using the inner perivitelline layer from laid chicken eggs. J Reprod Fertil 110:205–211
Robertson L, Horrocks AJ, Wishart GJ (1997b) Perivitelline N-linked glycans initiate sperm:egg interaction in the chicken. Biochem Soc Trans 25:S655
Robertson L, Horrocks AJ, Jackson L, Olusanya O, Wishart, GJ (1998) The role of oligosaccharides in avian sperm:egg interaction. Abstr 8th Int Symp on Spermatology, P3–22, p 89
Romanoff AL (1960) The avian embryo. MacMillan, New York
Sibley CG (1994) On the phylogeny and classification of living birds. J Avian Biol 25:87–92
Stambaugh R, Brackett BG, Mastroianni L (1969) Inhibition of in vitro fertilization of rabbit ova by trypsin inhibitors. Biol Reprod 1:223–227
Steele MG, Wishart GJ (1992a) Characterisation of a mechanism impeding sperm transport through the vagina of the chicken. In: Proc 12th Int Congr Anim Reprod, The Hague, pp 1593–1595
Steele MG, Wishart GJ (1992b) Evidence for a species-specific barrier to sperm transport within the vagina of the chicken hen. Theriogenol 38:1107–1114
Steele MG, Wishart GJ (1996a) Demonstration that the removal of neuraminic acid from the surface of chicken spermatozoa impedes their transvaginal migration. Theriogenology 46:1037–1044
Steele MG, Wishart GJ (1996b) Effect of removing surface-associated protein from viable chicken spermatozoa on sperm function in vivo and in vitro. Anim Reprod Sci 45:139–147
Steele MG, Meldrum W, Brillard J-P, Wishart GJ (1994) The interaction of avian spermatozoa with the perivitelline layer in vitro and in vivo. J Reprod Fertil 101:599–603
Takeuchi Y, Nishimura K, Aoki N, Adachi T, Sato C, Kitajima K, Matsuda T (1999) A 42-kDa glycoprotein from chicken egg-envelope, an avian homolog of the ZPC family glycoproteins in mammalian zona pellucida. Eur J Biochem 260:736–742
Tauber P (1875) Om hønseæggets befrugtning I æggelederen. Naturhist Tidskr (Copenh) 10:63–106

Thurston RJ, Hess RA (1987) Ultrastructure of spermatozoa from domesticated birds: comparative study of turkey, chicken and guinea fowl. Scanning Microsc 1:1829–1838

Van Krey H, Ogasawara F, Lorenz F (1966) Distribution of spermatozoa in the oviduct and fertility in domestic birds IV. Fertility of spermatozoa from the infundibular and uterovaginal glands. J Reprod Fertil 11:257–262

Vernon G, Woolley DM (1999) Three-dimensional motion of avian spermatozoa. Cell Mot Cytoskel 42:149–161

Waclawek M, Foisner R, Nimpf J, Schneider WJ (1998) The chicken homologue of zone pellucida protein 3 is synthesised by granulosa cells. Biol Reprod 59:1230–1239

Waddington D, Gribbin C, Sterling RJ, Sang H, Perry MM (1998) Chronology of events in the first cell cycle of the polyspermic egg of the domestic fowl (*Gallus domesticus*). Int J Dev Biol 42:625–628

Wassarman PM (1999) Mammalian fertilization: molecular aspects of gamete adhesion, exocytosis, and fusion. Cell 96:175–183

Wishart GJ (1987) Regulation of the length of the fertile period in the domestic fowl by numbers of oviducal spermatozoa. J Reprod Fertil 80:493–498

Wishart GJ (1988) Numbers of oviducal spermatozoa and the length of the fertile period in different avian species. In: Proc 11th Int Congr Anim Reprod AI, vol 3, Dublin, pp 362–364

Wishart GJ (1994) Vaginal sperm transport and sperm selection. In: Proc 9th Eur Poult Conf, Glasgow, pp 161–164

Wishart GJ (1995) New approaches to evaluating male and female fertility. In: Bakst MR, Wishart GJ (eds) Proc 1st Int Symp AI Poult, Poult Sci Assoc, Savoy, Illinois, pp 207–223

Wishart GJ (1997) Quantitative aspects of sperm:egg interaction in chickens and turkeys. Anim Reprod Sci 48:81–92

Wishart GJ, Fairweather JO (1999) The chicken 'fertilization window'. Poult Avian Biol Rev 9:124

Wishart GJ, Robertson L (1996) Chicken sperm:perivitelline interaction in vitro. J Reprod Fertil (Suppl) 17:113

Wishart GJ, Staines HJ (1999) Measuring sperm:egg interaction to assess breeding efficiency in chickens and turkeys. Poult Sci 78:428–436

Wishart GJ, Wilson YI (1999) Temperature-dependent inhibition of motility in spermatozoa from different avian species. Anim Reprod Sci 57:229–235

Woolley DM (1995) The structure of the spermatozoon of the Japanese quail, *Coturnix coturnix* L., var. *japonica*. Acta Zool (Stockh) 76:45–50

Yanagimachi R (1994) Mammalian fertilization. In: Knobil E, Neil JD (eds) The physiology of reproduction. Raven Press, New York, pp 189–317

Yanagimachi R, Teichman RJ (1972) Cytochemical demonstration of acrosomal proteinase in mammalian and avian spermatozoa by a silver proteinate method. Biol Reprod 6:87–97

Zanfeld LJD, Robertson RT, Williams WL (1970) Synthetic enzyme inhibitors as antifertility agents. FEBS Lett 11:345–350

Zavaleta D, Ogasawara F (1987) A review of the mechanism of the release of spermatozoa from storage tubules in the fowl and turkey oviduct. World Poul Sci J 43:132–139

CHAPTER 6

Fertilization in Monotreme, Marsupial and Eutherian Mammals

K.E. MATE[1], M.S. HARRIS[2], AND J.C. RODGER[3]

1 Introduction

1.1 The Three Groups of Mammals

There are three quite distinct evolutionary lineages of mammals which have been separated for many millions of years by time and place. The northern-origin (European, Asian, African and North American) eutherian mammals – often called placental mammals – and the southern-origin monotremes (Australasian) and marsupials (Australasian and American).

1.2 Monotremes

There are only three living species of monotreme – the Australian platypus and echidna and the larger New Guinean echidna (echidnas, or spiny anteaters, externally resemble porcupines). Monotremes are hair-bearing animals with a true cloaca for faeces, urine and reproduction exiting via a single external opening (monotreme means one hole). Monotremes lay relatively large yolky eggs approximately 3.5–4.0 mm in diameter (Fig. 1) and cleavage is meroblastic as in birds and reptiles. However, there is a small component of uterine support of embryo development and once hatched the embryo-like young is supported by milk as in all mammals. Male monotremes have abdominal testes and relatively simple sperm more similar to those of reptiles than the other mammals. For further information on monotreme biology see Grant (1989) and Griffiths (1989).

[1] Cooperative Research Centre for Conservation and Management of Marsupials, Department of Biological Sciences, Macquarie University, Sydney 2109, Australia
[2] Manaaki Whenua Landcare Research, Lincoln 8152, New Zealand
[3] Department of Biological Sciences, University of Newcastle, Newcastle 2308, Australia

Juan J. Tarín and Antonio Cano (Eds.)
Fertilization in Protozoa and Metazoan Animals

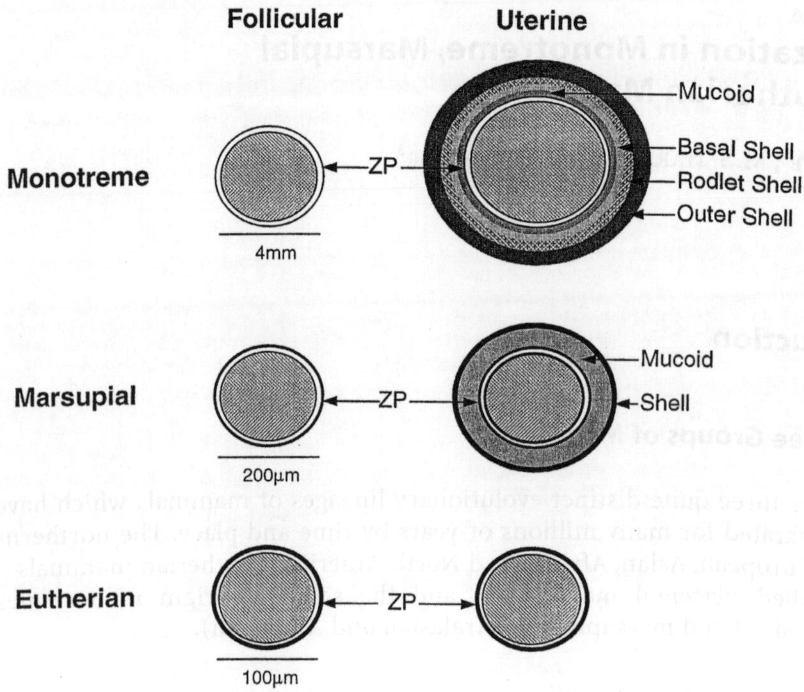

Fig. 1. Comparison of egg structure and vestments in monotreme, marsupial and eutherian mammals. *ZP* Zona pellucida

1.3 Marsupials

Marsupials are hair-bearing animals that give birth to embryo-like young which complete development supported by milk usually in a pouch or marsupium. Marsupials are placental but the period of placental support is remarkably short – only a few days towards the end of a 2- to 4-week gestation. A choriovitelline (yolk sac) placenta is the norm, although some develop a chorioallantoic placenta shortly before birth. Marsupial reproduction is otherwise similar to that of eutherian mammals in character and is regulated by the same suite of reproductive hormones. The marsupial ovary and oocyte are very similar to that of eutherians and even an expert would be hard pressed to distinguish between the two on basic morphology (Fig. 1). Marsupial sperm also share many structural features with those of eutherians although the manner of spermiogenesis means that the sperm head and acrosome are quite different to those of eutherians and are easily distinguishable. The reproductive tract of the female marsupial develops quite differently to that of eutherian mammals due to the relative positions of the embryonic ureters and Müllerian (female) and Wolffian (male) ducts (Fig. 2). In marsupials (plus

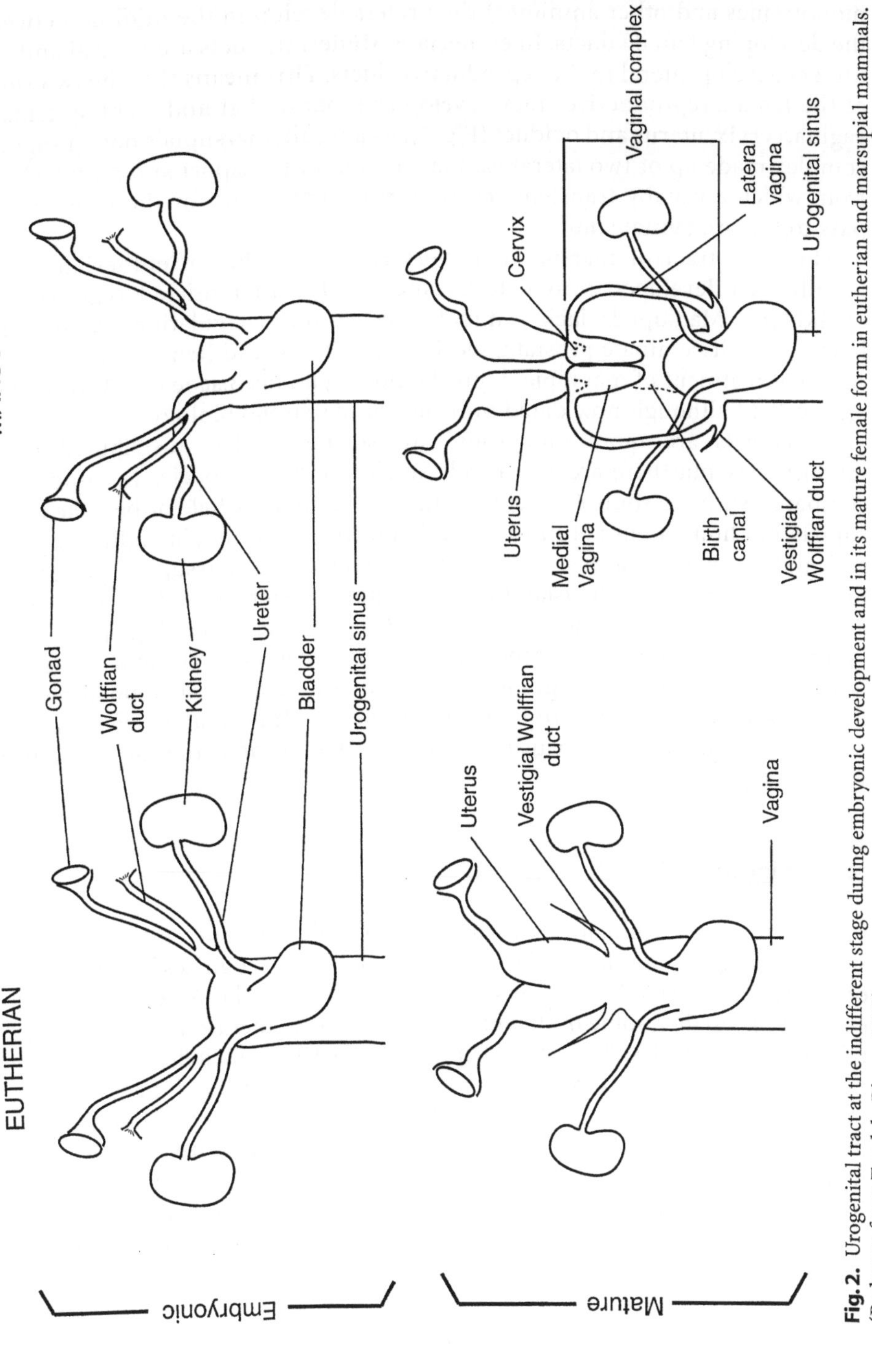

Fig. 2. Urogenital tract at the indifferent stage during embryonic development and in its mature female form in eutherian and marsupial mammals. (Redrawn from Tyndale-Biscoe 1973)

monotremes and other amniotes) the ureters develop in the midline between the developing female ducts. In eutherians Müllerian ducts are central and the ureters develop lateral to the reproductive ducts. This means that the two sides of the female reproductive tract develop as separate left and right systems – vagina, cervix, uterus and oviduct (Fig. 2). As a result, marsupials have a vaginal complex made up of two lateral vaginae for sperm transport and a central sac from which a usually transient central birth canal forms for the newborn to pass to the urogenital sinus.

The male tract of marsupials is very similar to that of eutherians with usually well-developed scrotal testis and highly differentiated regionalized epididymis. Marsupials lack seminal vesicles and coagulating glands and have only a disseminate prostate; see Tyndale-Biscoe and Renfree (1987) for a detailed treatment of marsupial reproduction; Tyndale-Biscoe (1973) provides an excellent although now dated overview of marsupial biology.

Marsupials are more numerous, diverse and widely distributed than monotremes, but there are a relatively small number of species (around 300) compared with the some 6000 eutherians. There are no whale or bat-like marsupials, but homologues of most eutherian forms are found, including aquatic and gliding specializations. Roughly three-quarters of marsupial species are found in Australia, on the island of New Guinea, eastern islands of Indonesia and West Pacific islands adjacent to New Guinea. The remaining species are found in Central and South America and one species, the Virginia opossum, is a relatively recent immigrant into North America. Feral populations of marsupials are found in Europe including the UK, Hawaii and New Zealand, where one, the common brushtail possum, has become the latter's major vertebrate pest.

1.3.1 Marsupial Diversity

Marsupials were originally thought to be equivalent to one eutherian order such as Carnivora or Rodentia. Current taxonomy recognizes seven orders of living marsupials, four in Australasia and three in the Americas (Kirsch et al. 1997). Further, some of the Australian orders are themselves diverse, containing many species. The order Diprotodontia which includes most of the more familiar Australian marsupials (kangaroos, possums and the koala) has been divided into three suborders, and these into five superfamilies containing ten families in all. The diversity of marsupials is emphasized here because there is a tendency to think that marsupials will display common biology. In this chapter we will thus be discussing and comparing not marsupial and eutherian fertilization but fertilization in diverse groups of marsupials and eutherians.

1.4 Eutherians

The eutherian mammals, which include all familiar domestic and laboratory animals and humans, are hair-bearing animals that give birth to well-developed often essentially independent young after a period of uterine development supported by a usually chorioallantoic placenta. We will not go into detail on the eutherians except to emphasize their remarkable diversity (111 families within 17 orders) and their success in essentially all habitats except the terrestrial Antarctic. Thus, as with the diverse marsupials, when considering the processes of fertilization one cannot assume that this will be the same in all species.

1.5 Mammalian Relationships and Evolution

The relationship between living and extinct mammal groups remains an area of considerable debate. A very important early and continuing influence is the Prototheria (monotreme), Metatheria (marsupial), Eutheria system of T.H. Huxley. This lead to a popular conception which saw the three groups of mammals as successive steps to 'true' therian (mammal) status and that marsupials represented a halfway position. As marsupial biology became better understood biologists working with marsupials have strongly argued that marsupials should be regarded as alternative rather than inferior therians. Currently, views are divided on whether marsupials are phylogenetically linked to monotremes (Marsupionta) or eutherians (Theria) (Janke et al. 1997; Kirsch et al. 1997; Kirsch and Mayer 1998). The renewed support for a marsupial monotreme link is not a return to stages of 'therianness' as in Huxley's system. Rather it argues that the placental support of embryo development in utero arose independently in marsupials and eutherians.

1.6 Fertilization in Mammals

Unfortunately the available information on Monotreme gametes is very limited and nothing is known about fertilization. Where some knowledge of gametes, or relevant reproductive processes, is available this will be mentioned. However, a current synthesis of fertilization in mammals must lack a significant monotreme perspective. It is encouraging that only three decades ago the situation was little better for marsupials; thus, provided appropriate access to monotreme gametes and early embryo stages can be gained, increased understanding of monotreme gametes and fertilization can be expected. Since the early 1970s there has been a rapid growth in knowledge of marsupial gametes and fertilization which would now equal, and in many cases exceed, that known about eutherians which are not laboratory or domestic animals. In this chapter, where possible, we will draw on what is known about a range of eutherian mammals not simply the best studied groups. An excellent survey of fertilization phenomena across the therian mammals is found in Dunbar and O'Rand (1991). One feature of reproduction which would appear to unite all mammals is the absolute dependence on fertilization as the start of a new individual. All mammalian oocytes require a sperm to initiate successful development. Parthenogenetic development can be initiated experimentally, or may occur naturally, but it does not result in full development and rarely proceeds beyond the blastocyst stage. Eutherian mammals, and presumably marsupials, do occasionally have a form of asexual reproduction where post-fertilization events result in the single fertilized oocyte giving rise to partially, or completely, separated individuals which are genetically identical clones. An unusual example of a mammal where post-fertilization asexual reproduction is the norm is the armadillo which routinely undergoes embryo splitting to produce a litter of up to eight identical individuals derived from a single fertilized oocyte.

2 Sperm Production

2.1 Marsupials and Eutherians

Although the processes of spermatogenesis are essentially similar in marsupials and eutherians, the spermatid nuclei flatten in different planes (Fig. 3). In eutherians nuclear flattening is lateral, parallel to the forming tail. As a result the acrosome and tail come to lie at opposite ends of the flattened head and the acrosome forms a cap over the leading edge of the head. In marsupials flattening occurs at 90° to the forming tail. The result is a sperm with a distinct dorsal ventral structure with the acrosome, which never forms a cap, restricted to the dorsal side of the head, and the tail on the ventral side attached opposite the acrosome. As discussed in more detail later, in many marsupials the sperm leaves the testis in a quite immature morphological form – the head and tail are not aligned and the acrosome is only partly formed (Fig. 3).

2.2 Monotremes

Although marsupial and eutherian spermatozoa show differences in the manner and timing of sperm formation the final product is very similar, with a compact head and acrosome, well-developed midpiece with mitochondrial spiral and a tail with a complex of accessory fibres. Indeed, it would require expertise to differentiate between marsupial and eutherian sperm at the ultrastructural level if one only examined the midpiece and tail. Monotreme sperm

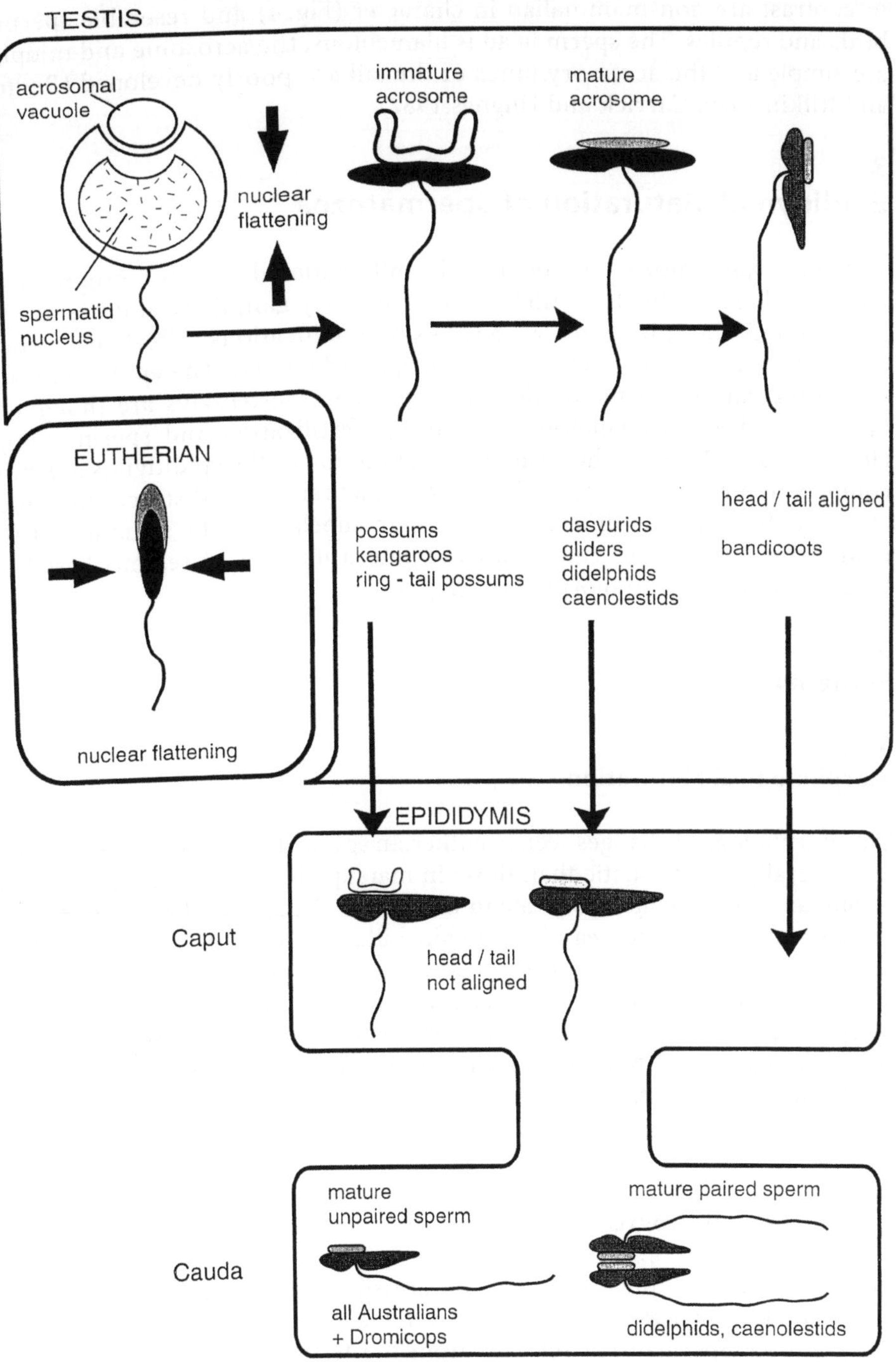

Fig. 3. Spermiogenesis in the testis and epididymal maturation in marsupials. (Redrawn from Rodger 1991)

in contrast are non-mammalian in character (Fig. 4) and resemble sperm of birds and reptiles. The sperm head is filamentous, the acrosome and midpiece are simple and the accessory fibres of the tail are poorly developed (Bedford and Rifkin 1979; Carrick and Hughes 1982).

3 Epididymal Maturation of Spermatozoa

Testicular spermatozoa, although fully differentiated, are not progressively motile and are unable to fertilize the oocyte. They gain these abilities as they pass through the epididymis via a series of modifications collectively termed epididymal maturation (reviewed in Yanagimachi 1994). The epididymis and post-testicular sperm maturation seen in higher vertebrates are thought to have evolved as an adaptation for internal fertilization and sperm storage (Jones 1998b). However, the structure and function of the epididymis are more complex in mammals compared to other vertebrates, and distinct differences also exist between the different mammalian subclasses. The structure of the mammalian epididymis can be divided into a number of regions based on function and epithelial morphology (Fig. 4).

3.1 Eutherians

3.1.1 Morphological Maturation

The morphological changes seen in eutherian sperm during epididymal transit are generally less dramatic than those in marsupials – ranging from minimal in humans and rodents, moderate in the rabbit, elephant and some primates, to substantial in the guinea pig and chinchilla (Bedford 1979). The cytoplasmic droplet, containing the residual spermatid cytoplasm, is initially located in the neck region of eutherian sperm and migrates distally to the posterior midpiece during epididymal transit. Loss of the cytoplasmic droplet accompanies ejaculation. Migration of the droplet is accompanied by midpiece modifications, whilst the sequestering of redundant enzymes in these vesicles may also protect the sperm from oxidative damage (Moore 1995).

3.1.2 Development of Progressive Motility

The development of the capacity for progressive motility is perhaps the most conspicuous change associated with epididymal maturation in all mammalian spermatozoa. Spermatozoa released from the testis and head of the mammalian epididymis are immotile or only very weakly motile. In contrast,

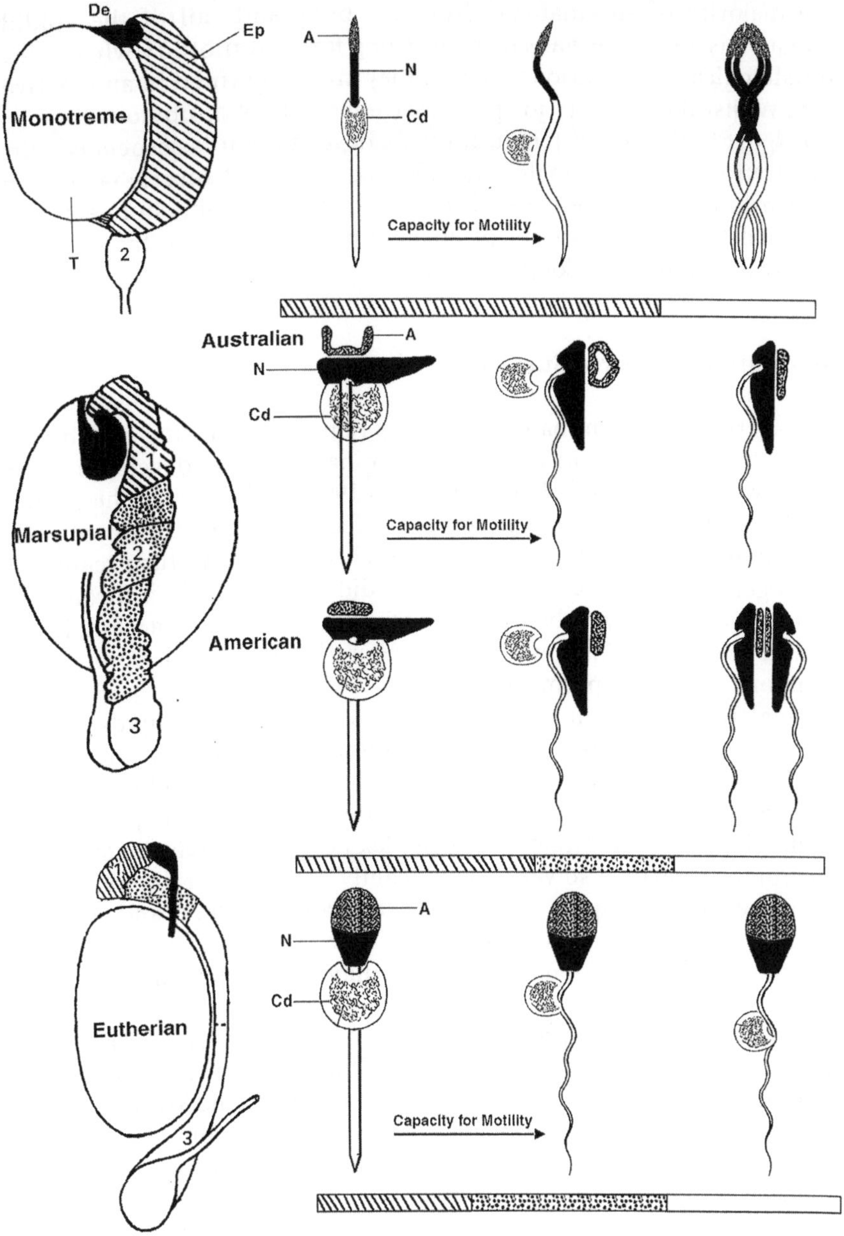

Fig. 4. Structural differentiation of efferent ducts and epididymis of monotreme, marsupial and eutherian mammals, showing relationship between histologically distinct zones and sperm maturation. The efferent duct is *black*, initial segment is *striped*, distal initial segment is *speckled* and terminal segment is *white*. *A* Acrosome ; *N* nucleus; *Cd* cytoplasmic droplet; *T* testis; *De* efferent duct; *Ep* epididymis. (Redrawn from Jones 1998b and Rodger 1991)

the majority of spermatozoa from the body and tail of the epididymis in marsupials and eutherians display strong forward motility when released into physiological salt solutions, though they are maintained in an inactive state *in situ*. Whilst the capacity for progressive motility is of obvious importance for transport to the site of fertilization, its development is associated with the maturation of the plasmalemma. The acquisition of several factors, including forward motility, protein, carnitine and glycerophosphocholine, from the epididymal fluid to the sperm plasmalemma may mediate the development of progressive motility (Cooper 1992).

3.1.3
Maturation of Sperm Surface

The sperm plasmalemma is extensively remodelled during epididymal maturation by a variety of mechanisms, culminating in the development of sperm fertility (Kirchhoff and Hale 1996; Jones 1998a). Changes in the configuration of the eutherian sperm surface are effected by the continually changing epididymal milieu of the luminal fluid. This milieu is in turn regulated by the androgen-dependent activity of the epididymal epithelium through selective and/or non-selective absorption, active secretion and accumulation from and in the blood. This results in the alteration of both the protein and lipid components of the sperm surface.

Selective changes in the amount and composition of lipids in the sperm membrane during epididymal maturation are thought to affect membrane fluidity and thus allow the reorganization of some membrane components (reviewed in Jones 1998a). There are differences between species, and much conflicting data. The situation is complex because the relative fluidity of each membrane domain also varies significantly as a result of differential densities of some phospholipids within the domains (Parkes and Hammerstedt 1985). The establishment of such lipid domains may precede, and perhaps direct, the reorganization of integral membrane components during epididymal transit.

Remodelling of the protein components of the sperm surface appears to be mediated by the movement and modification of existing proteins, the addition of proteins from the luminal secretions, or the loss of proteins from the spermatozoon. Approximately 50% of the glycoproteins added to the sperm surface are actively synthesized and secreted by the epithelium in different regions of the epididymis. Roles for these 'epididymal maturation antigens' have been proposed in the development of sperm fertility and the fertilization process itself (Martinez et al. 1995), acrosome stabilization, decapacitation and as motility regulators (Jones 1989; Cooper 1992). Some of these maturation antigens are only weakly associated with the sperm surface via ionic bonds. However, others are bound strongly to the membrane by a number of mechanisms (Jones 1998a) including direct covalent binding by enzymes such as transferases, and direct membrane interchange mediated by glycosylphosphatidylinositol lipid anchors (Kirchhoff and Hale 1996; Moore 1996).

Both pre-existing and epididymal maturation proteins on the sperm surface are also modified by post-translational processing and/or a move to different surface domains during epididymal maturation. Post-translational processing of sperm proteins is known to include glycosylation, methylation, disulphide bonding, transglutaminase-mediated cross-linking and phosphorylation (Jones 1998a). A number of proteins involved in fertilization such as fertilin and PH-20 undergo site-specific endoproteolytic cleavage during epididymal maturation. Such processing of these proteins may be a prerequisite for their movement from the plasmalemma from the whole sperm head to the post-acrosomal domain (Phelps et al. 1990). However, epididymal remodelling is not restricted to the sperm surface. Growing evidence suggests that significant processing of intracellular acrosomal and flagellar antigens also occurs (Phillips et al. 1991). For example, stabilization of a number of sperm tail and head structures, including the nucleus, during epididymal transit by increased disulphide bonding is likely to lead to more rigid tail movement for motility and stabilization of the head structures for zona penetration (Bedford 1996).

Together these epididymal modifications to the form and location of sperm components produce mature spermatozoa that have the ability to survive prolonged storage in the epididymis and are capable of both progressive motility and undergoing capacitation prior to fertilization. In most mammals this period of storage lasts days to a few weeks, but in certain bats fertile sperm may be stored in the tail of the epididymis for many months (Krutzsch and Crichton 1991).

3.2 Marsupials

3.2.1 Morphological Changes

Whilst eutherian spermatozoa undergo predominantly functional modifications during epididymal transit, marsupial spermatozoa also undergo significant morphological alterations (Figs. 3 and 4). By far the most extensive post-testicular reorganization of mammalian acrosomes occurs in phalangerid (possums), macropodid (kangaroos) and petaurid marsupials (Harding et al. 1979; Setiadi et al. 1997). The most detailed information on the process is available for the tammar wallaby and common brushtail possum. On leaving the testis the immature acrosome is a 'scoop'-shaped structure projecting from the sperm head which bends in on itself and fuses during transit through the body of the epididymis to produce the mature button-shaped acrosome (Fig. 5; Lin et al. 1997; Setiadi et al. 1997; Lin and Rodger 1999). The acrosome of American marsupial species does not undergo such marked morphological

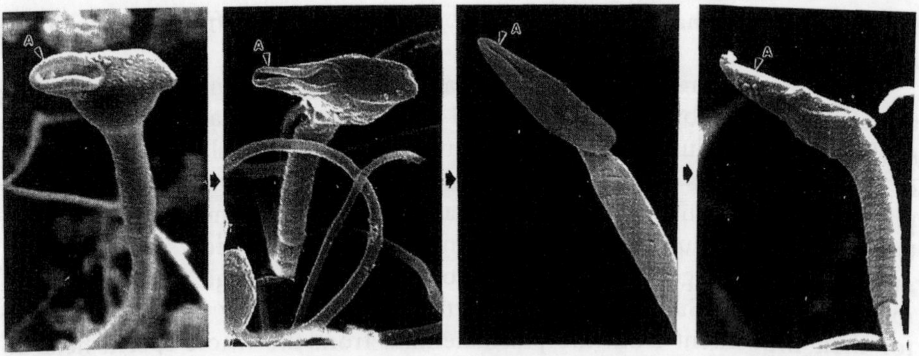

Fig. 5. Ultrastructure of extratesticular acrosome formation in the tammar wallaby. *A* Acrosome

modification but is instead involved in a pairing process that begins in the body of the epididymis. This pairing occurs via an association between the plasma membranes overlying the acrosome of two randomly selected spermatozoa. Sperm pairing does not occur in Australian marsupials nor in the American marsupial *Dromiciops australis* (Temple-Smith 1987).

The nucleus of the majority of marsupial species rotates during epididymal transit, to produce a streamlined spermatozoon (Fig. 3 and 4; Temple-Smith and Bedford 1980). In marsupials, the cytoplasmic droplet is initially located around the head and neck region, it condenses around the neck region of the spermatozoa and is lost directly into the lumen in the proximal epididymis (Temple-Smith and Bedford 1980). Other features of epididymal maturation in marsupials include the development of a unique cytoskeletal structure called the midpiece fibre network and, in phalangerids and macropodids, membrane stacks in the neck region (Temple-Smith and Bedford 1976; Harding et al. 1979).

3.2.2
Molecular Changes

Molecular modifications of the marsupial sperm surface are likely to rival those seen in eutherian species, but the evidence for such modification has been obtained using relatively non-specific probes. Freeze-fracture studies of marsupial spermatozoa have demonstrated the existence, and modification, of plasmalemma domains during epididymal transit (Olsen 1980; Cooper et al. 1998). Differences in the lectin-binding characteristics of sperm from the head and tail of the marsupial epididymis suggest modification of carbohydrate moieties during epididymal transit (Temple-Smith and Bedford 1976; Cooper et al. 1998). Electrophoretic studies provide some evidence for the acquisition and loss of proteins from the marsupial sperm surface during epididymal transit (Chaturapanich et al. 1992; Cooper et al. 1998). Recently it was found that tammar wallaby sperm acquired a 31-kDa maturation antigen during epididymal transit (Harris and Rodger unpublished observation). The protein was secreted by the epithelium of the epididymis and associated with the whole plasmalemma as sperm passed into region 2 (Fig. 4). Some evidence also exists for the maturation of intracellular antigens during epididymal transit in the marsupial. Sperm surface and intracellular structures are stabilized by disulphide bonding as in eutherians (Setiadi et al. 1997) and new antigenic epitopes are generated by the formation of the midpiece fibre network (Harris and Rodger 1998).

3.3
Monotremes

Epididymal maturation of monotreme spermatozoa involves the development of motility in the initial segment and movement of the cytoplasmic droplet, and there is evidence of changes in surface properties (Bedford and Rifkin 1979; Djakiew and Jones 1983). However, there are some notable differences. Monotreme sperm form into bundles with their heads embedded in electron-dense material, adsorbed to the periacrosomal plasmalemma, as they enter the terminal segment of the epididymis; formation of sperm bundles is associated with an increase in sperm velocity in vitro, which may be comparable to the change in the pattern of motility seen in other mammals (Djakiew and Jones 1983).

Whilst there is some evidence that the initial segment of the monotreme epididymis secretes specific proteins (Djakiew and Jones 1983), the molecular modification of the sperm plasmalemma has not been meaningfully examined. However, sperm maturation in these species does not appear to involve disulphide stabilization (Bedford and Rifkin 1979). The cytoplasmic droplet is initially located around the head of monotreme spermatozoa and migrates to the distal end of the midpiece in the initial segment. It is not discarded until sperm enter the terminal segment (Fig. 4; region 2) of the epididymis (Djakiew and Jones 1983). The terminal segment of the epididymis is involved in sperm storage, but the period of storage may be relatively short compared to that of other mammals (Djakiew and Jones 1983).

3.4
Common Themes and General Overview

In all mammalian species, epididymal maturation of spermatozoa results in the development of progressive motility, fertilizing capacity, and prepares the sperm for storage in the distal epididymis. These maturation changes are associated with morphological modifications that range from being minimal in monotremes, moderate (but variable) in eutherians to considerable in marsupials. A change in the location and loss of the cytoplasmic droplet is a consistent feature of epididymal maturation in all mammalian species, though there are differences in the location, migration and loss in the reproductive tract.

Molecular modification of the sperm surface and intracellular components, through the addition of proteins secreted by the epididymal epithelium and the alteration of others, is likely to occur in all mammals. Disulphide stabilization of sperm surface and intracellular organelles occurs in the epididymis of marsupials and eutherians, but apparently not in monotremes. Numerous studies on eutherian spermatozoa indicate that extensive molecular modification occurs during epididymal transit. Some evidence for a similar complexity of molecular maturation of marsupial sperm exists.

4 Sperm Deposition, Storage and Transport in the Female Tract

4.1 Marsupials and Eutherians

In the interval between mating and ovulation sperm are stored in the female reproductive tract awaiting the oocyte. Fertilization in most mammals occurs within 24h of mating (most eutherians, possums, kangaroos and opossums). Nevertheless, in a number of species sperm are stored for several days (dasyurid and didelphid marsupials, canid eutherians) to several months in insectivorous bats (see Mahi-Brown 1991; Taggart 1994). In most species the caudal isthmic oviduct serves as the sperm storage reservoir (Suarez 1998). Within this structure the spermatozoa are maintained either by interactions with the oviduct epithelial lining (e.g. pig, wallaby and possum) or in subluminal oviductal crypts lined with a specialized epithelium (reviewed by Taggart 1994). Oviductal storage crypts are peculiar to dasyurid and didelphid marsupials and only one group of eutherians, the insectivorous shrews (Bedford 1996). These storage crypts may allow a reduction in the number of sperm deposited into the female reproductive tract whilst retaining maximum reproductive efficiency. In comparison, other marsupials and eutherians deposit high concentrations of sperm into the female tract, far in excess of the number required for fertilization (Bedford 1996).

Sperm association with the oviduct epithelium is quite strong since repeated flushing is required to release the sperm in vivo, and enzymatic treatment of sperm bound to oviductal explants in vitro does not release the sperm (Raychoudhury and Suarez 1991; Smith and Yanagimachi 1991). Indeed, the sperm remain associated with the oviduct epithelium until ovulation, when a small number are released to meet the oocyte. There is evidence in both eutherian and marsupial species that association with the epithelium prolongs sperm motility and viability (Smith and Yanagimachi 1991; Sidhu et al. 1998). This interaction may also have a role in mediating capacitation-related changes in head orientation (marsupials) and hyperactivation (eutherians) which coincide with sperm release from the epithelium (Rodger and Bedford 1982; Yanagimachi 1994; Sidhu et al. 1998). Indeed, a change in the sperm surface at capacitation is proposed to trigger the release of spermatozoa from the epithelium (Smith and Yanagimachi 1991). Sperm transport from the lower isthmus to the site of fertilization in the upper ampulla is then achieved by the combined action of cilia-mediated fluid flow, oviductal contraction and sperm motility (Taggart 1994). Thus, sperm storage is ultimately responsible for the temporal coordination of fertilization by ensuring that the appropriate number of sperm arrive at the site of fertilization in the appropriate condition.

4.2 Monotremes

There is circumstantial evidence, based on the apparent variability of the gestation period, that monotremes may also store sperm in the oviduct for long periods of time (Griffiths 1978). A female echidna examined 34 days after segregation from males, had a 4.5-mm fertilized oocyte of normal appearance in one of her uteri and a recently formed corpus luteum, indicating that ovulation had occurred recently and that the oocyte had been apparently fertilized by a stored spermatozoon. Further evidence for sperm storage in monotremes is the presence of sperm in the lumina of the uterus and uterine glands of a platypus with oocytes that were not fully grown (Flynn and Hill 1939). These observations suggest that viable spermatozoa may be stored for up to 30 days in the female monotreme reproductive tract prior to fertilization.

5 Capacitation

Eutherian spermatozoa must undergo a period of 'capacitation' in the female reproductive tract before they can fertilize the oocyte. The capacitation of eutherian spermatozoa is a reversible phenomenon involving a complex array of post-ejaculatory functional modifications in the female reproductive tract. Only capacitated spermatozoa are able to undergo a regulated acrosome reaction in response to zona binding and subsequently fertilize the oocyte (reviewed in Yanagimachi 1994).

5.1 Eutherians

Capacitation is achieved in vivo during sperm interactions with the female reproductive tract, and takes from 6 h in the rabbit to as little as 0.5 h in the mouse (Yanagimachi 1994). Capacitation has also been achieved in vitro in a number of eutherian species, in defined synthetic medium containing energy sources, serum albumin, bicarbonate and calcium ions. In eutherians capacitation is associated with modifications of the sperm surface, metabolism and motility (hyperactivation – discussed separately) which provide the spermatozoon with the reaction pathways necessary to undergo a regulated acrosome reaction in response to signal induction by the zona pellucida (ZP; the extracellular coat surrounding the oocyte; Storey 1995).

Sperm surface modifications include the redistribution and removal of membrane proteins. For example, in the rabbit a surface-coating protein termed acrosome stabilizing factor added during epididymal maturation in preparation for sperm storage is removed during capacitation (Eng and Oliphant 1978). Cholesterol removal from the plasmalemma by oviductal fluid proteins leads to a concomitant decrease in the cholesterol/phospholipid ratio of the membrane (reviewed in Yanagimachi 1994; Cross 1998; Visconti et al. 1998).

Capacitation is also associated with a submicromolar rise in intracellular calcium ions, pH and cAMP, as well as hyperpolarization of the sperm plasmalemma. Though the mechanisms are poorly understood these changes are likely to be involved in transmembrane and intracellular signal transduction pathways which ultimately lead to the tyrosine phosphorylation of a specific subset of proteins essential for the attainment of capacitation (reviewed in Visconti and Kopf 1998).

5.1.1 Molecular Basis

The molecular basis of sperm capacitation is complex and controversial, though some consensus on a number of the controlling mechanisms is beginning to emerge (Aitken et al. 1998; Visconti and Kopf 1998; Visconti et al. 1998). Elevation of intracellular ions such as bicarbonate and calcium through ion channels, whose activity could be mediated by cholesterol efflux, may serve as transmembrane signals initiating intracellular signal transduction cascades. These reaction pathways involve the elevation of intracellular cAMP, brought about through bicarbonate (oviductal fluid), activation of sperm adenyl cyclase or redox regulation by reactive oxygen species. The elevation of cAMP activates protein kinase A (PKA) which acts on tyrosine kinases to induce protein tyrosine phosphorylation.

5.2 Marsupials

In the simplest sense, the fact that sperm flushed from the oviduct of the opossum after mating readily penetrated the ZP in vitro, whereas sperm from the vas deferens did not (Rodger and Bedford 1982), indicates that capacitation of marsupial spermatozoa in the female reproductive tract is a prerequisite for fertilization. The fact that many marsupial spermatozoa reside in the female tract for moderate to long periods (up to several days) prior to fertilization also suggests a requirement for capacitation (Selwood and McCullum 1987; Taggart and Temple-Smith 1991).

5.2.1 Morphological Correlates

To date there have been no definitive timing studies of marsupial capacitation, nor have any molecular markers of possible capacitation-related changes been identified. However, morphological changes that accompany sperm transit in the female tract have been suggested to be indicative of functional maturation in these mammals (see Fig. 6). Unpairing of American marsupial spermatozoa and reversion to a T-shaped (thumb tack) configuration in the female reproductive tract appear to be prerequisites for fertilization in these species, and are thus interpreted as being analogous to capacitation-associated changes in eutherian species (Rodger and Bedford 1982; Moore and Taggart 1993). Indeed, the spermatozoa of a number of Australian marsupials are also T-shaped after incubation in the female reproductive tract and approach the oocyte in this orientation, supporting the hypothesis of an involvement in capacitation (Mate and Rodger 1991; Bedford and Breed 1994). However, marsupial spermatozoa do not display hyperactivated motility as a result of capacitation.

5.2.2 Differences Between Marsupials

There also appear to be differences in the requirements for achieving capacitation between Australian and American marsupials. Capacitation in vitro is easily achieved in *Monodelphis domestica* by the incubation of epididymal spermatozoa in a simple culture medium containing bovine serum albumin (BSA). During incubation the spermatozoa unpair and are subsequently capable of fertilizing (Moore and Taggart 1993). In contrast, capacitation in Australian marsupials, such as the tammar wallaby and common brushtail possum, is not achieved in vitro in media conditions that capacitate monodelphid and many eutherian sperm, including addition of protein sources and increased temperature (Mate and Rodger 1996). *Sminthopsis crassicaudata* spermatozoa flushed from the isthmus of the oviduct displaying T-shape orientation (Bedford and Breed 1994) were still incapable of fertilizing oocytes in vitro, suggesting that further capacitative modifications were necessary (Bedford and Breed 1994). Thus, different mechanisms may regulate capacitation in Australian and American marsupials.

Recent experiments using various oviduct epithelial cell culture and co-culture systems using possum and wallaby sperm are providing insights into capacitation in Australian marsupials. Sperm–oviduct epithelial cell monolayer cultures and culture supernatants have been shown to prolong the viability and

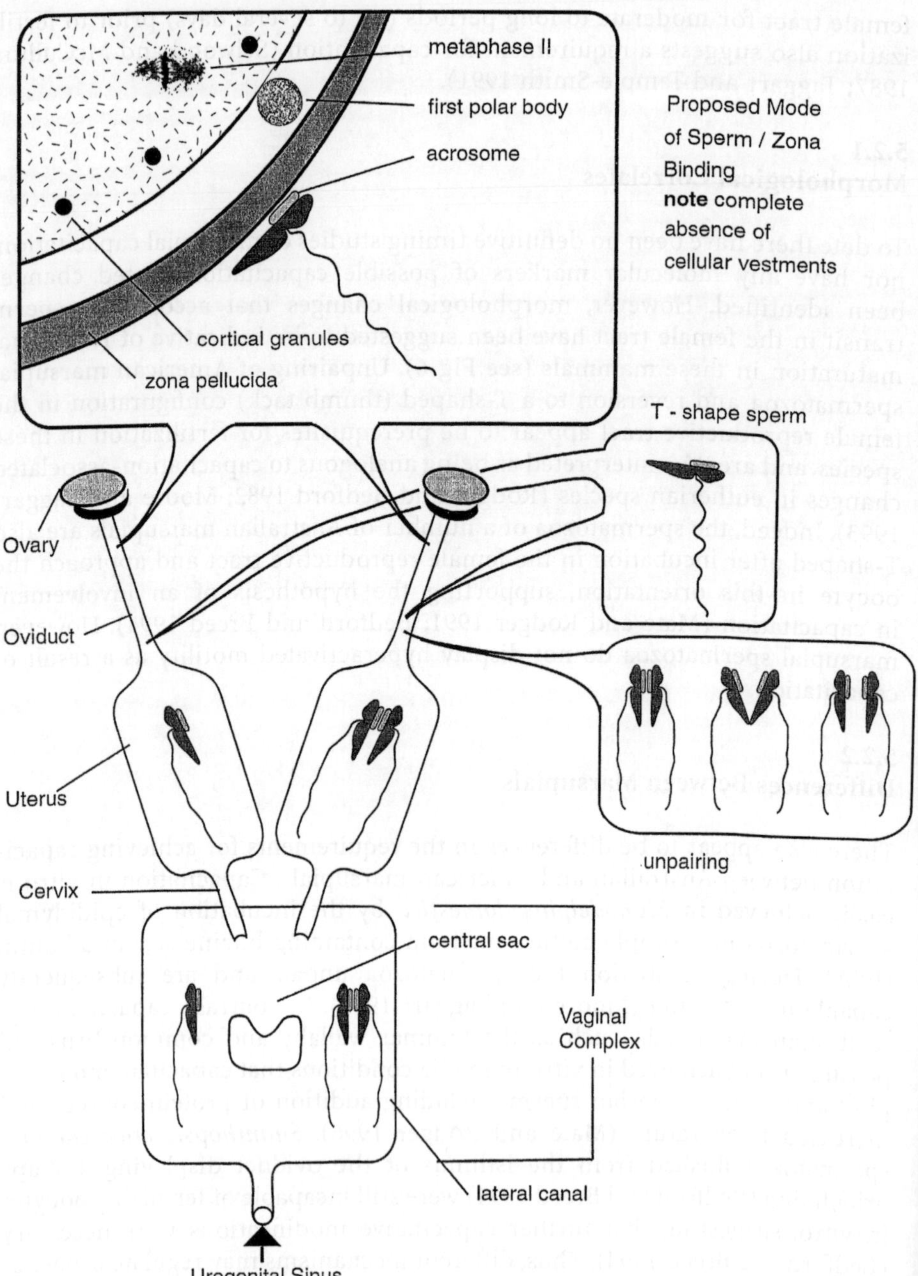

Fig. 6. Sperm transport, function and fertilization events in the marsupial female tract

motility of tammar wallaby and brushtail possum spermatozoa (Sidhu et al. 1998, 2000). Co-cultured spermatozoa attached to the ampulla and isthmus epithelial cells. Significantly, a percentage of these spermatozoa were transformed into the T-shape orientation upon release from the epithelium (10–15% wallaby, 60–70% possum). In IVF experiments using these in-vitro-produced T-shaped sperm, and in-vivo-capacitated T-shaped sperm flushed from the oviduct as a control, resulted in sperm binding to and penetration of the ZP (Mate et al. 2000). Thus, this study provides preliminary evidence for the necessity of sperm–oviduct epithelial cell interaction for capacitation in Australian species and lends further support to the suggestion that the T-shape head orientation is indicative of capacitation. Despite the occurrence of sperm–ZP binding and penetration, sperm–oocyte membrane fusion and oocyte activation were not observed in the IVF studies. Although it is possible that capacitation in these marsupials also involves sperm membrane alterations which facilitate sperm–oocyte membrane fusion it is more likely that cytoplasmic maturation and development of zona receptivity of the oocyte (discussed in ensuing sections) is a major factor.

Although the sperm of many eutherian species have been able to be capacitated in vitro using defined media and supplements, this has not been achieved for bats (Krutzsch and Crichton 1991). Perhaps some eutherian species also require unique interactions between the sperm and the oviduct and its products to achieve capacitation.

5.3 Monotremes

There have been no studies of sperm capacitation in monotremes. Phenotypically, monotreme spermatozoa more closely resemble those of reptiles and birds than their mammalian counterparts. As reptile and bird sperm do not require capacitation it is possible that monotreme spermatozoa also do not need to undergo maturational changes in the female reproductive tract (Bedford 1991). It is not known whether stabilizing factors/proteins are added to the monotreme sperm surface during epididymal maturation or ejaculation. However, prolonged storage of sperm in the female tract of preovulatory monotremes has been observed, suggesting that sperm are in some arrested state awaiting ovulation and thus presumably require a reactivation step equivalent to eutherian capacitation.

5.4 Common Themes and General Overview

The caudal isthmic oviduct is a site of sperm storage prior to fertilization in most eutherians and marsupials and probably monotremes. Sperm are stored in association with the oviduct epithelium in the majority of eutherians and in most marsupials other than dasyurids and didelphids. In these marsupials specialized oviductal crypts store sperm, perhaps reducing the number of sperm lost in the female tract prior to fertilization.

Coincident with release from the oviduct epithelium, both marsupial and eutherian sperm display capacitation associated changes. Capacitation is a prerequisite of fertilization in both marsupials and eutherians and enables sperm to undergo a regulated acrosome reaction in response to oocyte signals and thus fertilize. In eutherians, but not marsupials, capacitation is also associated with hyperactivated motility. However, reorientation of the sperm head to a T-shape appears to be a unique characteristic of capacitated marsupial sperm. The sperm of many eutherians and monodelphid marsupials can be capacitated in vitro under relatively simple media and conditions. In contrast, capacitation in Australian marsupials apparently requires interaction with the oviduct epithelium. Whilst it is likely that monotreme spermatozoa are stored for prolonged periods in the female tract it is not known if they require capacitation prior to fertilization. It is possible, given their similarity to avian spermatozoa that they may be fertile after epididymal maturation.

The molecular modifications associated with capacitation have only been investigated in eutherian species but appear to involve modification of sperm surface proteins and lipids, including tyrosine phosphorylation of specific proteins.

6 Hyperactivation of Spermatozoa

6.1 Eutherians

Capacitation in eutherian species is accompanied by a distinct change in the pattern of sperm motility, termed hyperactivity (reviewed in Yanagimachi 1994). In vivo, hyperactivation occurs in the oviduct and sperm approaching and interacting with the oocyte display the hyperactive pattern of motility. Hyperactivity has been observed in vitro in a wide variety of eutherian species. Hyperactivated motility is characterized by high-amplitude flagellar bending and permits increased penetration of viscous fluids, such as those encountered in the oviduct. However, when placed into low-viscosity medium, hyperactivated sperm, although vigorous, are erratic and generally non-progressive. A number of biological roles have been proposed for hyperactivated motility including aiding sperm detachment from the oviduct epithelium (Demott and Suarez 1992; Suarez 1996). However, the relative roles of the extra force generated by hyperactivated motility, and membrane changes associated with capacitation, in mediating sperm detachment have not been assessed. The ability of the hyperactivated spermatozoon to better penetrate viscous oviductal fluid increases the cell's chance of reaching the oocyte. The frequent direction changes associated with this asymmetric flagellar bending pattern may also help the sperm escape the grooves of the oviductal mucosa and increase its probability of encountering the oocyte (Suarez 1996). Hyperactivated sperm have also been observed to penetrate the zona far more effectively than progressively motile sperm, even when both had undergone the acrosome reaction (Stauss et al. 1995). Thus, the extra thrust generated by hyperactivated motility appears to be important for sperm penetration of the cumulus cells and zona in eutherian mammals (Bedford 1996).

In vitro studies have indicated that hyperactivity is a reversible event requiring a number of physiological factors including metabolic substrates, BSA and calcium, bicarbonate and potassium ions. Like other capacitation-related changes which enable the sperm to undergo a regulated acrosome reaction, hyperactivation appears to be activated via a signal transduction pathway involving cAMP-dependent phosphorylation of specific proteins via PKA (Si and Okuno 1999). Some of the proteins that are phosphorylated during capacitation and hyperactivation have been identified as A kinase anchor proteins which bind protein kinase A to the fibrous sheath (Carrera et al. 1996). The number of hyperactivated spermatozoa in the eutherian oviduct rises near the time of ovulation, suggesting that a specific oviductal signal may also be associated with hyperactivation in vivo (Cooper et al. 1979).

6.2 Marsupials and Monotremes

The resolution of whether marsupial or monotreme spermatozoa undergo any changes in motility associated with transport in the female reproductive tract or capacitation awaits thorough investigation of these processes in these species. Hyperactivity was not observed by Rodger and Bedford (1982) during in vitro fertilization in *Didelphis virginiana*, but Moore and Taggart (1993) suggested that hyperactivity was evident in unpaired *Monodelphis domestica* spermatozoa capacitated in vitro. However, these unpaired sperm exhibited only periods of erratic 'whiplash' tail movement with non-progressive motility after 4 h of incubation and may therefore represent degenerative effects. More recently, Taggart (1994) proposed that the sinusoidal or 'snake-like' sperm motility observed in the ejaculatory fluid of the dasyurid marsupial *Antechinus stuartii* (Taggart and Temple-Smith 1990) may be the marsupial equivalent of eutherian hyperactivation since it permits sperm penetration of viscous fluids.

6.3 Common Themes and General Overview

Marsupial sperm approaching and interacting with the oocyte have a T-shaped head, but no obvious change in motility or flagellar bending pattern similar to the hyperactivation observed in eutherian species. The extraflagellar force generated by hyperactivated motility appears to assist sperm to escape the oviduct epithelium and to possibly aid penetration of the relatively thick eutherian zona. Such a mechanism would seem unnecessary in marsupial species in which large holes in the zona are induced by the acrosomal contents (Taggart et al. 1993).

7
The Acrosome

7.1
Mammalian Acrosome Morphology and Enzymes

The acrosome is the sperm organelle specialized for sperm binding to, and penetration of, the ZP (Fig. 7). In eutherians and monotremes the acrosome forms a cap-like structure over the anterior portion of the head. However, in marsupial spermatozoa the acrosome is restricted to the dorsal side of the nucleus (Rodger 1991). Nevertheless, in all species the acrosome is a species-specific, membrane-enclosed organelle containing an array of hydrolytic enzymes (Yanagimachi 1994). The eutherian sperm acrosome is a relatively fragile structure (Dudenhausen and Talbot 1982), while in marsupial sperm it is remarkably stable and resists a variety of disruptive treatments (Cummins 1980; Mate and Rodger 1991). This stability is apparently explained by disulphide stabilization of the acrosomal membranes and/or matrix (Mate et al. 1994; Setiadi et al. 1997).

7.2
Acrosome Reaction in Eutherians

7.2.1
Morphology and Functional Significance in Vivo

Only capacitated spermatozoa are capable of undergoing a regulated, physiological acrosome reaction in response to ZP binding, and only acrosome-reacted sperm can penetrate the zona and fuse with the oocyte. In eutherian sperm the acrosome reaction (AR) involves multiple-point fusions between the outer acrosomal membrane and the overlying plasmalemma, vesiculation and the exocytosis of the acrosomal contents (Fig. 7; Yanagimachi 1994). The hydrolytic acrosomal contents released by the AR have long been thought to facilitate the enzymatic digestion and subsequent penetration of the eutherian ZP. However, recent contradictory evidence such as the fertility of 'knockout' mice which do not produce the putative zona protease, acrosin, has brought into question the relative importance of zona digestion in eutherian fertilization (reviewed in Bedford 1999). The physical thrust generated by the hyperactivated sperm tail may be more important for ZP penetration in eutherian species (Bedford 1991). The acrosomal contents have been implicated in other roles including the development of the AR (Meizel 1984)

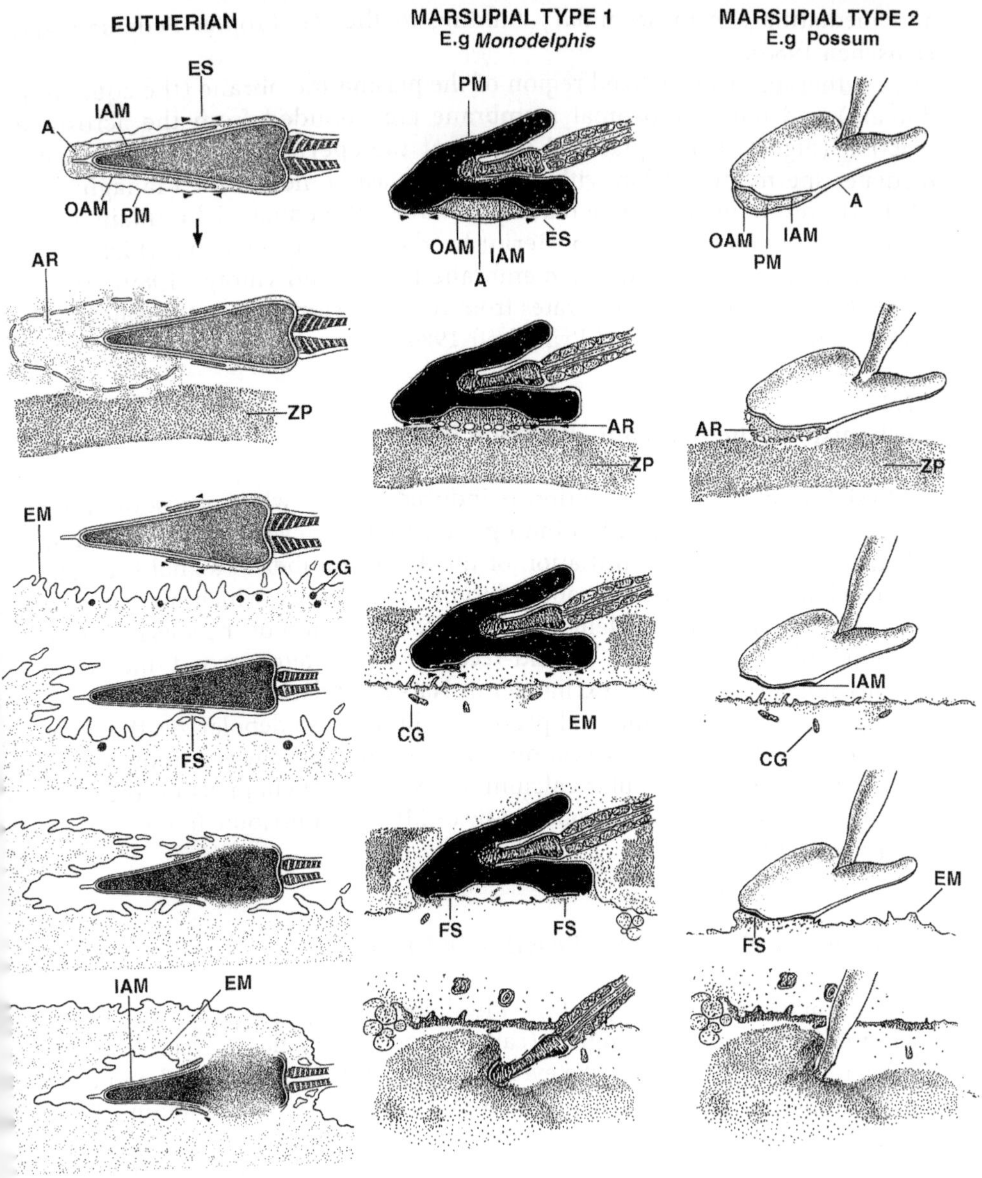

Fig. 7. Morphology of the mammalian acrosome reaction and sperm–egg fusion. *A* Acrosome; *PM* plasma membrane; *OAM* outer acrosomal membrane; *IAM* inner acrosomal membrane; *ES* equatorial segment; *ZP* zona pellucida; *EM* egg plasma membrane; *CG* cortical granules; *FS* sperm–egg fusion. (Adapted from Moore 1996 and Bedford 1982)

and mediating sperm secondary binding to the ZP (Topfer-Pederson and Henschen 1988).

In eutherians a specialized region of the plasma membrane (the equatorial domain) and outer acrosomal membrane are excluded from the acrosome reaction (Fig. 7). This equatorial region of the sperm surface subsequently mediates sperm interaction with the oocyte plasma membrane (Yanagimachi 1994). However, the acrosome reaction modifies the equatorial domain, inner acrosomal membrane and/or posterior head of the spermatozoa which facilitates attachment to the oocyte membrane (Sidhu and Guraya 1989). Mouse sperm PH-20, for example, migrates from the posterior head to the inner acrosomal membrane (Myles and Primakoff 1984).

7.2.2 Mechanism of the Acrosome Reaction in Vivo

A physiological acrosome reaction is induced by ZP3 when acrosome-intact sperm bind to the zona. ZP3 binding triggers the elevation of intracellular calcium ions essential for initiation of the AR. A number of signal transduction mechanisms have been proposed to mediate the AR, which may turn out to be dependent on multivalent interactions. Interaction of the sperm surface receptors with GI-proteins (ligand-receptor–GI-protein second messenger system) activates an Na^+/H^+ exchanger or Na^+-dependent Cl^-/HCO_3 exchanger leading to an increase in internal pH. Zona binding also depolarizes the membrane by activating a selective cation channel which produces a transient and focal elevation of intracellular calcium levels. Calcium channels on the outer acrosomal membrane are also opened by cAMP and contribute to the increase in Ca^{2+} concentration.

7.2.3 Mechanism of the Acrosome Reaction in Vitro

In addition to ZP3, a long list of physiological (e.g. progesterone, prostaglandins) and non-physiological (calcium ionophore, diacylglycerol) agents has been used to induce AR-like events in vitro in both epididymal and ejaculated eutherian spermatozoa (reviewed in Breitbart and Spungin 1997). However, the mechanisms underlying AR induction by the physiological inducers are not clearly understood and the putative sperm receptors are unidentified. Non-physiological inducers such as calcium ionophore and diacylglycerols (DAG) can be used to effectively bypass the receptor-mediated signalling pathways described above and explore the down stream mechanisms involved in AR induction.

7.3
Acrosome Reaction in Marsupials and Monotremes

There is no information about the acrosome reaction in monotreme spermatozoa, though it is undoubtedly necessary to facilitate ZP penetration. The physiological acrosome reaction in *Monodelphis* has been shown to involve multiple point fusions as in eutherians (Taggart et al. 1993). Similar multiple point fusions are induced by the diacylglycerol DiC8 and arachidonic acid in brushtail possum and tammar wallaby spermatozoa (Sistina et al. 1993; Sistina and Rodger 1998) suggesting that similar acrosome reaction mechanisms function in marsupials and eutherians. A major difference between the acrosomes of marsupial and eutherian sperm is that marsupial sperm are unresponsive to calcium ionophores and that diacylglycerol-induced AR occurs in the absence of extracellular calcium, suggesting a role for internal calcium stores (Mate and Rodger 1993). A region of the acrosome equivalent to the eutherian equatorial segment has been observed to persist in *Monodelphis domestica* and *Sminthopsis crassicaudata* sperm penetrating the zona (Taggart et al. 1993; Breed 1994b). This equatorial segment-like region subsequently appeared to participate in sperm–oocyte interaction. However, no region of the brushtail possum or tammar wallaby acrosome remained intact after AR induction by diacylglycerol or arachidonic acid treatment in vitro (Sistina et al. 1993; Sistina and Rodger 1998) or in possum sperm interacting with oocytes in vivo (Jungnickel et al. 1999).

7.4
Common Themes and General Overview

The acrosome is similar in all mammalian species in that it is a membrane-bounded organelle containing a variety of hydrolytic enzymes. However, characteristic differences in the stability of the marsupial versus the eutherian acrosome exist. The physiological significance of this difference is not known. The acrosome reaction facilitates sperm penetration of the zona in all mammalian spermatozoa. In eutherian species, and some marsupials, an unreacted equatorial region persists after the AR to participate in sperm–oocyte interaction, but no equivalent equatorial region has been observed in diacylglycerol-treated wallaby and possum spermatozoa or during in vivo fertilization in the possum. It is possible that two different AR systems exist in marsupials (Fig. 7; Mate and Rodger 1996).

8 Sperm Interaction with the Zona Pellucida

8.1 Eutherians

8.1.1 Structure and Function of the ZP

All growing oocytes, ovulated oocytes and pre-implantation embryos are surrounded by a ZP, an extracellular coat responsible for vital functions during the fertilization process including specific binding of spermatozoa, induction of the acrosome reaction, prevention of polyspermy and physical protection of the embryo.

The ZP of eutherian mammals consists of three sulphated glycoproteins (designated ZP1, ZP2 and ZP3) that serve specific functions during fertilization and early development, which are well defined in the mouse. ZP1 is a structural protein that cross-links filaments made of ZP2 and ZP3 heterodimers (Greve and Wassarman 1985; reviewed by Green 1997). During fertilization sperm initially bind to ZP3 (Bleil and Wassarman 1980), which induces the sperm acrosome reaction (Leyton and Saling 1989; Ward and Kopf 1993; Tomes et al. 1996). After induction of the acrosome reaction, ZP2 acts as a secondary sperm receptor (Bleil et al. 1988) which is necessary for penetration of the sperm through the ZP. Following fertilization the proteolytic cleavage of ZP2 and modification of ZP3 are thought to be involved in the prevention of polyspermy and protection of the early embryo.

Eutherian *ZP* genes are expressed primarily in the oocytes (reviewed by Liang and Dean 1993) and also in the granulosa cells of some species (Grootenhuis et al. 1996). The ZP genes are conserved among diverse vertebrate species (Fig. 8). The *ZP3* gene is the most highly conserved of the three with amino acid identities between eutherian mammals (human, mouse, rabbit, pig) ranging from 66 to 74% (Fig. 9; reviewed by Epifano and Dean 1994). The *ZP2* gene is also well conserved with amino acid identities between eutherian mammals (human, mouse, rabbit, pig) ranging from 54 to 72% (Fig. 9; Epifano and Dean 1994). Although the amino acid sequence of the ZP proteins is conserved between species, variations in the post-translational modifications of the ZP proteins result in significant differences in the molecular weights of the proteins between species (reviewed by Dunbar et al. 1991; Castle and Dean 1996; McLeskey 1998). These extensive post-translational modifications, which include glycosylation and sulphation, are critical to the structural and functional integrity of the ZP (Greve et al. 1982; Salzmann et al. 1983). In addition to the conservation of ZP proteins between species, specific protein domains are conserved among the three classes of zona proteins. A 277-amino acid domain exists in mouse ZP1 and ZP2 in which the two polypeptides are 47% similar and 34% identical (Epifano et al. 1995).

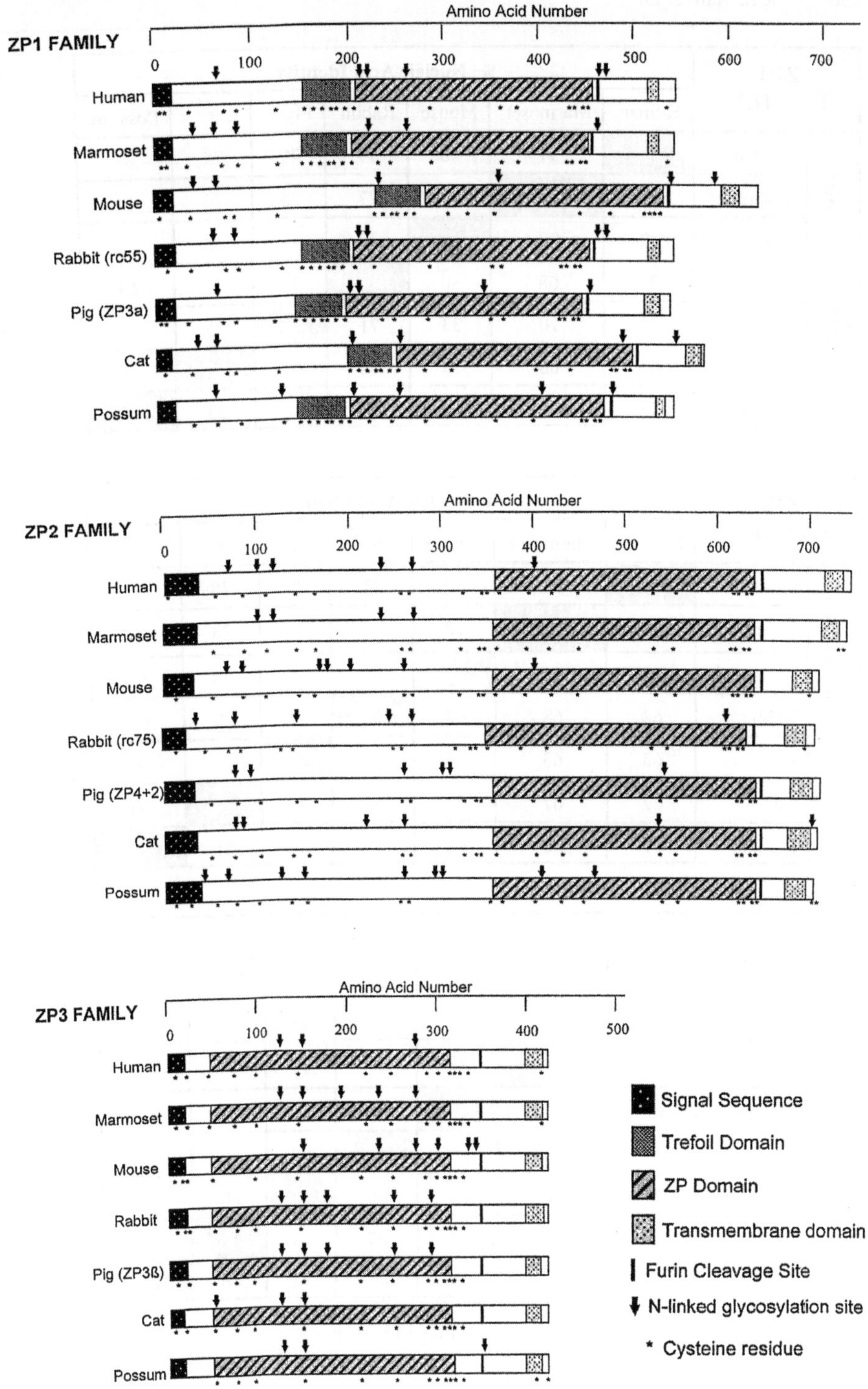

Fig. 8. Structure of the zona pellucida 1 (*ZP1*), 2 (*ZP2*) and 3 (*ZP3*) protein families from a representative group of mammals. (Redrawn from McLesky et al. 1998)

ZP1 FAMILY		% Nucleic Acid Identity						
		Human	Marmoset	Mouse	Rabbit	Pig	Cat	Possum
% Amino Acid Identity	Human		91	40	79	79	74	64
	Marmoset	87		48	77	75	76	64
	Mouse	32	39		39	41	43	48
	Rabbit	73	68	30		78	79	64
	Pig	72	70	33	71		75	46
	Cat	67	68	33	64	67		62
	Possum	54	53	40	54	30	53	

ZP2 FAMILY		% Nucleic Acid Identity						
		Human	Marmoset	Mouse	Rabbit	Pig	Cat	Possum
% Amino Acid Identity	Human		92	69	77	75	78	63
	Marmoset	86		72	77	77	78	62
	Mouse	57	59		73	68	70	60
	Rabbit	68	68	62		76	79	65
	Pig	63	63	55	64		82	62
	Cat	67	67	58	66	72		63
	Possum	51	50	48	54	50	51	

ZP3 FAMILY		% Nucleic Acid Identity						
		Human	Marmoset	Mouse	Rabbit	Pig	Cat	Possum
% Amino Acid Identity	Human		92	72	70	78	78	52
	Marmoset	91		71	70	76	76	54
	Mouse	67	67		64	69	71	52
	Rabbit	69	68	65		69	69	56
	Pig	74	72	66	68		80	55
	Cat	64	68	64	65	73		54
	Possum	47	46	46	47	47	46	

Fig. 9. Zona pellucida family cDNA and amino acid identities

8.1.2 Primary Sperm–ZP Binding

Ultrastructural Observations. Primary sperm–ZP binding is a rapid transitional event, and it is very rare to observe the fertilizing spermatozoon at the point of initial contact with the ZP, as penetration follows almost immediately. Saling et al. (1979) observed that mouse sperm bound to the outside of the ZP were acrosome intact; however, changes observed in the structure of the acrosomal cap of rat and rabbit sperm bound to the ZP (Esaguy et al. 1988; Shalgi et al. 1989) suggest that there may be some leakage of acrosomal enzymes to aid progress through the cumulus (a complex of former follicular cells in an acellular matrix) enclosing the ZP and oocyte.

Molecular Mechanisms. Acrosome-intact mouse sperm bind to the ZP via O-linked oligosaccharide residues that are located on a cluster of serine residues near the carboxy terminus of the ZP3 glycoprotein (Litscher and Wassarman 1996). Evidence suggests that a terminal α-galactose and β-N-acetylglucosamine, as well as the presence of a fucosyl residue, are important functional components of these oligosaccharides (Bleil and Wassarman 1988; Miller et al. 1993; Johnston et al. 1998). The importance of heterogeneity in the post-translational glycosylation of the ZP proteins for specificity of sperm binding is well illustrated by the restoration of fertility in Zp3 null mice by breeding with transgenic mice expressing human ZP3 (Rankin et al. 1998). The resultant chimeric ZP consisting of mouse ZP1, mouse ZP2 and human ZP3 did not bind human sperm but bound mouse sperm, resulting in restoration of fertility.

Identification of a single complementary ZP3-binding protein on the sperm surface has been more difficult and remains controversial, but candidates include (1) zonadhesin, a large membrane protein containing multiple types of domains (Gao and Garbers 1998), and (2) spermadhesins, a family of small sperm-associated proteins, which bind solubilized ZP (Topfer-Petersen and Calvete 1995). Reviews of the molecules involved in sperm–oocyte interaction in eutherian mammals can be found in McLesky et al. (1998) and Wassarman (1999).

8.1.3 Penetration of the ZP and Secondary Sperm–ZP Binding

Ultrastructural Observations. The eutherian spermatozoon undergoes the acrosome reaction soon after binding to the ZP, releasing its acrosomal contents at the ZP surface. The penetrating sperm leaves a narrow, sharply defined penetration slit in the ZP which appears more indicative of a mechanical cutting action than enzymatic digestion (Bedford 1998). Also supporting this view are ultrastructural observations of distortion and displacement of the zona's fibrous matrix around the head of the penetrating sperm (Phillips 1991). The relative contribution of mechanical versus enzymatic penetration of the

ZP in eutherian mammals was reviewed by Yanagimachi (1994), who suggested that acrosin and perhaps other acrosomal enzymes may be important for binding of the acrosome-reacted sperm and insertion of the sperm head into the ZP matrix. When the sperm head is within the zona, acrosomal enzymes may no longer be required and mechanical forces may be sufficient for progression of the sperm through the ZP.

Molecular Mechanisms. Sperm binding to ZP3 initiates the acrosome reaction, a fusion of the sperm plasma membrane and outer acrosomal membrane resulting in the release of hydrolytic enzymes. Following the AR, a secondary binding event takes place which again adheres the sperm to the ZP and facilitates transit through the ZP. Candidate secondary ZP binding proteins on sperm have been reviewed recently by McLesky et al. (1998) and include: (1) PH-20, a glycosyl phosphatidylinositol-anchored membrane protein located on the post-acrosomal surface of the sperm head as well as within the acrosome (Phelps et al. 1988), which has both ZP-binding and hyaluronidase activity (Primakoff et al. 1985; Hunnicutt et al. 1996); (2) proacrosin, which is located within the acrosome of all mammalian sperm and has ZP-binding and proteolytic properties (Jones and Brown 1987); and (3) Sp17, a 17-kDa protein that becomes exposed during the AR and migrates posteriorly to the equatorial region of the head of acrosome-reacted sperm (Richardson et al. 1994).

8.2 Monotremes

The ZP of the platypus reaches a maximum thickness of approximately 11 µm at the onset of vitellogenesis but is less than 0.5 µm in the preovulatory oocyte (Hughes 1977). There is no information about sperm–ZP interaction in monotremes. It has been suggested that the filiform sperm head and modest acrosome of echidna sperm may be related to the fragile nature of the ZP (Bedford and Rifken 1979).

8.3 Marsupials

8.3.1 Structure and Function of the ZP

The marsupial ZP is thought to perform similar functions to that of eutherian mammals but is considerably thinner, varying from 6 to 8 µm in the brushtail possum (*Trichosurus vulpecula*) and tammar wallaby (*Macropus eugenii*) (Tyndale-Biscoe and Renfree 1987; Mate 1998) to only 1–2 µm in the didelphid marsupials (Talbot and DiCarlantonio 1984; Falconnier and Kress 1992). The marsupial ZP also appears to be more easily digested than the zona of eutherian oocytes (Rodger and Young 1981; Baggott and Moore 1990; Rodger

1991). Like that of eutherians, the marsupial ZP is composed of a matrix of three proteins which have approximately 50% homology to eutherian ZP1, ZP2 and ZP3 (Fig. 9; Mate and McCartney 1998; Haines et al. 1999; McCartney and Mate 1999) and are antigenically related to porcine ZP (Mate 1998). Although the homology of the marsupial ZP genes to their eutherian counterparts is relatively low, the structure of the ZP proteins is highly conserved (Fig. 8).

There is evidence of maturational changes to the ZP, affecting both structure and function, during the latter stages of follicular development and even following ovulation in some marsupial species. There are prominent structural differences between the ZP of follicular and ovulated oocytes from the brushtail possum and *Monodelphis* (Mate 1996). The ZP surrounding follicular oocytes in these species is a wide diffuse structure, but in ovulated oocytes it is narrow and compact (Mate 1996). These periovulatory changes are suggestive of a physical compression of the ZP, perhaps as a consequence of late preovulatory growth of the oocyte "stretching" the ZP. An in vitro fertilization study of *Monodelphis* suggests that the ZP also undergoes maturational changes vital for acquisition of sperm-binding ability; germinal vesicle (GV)-stage oocytes collected within 15 h after mating were incapable of binding sperm; however , PB1-stage oocytes collected 15–18 h after mating were capable of sperm–ZP binding and penetration (Moore and Taggart 1993). As secretion of the ZP proteins and formation of the ZP is essentially complete by this stage of follicular development, it is likely that these functional changes in the ZP are due to modifications to the carbohydrate residues on the ZP proteins.

8.3.2
Sperm Interaction with the ZP

Marsupial oocytes are ovulated free of any surrounding granulosa cells; thus a naked ZP forms the only barrier to the fertilizing spermatozoon. Our current knowledge of sperm–ZP interaction in marsupials is limited to ultrastructural observations, but suggests that quite different modes of zona penetration may operate. Prior to binding to the ZP, marsupial sperm undergo reorientation of the sperm head to form a T-shape so that the flat surface of the intact acrosome can be applied to the ZP (Bedford and Breed 1994; Molinia et al. 1998) perhaps to maximize the subsequent effects of acrosomal enzymes on the ZP. *Monodelphis* spermatozoa have been observed to undergo the acrosome reaction at the zona surface, suggesting that it may be triggered by zona components (Taggart et al. 1993).

In the opossums *Didelphis* and *Monodelphis* (Rodger and Bedford 1982; Taggart et al. 1993) and the brushtail possum (Jungnickel et al. 1999; Mate et al. 2000), a large hole is digested in the ZP as a result of the AR. However, in *Sminthopsis*, a narrow penetration slit is formed by the penetrating sperm, and electron micrographs suggest the involvement of both physical pressure and localized enzymic digestion in penetration of the ZP (Breed and Leigh 1988, 1992; Breed 1994a).

8.4 Common Themes and General Overview

Despite significant variation in the thickness of the ZP amongst the three mammalian groups, its structural components are well conserved. The eutherian and marsupial ZP proteins share several structural features including short 5′ and 3′ untranslated regions, N-terminal hydrophobic signal peptides, potential and O-linked glycosylation, a hydrophobic transmembrane domain near the C terminus, a furin or furin-related cleavage site upstream of the transmembrane domain and a ZP signature domain. As most of these structural features are also present in non-mammalian vertebrate ZP proteins, it is likely that they are also conserved in monotremes. However, our current understanding of the function of the ZP glycoproteins is based almost entirely on the mouse and may not be representative of gamete interactions in all eutherian mammals, marsupials or monotremes.

It appears that in some marsupials at least, the completion of oocyte maturation and acquisition of fertilizability, including changes to the ZP, may not occur until late in the periovulatory period. In contrast, the ZP appears penetrable in eutherian GV oocytes as well as in subsequent stages of preovulatory maturation (reviewed by Bedford 1991).

9 Sperm–Oocyte Fusion

9.1 Eutherians

Only acrosome-reacted spermatozoa are capable of fusing with the oocyte. Sperm binding and fusion with the oocyte plasma membrane are independent events which, together, result in the formation of the zygote. In eutherian mammals, sperm binding and subsequent membrane fusion is initiated in the equatorial region (Yanagimachi 1994). The stabilized inner acrosomal membrane is excluded from the fusion process and is instead encapsulated in a vacuole within the oocyte cytoplasm (Bedford 1991). In species in which the tail is incorporated into the vitellus, membrane fusion spreads posteriorly to the flagellar membrane and the sperm–oocyte membranes form a continuum.

The sperm proteins proposed to participate in sperm–oocyte binding and fusion include DE/AEG (Rochwerger at al. 1992) and fertilin (PH-30) (Primakoff et al. 1987). DE/AEG is an epididymal maturation antigen that migrates from the post-acrosomal region to the equatorial region during capacitation (Rochwerger and Cuasnicu 1992). Antibodies raised against

DE/AEG inhibit sperm–oocyte fusion (Rochwerger at al. 1992). Fertilin arises in the testis and undergoes endoproteolytic cleavage during testicular and epididymal maturation. Both the a and b subunits of this heterodimeric protein belong to a family of proteins containing a disintegrin and metalloprotease domain (ADAM) (Wolfsberg et al. 1995). Both the a and b subunits bind to the oocyte plasmalemma via their disintegrin domains through an interaction with integrins on the oolema (Myles et al. 1994). A second ADAM protein, cyritestin, functions with fertilin b in mediating sperm–oocyte plasma membrane adhesion leading to fusion (Yuan et al. 1997).

9.2 Monotremes

The only relevant information concerning monotremes is that the oolemma of monotreme oocytes has an abundance of microvilli (Hughes and Carrick 1978).

9.3 Marsupials

As discussed earlier, a narrow peripheral region of the *Monodelphis* sperm acrosome remains intact following the acrosome reaction, suggesting that it may be structurally and functionally similar to the stabilized equatorial region of eutherian spermatozoa (Taggart et al. 1993). Ultrastructural analysis of fertilization in vitro in *Monodelphis* indicates that sperm fusion with the oolemma occurs via this persistent region of the acrosome (Taggart et al. 1993), suggesting that it is the sperm plasma membrane overlying this region of the acrosome that actually fuses with the oolemma. An earlier study of fertilization in *Didelphis* suggested that it was the inner acrosomal membrane of the acrosome-reacted spermatozoon that fused with the oolemma. This conclusion was based on light microscopy observations and electron microscopy which indicated that there was no evidence of incorporated oocyte or sperm membranes surrounding the decondensing sperm head (Rodger and Bedford 1982). This is an important point because fertilization in eutherians is not only distinguished by the role of the equatorial segment as the site of sperm–oocyte membrane fusion but by the unusual phagocytic-like mode of sperm engulfment by the oocyte. Although an equatorial segment-like region may be present in the acrosomes of some marsupials it does not necessarily follow that the sperm is incorporated in a manner similar to that of eutherians.

9.4 Common Themes and General Overview

Until recently it was thought that the mechanism of sperm–oocyte fusion in eutherian mammals, involving the plasma membrane over the equatorial segment of the acrosome, was unique to this group. It was hypothesized that this departure from the conventional mode of fusion (via the inner acrosomal membrane of the sperm head), observed in all non-mammalian vertebrates and invertebrates, evolved in eutherian mammals from the need to stabilize the inner acrosomal membrane to withstand penetration through a relatively thick and resilient ZP (Moore and Bedford 1983; Bedford 1991). The observation that sperm–oocyte fusion in *Monodelphis* also occurs via the plasma membrane over an equatorial segment-like region (Taggart et al. 1993) indicates that this process in marsupials may be more akin to the eutherian situation (Fig. 7) than the lower vertebrates.

In contrast to *Monodelphis*, there is no evidence that a region of the acrosome persists after the acrosome reaction in brushtail possum and tammar wallaby sperm (Sistina et al. 1993; Sistina and Rodger 1998; Jungnickel et al. 1999), suggesting that there may be significant differences between marsupial species in the sperm membranes involved in fusion with the oolemma (see types 1 and 2 in Fig. 7). Observations of sperm–oocyte binding, fusion and incorporation in macropods and possums are required to establish definitively which region of the spermatozoon fuses with the oocyte so that a representative model(s) of marsupial sperm–oocyte fusion can be established.

10 Oocyte Activation

10.1 Eutherians

The fusion of sperm and oocyte triggers a cascade of cellular events in the oocyte that result in the exocytosis of cortical granules, recruitment and translation of maternal mRNAs and the resumption of meiosis known as activation. Many of the events of oocyte activation are triggered by a transient increase in intracellular calcium – $[Ca^{2+}]_I$ – at the time of fertilization. In the mouse there is a large initial increase in $[Ca^{2+}]_I$ followed by a series of cytosolic Ca^{2+} oscillations which last for several hours after sperm–oocyte interaction (reviewed by Ben-Yosef and Shalgi 1998). Unlike fish and amphibians, the intracellular pH of eutherian oocytes is not affected by fertilization and thus is not associated with activation (Kline and Zagray 1995; Ben-Yosef et al. 1996). Two pathways in the cascade of events leading to oocyte activation can be segregated: early events (including the cortical reaction) can be triggered by PKC which is activated by the initial rise in Ca^{2+}, whereas later events (such as resumption of meiosis) require higher Ca^{2+} elevation but are PKC independent (Raz et al. 1998a).

The increase in $[Ca^{2+}]_I$ is generated by IP_3-mediated release of Ca from intracellular stores (Miyazaki et al. 1992) but also requires Ca^{2+} entry from outside the oocyte to refill the Ca^{2+} pools and maintain Ca^{2+} oscillations (Miyazaki 1995). It is therefore likely that PLC is activated during fertilization as it is the enzyme responsible for hydrolysis of phosphatidyl inositol 4,5-bisphosphate (PIP_2) to IP_3 and DAG, leading to activation of PKC. One of the ways that the fertilizing sperm could initiate these events is via binding of a sperm ligand with a receptor on the oolemma and subsequent activation of PLC. It has been demonstrated in the mouse that this pathway does not involve G proteins (Williams et al. 1998) and therefore most probably involves receptor tyrosine kinases which are known to be involved in invertebrate oocyte activation. Another possibility is that activation of the oocyte may be initiated by a sperm factor(s) that is introduced into the oocyte during sperm–oocyte fusion. The candidates for this role include a component of the sperm cytosol (Wolny et al. 1999), and a relatively insoluble sperm head component(s) known as SOAF (Perry et al. 1999).

Development of activation competence by the oocyte occurs gradually after GV breakdown (Ducibella et al. 1990; Ducibella and Buetow 1994; Wang et al. 1997; Raz et al. 1998b) concomitant with migration of cortical granules from a subcortical to cortical location (Cran and Cheng 1985), accumulation of clusters of endoplasmic reticulum in the oocyte cortex (Ducibella et al. 1988) and a four-fold increase in the intracellular stores of Ca^{2+} (Tombes et al. 1992). Maximal activation competence is not fully developed until MII (Wang et al. 1997), and this suggests that there is likely to be a temporal window in which normal oocytes have the highest likelihood of undergoing a normal cortical reaction and activation of development (reviewed by Ducibella 1996). The general pattern for eutherian mammals is that oocytes are ovulated and fertilized at MII of meiosis. Fertilization reactivates meiosis and a second polar body is produced. A major exception to this rule is found in wild and domestic canids. Canid oocytes are ovulated with a germinal vesicle in meiotic arrest. The first meiotic division occurs in the oviduct around 2 days later and then meiosis arrests, awaiting fertilization (Mahi-Brown 1991).

10.2 Monotremes

The fully grown monotreme oocyte is surrounded by a follicular epithelium that is two cell layers thick. The follicle cells contain secretion granules that contain the precursors of a fluid which forms a continuous layer around the oocyte at the time of maturation. Flynn and Hill (1939) suggested that this secretion is the homologue of the follicular fluid of the Graafian follicle of the eutherians and marsupials. The first polar body is extruded from the monotreme oocyte before ovulation and like most of the mammals it is fertilized at MII of meiosis. Activation of the monotreme oocyte following fertilization results in the resumption of meiosis and the extrusion of the second polar body (reviewed by Griffiths 1968, 1978).

10.3 Marsupials

Activation of marsupial oocytes following fertilization is marked by the resumption of meiosis and extrusion of the second polar body (Rodger and Bedford 1982; Selwood 1982; Breed and Leigh 1990; Taggart et al. 1993). A protrusion of oocyte cytoplasm through the hole created in the ZP by the fertilizing sperm, somewhat reminiscent of a fertilization cone in invertebrates, has been observed in vitro in *Didelphis* and *Monodelphis* (Rodger and Bedford 1982; Moore and Taggart 1993; Taggart et al. 1993). However, the cytoplasmic protrusion in these oocytes did not display the organization of filamentous actin typical of fertilization cones in lower vertebrates (Schatten and Schatten 1987), suggesting that it is most likely an artefact of in vitro fertilization and the absence of the oviduct-deposited mucoid coat.

There is some ambiguity regarding the association of cortical granule exocytosis with activation in marsupials as an absence of cortical granules has been observed in both unfertilized and fertilized oocytes recovered from the oviducts of *Monodelphis* (Taggart et al. 1993) and brushtail possum (Jungnickel et al. 1999).

10.4 Common Themes and General Overview

Fusion of the membranes of sperm and oocyte sets off a chain of events which has a profoundly stimulatory effect on the oocyte. Although little is known about these events and their regulation in monotremes and marsupials the general pattern appears likely to be similar in all mammals. In eutherians activation is associated with a transient increase in intracellular calcium and it will be very surprising if similar events do not occur in the oocytes of marsupials and monotremes. A common feature of eutherian and marsupial, and presumably monotreme, oocytes is that the immature oocyte is not competent to undergo activation. This competence appears to develop gradually in the hours following GV breakdown. Even in canid oocytes which are ovulated at the GV stage fertilization appears not to occur until these cytoplasmic maturation events and first meiotic division are completed.

11 Cortical Reaction and Block to Polyspermy

11.1 Eutherians

The major block to polyspermy is provided by alterations to the ZP although there is at least some role played by the plasma membrane in some species (e.g.

rabbit; O'Rand and Nikolajczyk 1991). The ZP of eutherian oocytes becomes "hardened" following cortical granule release induced by sperm–oocyte fusion (reviewed by Ducibella 1996).

11.1.1 Cortical Granules

Cortical granules (CG) are small membrane-bound vesicles (0.2–0.6 µm in diameter, depending upon the species) located directly beneath the plasma membrane of mature unfertilized eutherian oocytes. Their formation in eutherian mammals is related to the onset of follicular growth (Gulyas 1980) and is associated with many active Golgi complexes in the peripheral cytoplasm of the oocyte (Zamboni 1970). Condensation and accumulation of the luminal contents of the fusing vesicles results in formation of the mature dense granules (Cran and Moor 1990) Migration of cortical granules to their final position just below the oocyte membrane occurs around the same time as the resumption of meiosis (Cran and Moor 1990). A marked polarity develops in the distribution of cortical granules after extrusion of the first polar body and formation of the second meiotic spindle; in the mouse and hamster up to 24% of the total cortex of the oocyte in the area of the second meiotic spindle is totally devoid of cortical granules (Cran 1989; Ducibella et al. 1990).

11.1.2 Cortical Reaction

Initiation of the cortical reaction is correlated with an increase in intracellular Ca^{2+} concentration and activation of protein kinase C, resulting in a block to polyspermy within 10 min of fertilization (reviewed by Ducibella 1996). A cortical granule protease released into the perivitelline space during the cortical reaction partially hydrolyses ZP2 (120 kDa), the secondary ZP ligand that preferentially binds acrosome-reacted spermatozoa to a 90-kDa form known as $ZP2_f$ (Bleil et al. 1981). The smaller peptides generated during this process remain covalently linked to the glycoprotein by intramolecular disulphide bonds (Bleil et al. 1981). ZP3 is also altered so that sperm are no longer able to bind or penetrate the ZP, thus establishing a barrier to supernumerary spermatozoa (Wassarman et al. 1985). Although the physical and chemical characteristics of ZP3 and $ZP3_f$ are virtually indistinguishable, $ZP3_f$ represents an inactive form of ZP3 perhaps due to modification of terminal carbohydrate residues (Aviles et al. 1997).

11.2 Monotremes

Shortly after fertilization, during passage through the oviduct, monotreme oocytes become surrounded by two additional coats. The mucoid coat and oocyte shell of monotremes are tertiary oocyte membranes, typical of the cleidoic oocytes of sauropsid vertebrates. The thickness of the mucoid coat of the echidna oocyte varies from 17 to 30 μm (Flynn 1930) and is composed of sulphated acidic glycoprotein derived from membrane-bound precursors within the cytoplasm of non-ciliated oviduct epithelial cells (reviewed by Hughes and Carrick 1978). There is no information on the functional role of the mucoid layer in monotremes; however, sperm have been observed trapped in the mucoid (Griffiths 1968) suggesting that it may play a role in the block to polyspermy. Cortical granules have not been observed in monotreme oocytes.

11.3 Marsupials

11.3.1 Cortical Granules

The cortical cytoplasm of marsupial oocytes contains spherical membrane-bound organelles with similar morphology to eutherian cortical granules (CG) (Breed and Leigh 1992; Mate et al. 1992). Their size of 0.2–0.4 μm diameter is comparable to pig and rabbit CG (Oh and Brackett 1975; Cran and Cheng 1985) but slightly smaller than most eutherian CG which generally range from 0.2 to 0.8 μm in diameter (Sathananthan and Trounson 1982; Cherr et al. 1988; Ducibella et al. 1988).

Formation of CG in the marsupial species studied thus far is closely associated with the Golgi apparatus and smooth endoplasmic reticulum (SER) (Mate et al. 1992), similar to that described in long-cycling domestic species (Cran and Cheng 1985; Hyttel et al. 1986). The timing of CG formation in some marsupial species represents a significant departure from the timing of CG formation in eutherian mammals. An apparently full complement of mature peripherally positioned CG is only observed in ovulated oocytes which have undergone the first meiotic division in the brushtail possum, tammar wallaby and grey short-tailed opossum (Falconnier and Kress 1992; Mate et al. 1992). In contrast, preovulatory oocytes from *Didelphis virginiana* contain a large number of CG indicating that the timing of their formation differs among different marsupial species. Cortical granules have also been observed in germinal vesicle stage oocytes from *Sminthopsis crassicaudata* (Breed and Leigh 1990, 1992; Breed 1994a), the brush-tailed bettong and northern short-nosed bandicoot (Breed 1994b), although there was no indication of their number or distribution.

11.3.2 Cortical Reaction

The efficiency of the cortical granule-induced block to polyspermy in marsupials has been questioned because of the large hole that the penetrating spermatozoa leaves in the zona in the opossum (Rodger and Bedford 1982); the occasional observation of polyspermy in *Antechinus* (Selwood 1982) and *Sminthopsis* (Breed and Leigh 1990) *in vivo*; and multiple sperm penetration of *Didelphis* and *Monodelphis* oocytes during *in vitro* culture (Rodger and Bedford 1982; Moore and Taggart 1993). The block to polyspermy in both *Monodelphis* and *Didelphis* appears to be efficient *in vivo*, with all supernumerary sperm restricted to the surface of the ZP or within the mucoid layer, secreted by the oviduct. Mucoid deposition prevents sperm from making contact with the ZP and may form the most important block to polyspermy *in vivo* as it is closely correlated temporally with the early stages of gamete interaction and penetration (Rodger and Bedford 1982; Baggott et al. 1987; Breed and Leigh 1990). Certainly, the mucoid layer traps a large number of sperm, which are clearly visible under phase contrast.

There is evidence, however, that the cortical reaction does play a role in the block to polyspermy in marsupials. The fusion of the membrane of cortical granules with the oolemma and exocytosis of the cortical granule contents have been described in *Sminthopsis* (Breed 1994b). Also, fertilized oviductal oocytes from the laboratory opossum lack cortical granules and exhibit a new coat within the perivitelline space between the ZP and the tips of the oocyte microvilli (Dandekar et al. 1995). This new coat, termed the cortical granule envelope, is more electron dense than the ZP and causes a significant increase in the width of the perivitelline space. A similar matrix has been described in the perivitelline space of fertilized oocytes from naturally mated females of the fat-tailed dunnart (Breed and Leigh 1992) as well as in eutherian species such as the human, mouse and hamster (Talbot and DiCarlantonio 1984; Dandekar and Talbot 1992) and is thought to arise, at least in part, from the cortical granule exudate. The function(s) of the cortical granule envelope is not known but may include a block to polyspermy at the level of the perivitelline space or oolemma.

11.4 Common Themes and General Overview

It remains uncertain whether the cortical reaction functions as an efficient block to polyspermy in all mammals, or whether it has evolved eutherian mammals concomitant with the disappearance of the mucoid layer. The study of cortical granule formation and function in marsupial oocytes remains limited as they cannot be visualized at the light microscope level using the lectin *Lens culinaris*, used to specifically label cortical granules in eutherian species (e.g. Ducibella et al. 1988).

12
Decondensation of Sperm Nucleus

12.1
Eutherians

The nuclear envelope rapidly disintegrates when the sperm nucleus is incorporated into the cytoplasm of eutherian oocytes, regardless of their maturational status (Usui et al. 1997). Factors within the oocyte cytoplasm then begin decondensation of the disulphide-stabilized "sperm-specific" protamines in the sperm nucleus. The reduced form of glutathione (GSH) has been proposed as one of the factors in the oocyte cytoplasm involved in decondensation by mediating the reduction of disulphide bonds (Calvin et al. 1986). Following reduction of SS bonds, sperm protamines are replaced by oocyte histones and other proteins allowing packaging of the sperm DNA into nucleosomal chromatin. The removal of sperm protamines may be mediated by high-affinity binding of the reduced protamines for a cytoplasmic nucleoplasmin-like protein (reviewed by Yanagimachi 1994).

Nucleoplasmin is released from the GV into the ooplasm of frog oocytes at the time of GV breakdown; it then must undergo massive hyperphosphorylation before it can replace sperm protamines with histones and enable chromatin decondensation (Leno et al. 1996). Frog nucleoplasmin can also bind human protamines and decondense DTT-treated human spermatozoa in vitro (Ito et al. 1993). A nucleoplasmin-like protein may also be involved in decondensation of eutherian spermatozoa, and its release into the ooplasm during GV breakdown, and a requirement for hyperphosphorylation, dependent upon oocyte maturation, would correlate well with the acquisition of the oocytes' ability to decondense sperm nuclei (McLay and Clarke 1997).

12.2
Monotremes

There is no information on the process of decondensation of the sperm nucleus in monotremes.

12.3
Marsupials

Sperm incorporation in marsupials seems to follow the conventional mammalian pattern. The sperm nucleus begins decondensation rapidly after entering the oocyte cytoplasm in *Monodelphis*, *Didelphis* and the brushtail possum (Rodger and Bedford 1982; Taggart et al. 1993; Jungnickel et al. 1999); however, in *Sminthopsis* the sperm head appears to travel a considerable distance within the oocyte cytoplasm before decondensation is initiated (Breed and Leigh 1988, 1992). The longer time taken for decondensation in *Sminthopsis* may be related to the large size of the sperm tail (approx. 100 μm), which is visible by phase contrast microscopy in the cytoplasm of recently fertilized oocytes (Rodger 1991). Rapid decondensation of the marsupial sperm nucleus is not surprising considering its lack of disulphide stabilization. The protamines of all marsupials were believed to lack cysteine and thus disulphide stabilization (Calvin and Bedford 1971; Balhorn et al. 1988; Fifis et al. 1990). Recently it has been demonstrated that in one genus of Australian dasyurid marsupial – *Planigale* – but not other dasyurids, sperm protamines contain cysteine and presumably the sperm heads are disulphide stabilized (Retief et al. 1995). Fertilization events in planigales have not been studied.

In *Sminthopsis*, expansion and decondensation of chromatin began in the ventral region of the sperm head, whilst chromatin along the dorsal region of the nucleus remained condensed for a longer period (Breed 1996). Membranous material was found associated with the decondensing sperm head in recently fertilized *Monodelphis* and *Sminthopsis* oocytes (Breed and Leigh 1988; Taggart et al. 1993) but not *Didelphis* oocytes (Rodger and Bedford 1982). As discussed earlier, the precise manner of sperm incorporation by the oocyte awaits direct evidence.

12.4 Common Themes and Overview

In both eutherians and marsupials (there is no information for monotremes), the sperm head begins to decondense rapidly on incorporation into the oocyte cytoplasm. In some cases this is extremely rapid and is quite advanced before the sperm tail is fully incorporated. Once decondensed the protamines of the sperm nucleus are transformed into the histones of somatic cell nuclei. An intriguing but as yet poorly understood difference between the sperm of apparently all but one genus of marsupials and all eutherians is the absence of cysteine in the protamines, the result being a lack of disulphide stabilization of the marsupial sperm head, which is far more fragile than those of eutherians and more similar to monotremes and non-mammalian vertebrates. Despite this in vitro fragility there is no evidence that the sperm head is particularly fragile in vivo because the heads of sperm injected into immature tammar wallaby GV oocytes remain condensed for many hours in the oocyte cytoplasm (K.E. Mate and G. Magarey, unpubl. observ.).

13 Completion of Meiosis, Pronuclei Development and Syngamy

13.1 Eutherians

13.1.1 Completion of Oocyte Meiosis

Meiotic maturation in eutherian oocytes involves the activation and inactivation of maturation-promoting factor (MPF) which is a serine/threonine protein kinase composed of a catalytic subunit, $p34^{cdc2}$, and a regulatory subunit, cyclin B. In order to be activated, MPF requires dephosphorylation of Thr14 and Tyr15 and phosphorylation of Thr161. The activity of MPF is low in the GV-intact oocyte, increases to a peak level by MI, transiently decreases following exit from MI, and then increases and remains elevated during MII (Choi et al. 1991). Activation of MPF induces chromosomal condensation and GV breakdown during the first meiotic division (Motlik and Kubelka 1990). The arrest of oocytes at MII of the second meiotic division is thought to be due to maintenance of elevated MPF levels by a cytostatic factor (CSF or p39mos). The rise in intracellular Ca^{2+} following sperm–oocyte fusion leads to inactivation of MPF due to the destruction of the cyclin component, which correlates with the resumption of meiosis and the emission of the second polar body (Moos et al. 1996).

Mitogen-activated protein kinase (MAP kinase) is another protein kinase involved in signalling pathways regulating the cell cycle of the oocyte (Moos et al. 1995; Sun et al. 1999). MAP kinase activation increases subsequent to that of MPF, and remains elevated during the transition from MI to MII and in the MII-arrested oocyte. Following fertilization there is a time-dependent decline in the activity of MAP kinase that is temporally correlated with the formation of the male and female pronuclear envelopes (Moos et al. 1995).

13.1.2 Pronuclei Development and Syngamy

A new nuclear envelope forms around the decondensed sperm head when the oocyte reaches telophase II. In contrast to disassembly of the sperm nuclear membrane which is independent of the cell cycle stage of the oocyte, the ability of the oocyte cytoplasm to assemble this new envelope is restricted to telophase of meiosis, correlated with the absence of active MAP kinase (Moos et al. 1996) and MPF (Usui et al. 1997) in the ooplasm. After incorporation, the structures of the sperm tail connecting piece are disassembled, exposing the single sperm centriole to the zygotic cytoplasm. Following accumulation of maternal centrosomal components around the sperm, cytoplasm-derived components around the sperm centriolar complex form the microtubule-based sperm aster, a structure responsible for union of the male and female pronuclei (Schatten 1994). With the exception of rodents, this scheme of centrosome assembly seems valid for all eutherian mammals. Concomitantly, the major structure of the sperm principal piece, the fibrous sheath, disappears. Clusters of sperm-derived mitochondria can be detected in early stage four-cell cow embryos (Sutovsky et al. 1996) but not by the beginning of the third mitotic cycle.

Integrated assembly of microtubules and microfilaments is required for migration of the male and female pronuclei towards the centre of the oocyte where they become closely apposed in preparation for the first mitotic division (Schatten et al. 1985; Kim et al. 1997; Palermo et al. 1997). Microtubules are primarily associated with the arrested meiotic spindle in the unfertilized oocyte. In most animals, the microtubules of the sperm aster are responsible for moving the male and female pronuclei from the oocyte cortex to the centre of the oocyte (Le Guen et al. 1989; Long et al. 1993). However, in the mouse numerous cytoplasmic asters assemble after incorporation of the sperm, and begin to associate with the developing pronuclei (Schatten et al. 1985). As the pronuclei migrate to the centre of the oocyte, these asters develop into a dense array, which is subsequently replaced by sheaths of microtubules surrounding the adjacent pronuclei, which give rise to the mitotic apparatus. Pronuclear fusion is not observed, but rather the envelopes of the sperm and oocyte pronuclei disintegrate and the maternal and paternal chromosome sets meet at metaphase.

13.2
Completion of Fertilization in Monotremes

The second polar body is extruded from the monotreme oocyte in response to fertilization. The male pronucleus comes to lie in the germinal disc approximately 1 mm away from the female pronucleus and the two polar bodies migrate to the margin of the disc (Griffiths 1968). The two pronuclei migrate towards each other and become closely associated, forming an almost spherical body divided at the equator by the limiting membrane. Shortly after, it is not possible to detect the membrane between the two pronuclei (Griffiths 1968).

13.3
Completion of Fertilization in Marsupials

Following sperm incorporation in both *Monodelphis* and *Sminthopsis*, a microtubule-based sperm aster is formed, indicating that the centrosome is paternally inherited in marsupials (Breed et al. 1994; Merry et al. 1995). The sperm aster is restricted to the non-yolky region of the oocyte cytoplasm, resulting in a cytoplasmic polarization consisting of a microtubule-rich region and an eccentrically located microtubule-free yolk mass (Breed et al. 1994; Merry et al. 1995). Once formed, the male and female pronuclei move toward each other and lie in close apposition (Breed and Leigh 1990), surrounded by abundant microtubules (Breed et al. 1994). The manner of syngamy has not been directly observed for any marsupial.

13.4
Common Themes and Overview

The cytoskeleton plays an integral role during the latter stages of fertilization in all mammals; the assembly of microtubules and microfilaments is vital for pronuclear formation and migration, as well as cleavage after fertilization. In those marsupials examined, as in most eutherian mammals, it is the microtubules of the sperm aster that are responsible for moving the male and female pronuclei from the inner face of the oocyte cortex in the oocyte cytoplasm. Thus, it appears that the observation of numerous cytoplasmic microtubule foci in the cytoplasm of fertilized mouse oocytes is atypical of all other mammals.

14
Concluding Remarks

We hope that it is now clear that despite many common themes running through fertilization in mammals, there are major differences between the three groups – monotremes, marsupials and eutherians. The greatest differences are between monotremes and the other two groups based on the character of the oocyte and sperm; however, our knowledge of monotreme fertilization is so limited that it is probably unwise to speculate more. Eutherian and marsupial fertilization clearly have much in common. Indeed, if there are two distinct AR and sperm–oocyte fusion systems in the American and Australian marsupials the major comparison perhaps is between marsupials rather than between marsupials and eutherians. Major issues that require closer study are the process of capacitation in marsupials, regulation of the AR, the manner of sperm–oocyte fusion and sperm incorporation. A better understanding of these dynamic aspects of marsupial fertilization and their molecular biology, together with the large body of knowledge of eutherian fertilization, will no doubt allow the development of a broad understanding of and perhaps a unified conceptual analysis of mammalian fertilization. The place of monotremes in this scenario is intriguing but awaits a new generation of intrepid gamete biologists.

Acknowledgements. The authors thank Janine Buist for assistance with the figures and Prof. Mike O'Rand and Prof. Harry Moore for comments on the manuscript.

References

Aitken RJ, Harkiss D, Knox M, Paterson M, Irvine DS (1998) A novel signal transduction cascade in capacitating human spermatozoa characterised by a redox-regulated, cAMP-mediated induction of tyrosine phosphorylation. J Cell Sci 111:645–656

Aviles M, Jaber L, Castells MT, Ballesta J, Kan FWK (1997) Modifications of carbohydrate residues and ZP2 and ZP3 glycoproteins in the mouse zona pellucida after fertilization. Biol Reprod 57:1155–1163

Baggott LM, Moore HDM (1990) Early embryonic development of the grey short-tailed opossum, *Monodelphis domestica*, in vivo and in vitro. J Zool (Lond) 222:623–629

Baggott LM, Davis-Butler S, Moore HDM (1987) Characterization of oestrus and timed collection of oocytes in grey short-tailed opossum, *Monodelphis domestica.* J Reprod Fertil 79:105–114

Balhorn R, Mazrimas JA, Corzett M, Cummins J, Fadem B (1988) Analysis of protamines isolated from two marsupials, the ringtailed wallaby and gray short-tailed opossum. J Cell Biol 107:167

Bedford JM (1979) Evolution of the sperm maturation and sperm storage function of the epididymis. In: Fawcett DW, Bedford JM (eds) The spermatozoon. Urban and Schwarzenberg, Munich, pp 7–21

Bedford JM (1982) Fertilization. In: Austin CR, Short RV (eds) Reproduction in mammals: 1. Germ cells and fertilization. Cambridge University Press, Cambridge, pp 128–163

Bedford JM (1991) The coevolution of mammalian gametes. In: Dunbar BS, O'Rand MG (eds) A comparative overview of mammalian fertilization. Plenum Press, New York, pp 3–35

Bedford JM (1996) What marsupial gametes disclose about gamete function in eutherian mammals. Reprod Fertil Dev 8:569–580

Bedford JM (1999) Mammalian fertilization misread? Sperm penetration of the eutherian zona pellucida is unlikely to be a lytic event. Biol Reprod 59:1275–1287

Bedford JM, Breed WG (1994) Regulated storage and subsequent transformation of spermatozoa in the fallopian tube of Australian marsupial, *Sminthopsis crassicaudata.* Biol Reprod 50:845–54

Bedford JM, Nicander L (1971) Ultrastructural changes in the acrosome and sperm membranes during maturation of spermatozoa in the testis and epididymis of the rabbit and monkey. J Anat 108:527–544

Bedford JM, Rifkin JM (1979) An evolutionary view of the male reproductive tract and sperm maturation in a monotreme mammal – the echidna, *Tachyglossus aculeatus.* Am J Anat 156:207–230

Ben-Yosef D, Shalgi R (1998) Early ionic events in activation of the mammalian egg. Rev Reprod 3:96–103

Ben-Yosef D, Oron Y, Shalgi R (1996) Intracellular pH of rat eggs is not affected by fertilization and the resulting calcium oscillations. Biol Reprod 55:461–468

Bleil JD, Wassarman PM (1980) Mammalian sperm–egg interaction: identification of a glycoprotein I mouse egg zonae pellucidae possessing receptor activity for sperm. Cell 20:873–882

Bleil JD, Wassarman PM (1988) Galactose at the nonreducing terminus of O-linked oligosaccharides of mouse egg zona pellucida glycoprotein ZP3 is essential for the glycoprotein sperm receptor activity. Proc Natl Acad Sci USA 85:6778–6782

Bleil JD, Beall CF, Wassarman PM (1981) Mammalian sperm–egg interaction: fertilization of mouse eggs triggers modification of the major zona pellucida glycoprotein, ZP2. Dev Biol 89:189–197

Bleil JD, Greve JM, Wassarman PM (1988) Identification of a secondary sperm receptor in the mouse egg zona pellucida: role in maintenance of binding of acrosome-reacted sperm to eggs. Dev Biol 128:376–385

Breed WG (1994a) How does sperm meet egg? – In a marsupial. Reprod Fertil Dev 6:485–506

Breed WG (1994b) Sperm–egg interactions in an Australian dasyurid marsupial with special reference to changes in acrosomal morphology. Zygote 2:201–211

Breed WG (1996) Egg maturation and fertilization in marsupials. Reprod Fertil Dev 8:617–643

Breed WG, Leigh CM (1988) Morphological observations on sperm-egg interactions during in vivo fertilization in dasyurid marsupial *Sminthopsis crassicaudata.* Gamete Res 19:131–149

Breed WG, Leigh CM (1990) Morphological changes in the oocyte and its surrounding vestments during in vivo fertilization in the dasyurid marsupial, *Sminthopsis crassicaudata.* J Morphol 204:177–196

Breed WG, Leigh CM (1992) Marsupial fertilization: some further ultrastructural observations on the dasyurid, *Sminthopsis crassicaudata.* Mol Reprod Dev 32:277–292

Breed WG, Simerly C, Navara CS, Vandeberg JL, Schatten G (1994) Microtubule configurations in oocytes, zygotes and early embryos of a marsupial, *Monodelphis domestica.* Dev Biol 164:230–240

Breitbart H, Spungin B (1997) The biochemistry of the acrosome reaction. Mol Hum Reprod 3:195–202

Calvin HI, Bedford JM (1971) Formation of disulphide bonds in the nucleus and accessory structures of mammalian spermatozoa during maturation in the epididymis. J Reprod Fertil (Suppl) 13:65–75

Calvin HI, Grosshans K, Blake EJ (1986) Estimation and manipulation of glutathione levels in prepubertal mouse ovaries and ova: relevance to sperm nucleus transformation in the fertilized egg. Gamete Res 14:265–275

Carrera A, Moos J, Ning XP, Gerton GL, Terarik J, Kopf GS, Moss SB (1996) Regulation of protein tyrosine phosphorylation in human sperm by a calcium/calmodulin-dependent mechanism: identification of A kinase anchor proteins as major substrates for tyrosine phosphorylation. Dev Biol 180:284–296

Carrick FN, Hughes RL (1982) Aspects of the structure and development of monotreme spermatozoa and their relevance to the evolution of mammalian sperm morphology. Cell Tissue Res 222:127–141

Castle PE, Dean J (1996) Molecular genetics of the zona pellucida: implications for immunocontraceptive strategies. J Reprod Fertil (Suppl) 50:1–8

Chaturapanich G, Jones RC, Clulow J (1992) Protein synthesis and secretion by the epididymis of the tammar wallaby *Macropus eugenii* (Macropodidae: Marsupialia). Reprod Fertil Dev 4:533–545

Cherr GN, Drobnis EA, Katz DF (1988) Localization of cortical granule constituents before and after exocytosis in hamster egg. J Exp Zool 246:81–93

Choi T, Aoki F, Mori M, Yamashita M, Nagahama Y, Kohmoto K (1991) Activation of $p34^{cdc2}$ protein kinase activity in meiotic and mitotic cell cycles in mouse oocytes and embryos. Development 113:789–795

Cooper GW, Overstreet JW, Katz DF (1979) The motility of rabbit spermatozoa recovered from the female reproductive tract. Gamete Res 2:35–42

Cooper NJ, Holland MK, Breed WG (1998) Extratesticular sperm maturation in the brush-tail possum, *Trichosurus vulpecula.* J Reprod Fertil (Suppl) 53:221–226

Cooper TG (1992) Epididymal proteins and sperm maturation. In: Nieschlag E, Habenicht UF (eds) Spermatogenesis, fertilization, contraception, molecular, cellular and endocrine events in male reproduction. Springer, Berlin Heidelberg New York, pp 285–318

Cran DG (1989) Cortical granules during oocyte maturation and fertilization. J Reprod Fertil (Suppl) 38:49–62

Cran DG, Cheng WTK (1985) Changes in cortical granules during porcine oocyte maturation. Gamete Res 11:311–319

Cran DG, Moor RM (1990) Programming the oocyte for fertilization. In: Bavister BD, Cummins J, Roldan ERS (eds) Fertilization in mammals. Serono Symposia USA, Norwell, Massachusetts, pp 35–47

Cross NL (1998) Role of cholesterol in sperm capacitation. Biol Reprod 59:7–11

Cummins JM (1980) Decondensation of sperm nuclei of Australian marsupials: effects of air drying and of calcium and magnesium. Gamete Res 3:351–367

Dandekar P, Talbot P (1992) Perivitelline space of mammalian oocytes: extracellular matrix of unfertilized oocytes and formation of a cortical granule envelope following fertilization. Mol Reprod Dev 31:135–143

Dandekar P, Mate KE, Talbot P (1995) Perivitelline space of marsupial oocytes – extracellular matrix of the unfertilized oocyte and the formation of a cortical granule envelope following the cortical reaction. Mol Reprod Dev 41:368–373

Demott RP, Suarez SS (1992) Hyperactivated sperm progress in the mouse oviduct. Biol Reprod 46:779–785

Djakiew D, Jones JC (1983) Sperm maturation, fluid transport, and secretion and absorption of protein in the epididymis of the echidna, *Tachyglossus aculeatus*. J Reprod Fertil 68:445–456

Ducibella T (1996) The cortical reaction and development of activation competence in mammalian oocytes. Hum Reprod Update 2:29–42

Ducibella T, Buetow J (1994) Competence to undergo normal, fertilization-induced cortical activation develops after metaphase I of meiosis in mouse oocytes. Dev Biol 165:95–104

Ducibella T, Anderson E, Albertini DF, Aalberg J, Ranjgarajan S (1988) The development of mouse oocyte cortical reaction competence is accompanied by major changes in cortical vesicles and not cortical granule depth. Dev Biol 130:789–792

Ducibella T, Duffy P, Reindollar R, Su B (1990) Changes in the distribution of mouse oocyte cortical granules and ability to undergo the cortical reaction during gonadotrophin-stimulated meiotic maturation and aging in vivo. Biol Reprod 43:870–876

Dudenhausen E, Talbot P (1982) Detection and kinetic of the normal acrosome reaction of mouse sperm. Gamete Res 6:257–265

Dunbar BS, O'Rand MG (eds) (1991) A comparative overview of mammalian fertilization. Plenum Press, New York

Dunbar BS, Prasad SV, Timmons TM (1991) Comparative structure and function of mammalian zonae pellucidae. In: Dunbar BS, O'Rand MG (eds) A comparative overview of mammalian fertilization. Plenum Press, New York, pp 97–114

Eng LA, Oliphant G (1978) Rabbit sperm reversible decapacitation stabilization with a highly purified glycoprotein from seminal plasma. Biol Reprod 19:1083–1094

Epifano O, Dean J (1994) Biology and structure of the zona pellucida: a target for immunocontraception. Reprod Fertil Dev 6:319–330

Epifano O, Liang L-F, Familari M, Moos MC, Dean J (1995) Coordinate expression of the three zona pellucida genes during mouse oogenesis. Development 121:1947–1956

Esaguy N, Welch JE, O'Rand MG (1988) Ultrastructural mapping of a sperm plasma membrane autoantigen before and after the acrosome reaction. Gamete Res 19:387–399

Falconnier C, Kress A (1992) Ultrastructural aspects of oocyte growth in the marsupial *Monodelphis domestica* (grey short tailed opossum). J Anat 181:481–498

Fawcett DW (1975) The mammalian spermatozoon. Dev Biol 44:394–436

Fifis T, Cooper DW, Hill RJ (1990) Characterization of the protamines of the tammar wallaby (*Macropus eugenii*). Comp Biochem Physiol 95B:571–575

Flynn TT (1930) On the unsegmented ovum of echidna (*Tachyglossus*). Q J Microsc Sci 74:119–131

Flynn TT, Hill JP (1939) The development of the Monotremata. IV. Growth of the ovarian ovum, fertilization and early cleavage. Trans Zool Soc Lond 24:445–622

Gao Z, Garbers DL (1998) Species diversity in the structure of zonadhesin, a sperm-specific membrane protein containing multiple cell adhesion molecule-like domains. J Biol Chem 273:3415–3421

Grant TR (1989) 16. Ornithorhynchidae. In: Walton DW, Richardson BJ (eds) Fauna of Australia, vol 1B: Mammalia. Australian Government Publishing Service, Canberra, pp 436–450

Green DPL (1997) Three-dimensional structure of the zona pellucida. Rev Reprod 2:147–156

Greve JM, Wassarman PM (1985) Mouse egg extracellular coat is a matrix of interconnected filaments possessing a structural repeat. J Mol Biol 181:253–264

Greve JM, Salzmann GS, Roller RJ, Wassarman PM (1982) Biosynthesis of the major zona pellucida glycoproteins secreted by oocytes during mammalian oogenesis. Cell 31:749–759

Griffiths M (1968) Echidnas. Pergamon Press, Oxford

Griffiths M (1978) The biology of the monotremes. Academic Press, New York

Griffiths M (1989) 15. Tachyglossidae. In: Walton DW, Richardson BJ (eds) Fauna of Australia, vol 1B: Mammalia. Australian Government Publishing Service, Canberra, pp 407–435

Grootenhuis AJ, Philipsen HLA, de Breet-Grijsbach JTM, van Duin M (1996) Immunocytochemical localisation of ZP3 in primordial follicles of rabbit, marmoset, rhesus monkey, and human ovaries using antibodies to human ZP3. J Reprod Fertil (Suppl) 50:43–54

Gulyas BJ (1980) Cortical granules of mammalian eggs. Int Rev Cytol 63:357–392

Haines BP, Rathjen PD, Hope RM, Whyatt LM, Holland MK, Breed WG (1999) Isolation and characterisation of a zona pellucida protein (ZPB) from the marsupial *Trichosurus vulpecula* (brushtail possum). Mol Reprod Dev 52:174–182

Harding HR, Carrick FN, Shorey CD (1979) Special features of sperm structure and function in marsupials. In: Fawcett DW, Bedford JM (eds) The spermatozoon. Maturation, motility, surface properties and comparative aspects. Urban and Schwarzenberg, Baltimore, pp 289–303

Harris M, Rodger JC (1998) Characterisation of fibrous sheath and midpiece fibre network polypeptides of marsupial spermatozoa with a monoclonal antibody. Mol Reprod Dev 50:461–473

Hughes RL (1977) Egg membranes and ovarian function during pregnancy in monotremes and marsupials. In: Calaby JH, Tyndale-Biscoe CH (eds) Reproduction and evolution. Australian Academy of Science, Canberra, pp 281–291

Hughes RL, Carrick FN (1978) Reproduction in female monotremes. In: Augee ML (ed) Monotreme biology. Royal Zoological Society New South Wales, Taronga Zoo, Mosman, pp 233–253

Hunnicutt GR, Primakoff P, Myles DG (1996) Sperm surface protein PH-20 is bifunctional – one activity is a hyaluronidase and a second distinct activity is required in secondary sperm–zona binding. Biol Reprod 55:80–86

Hyttel P, Callesen H, Greve T (1986) Ultrastructural features of preovulatory oocyte maturation in superovulated cattle. J Reprod Fertil 76:645–656

Ito T, Ohsumi K, Katagiri C (1993) Remodeling of human sperm chromatin mediated by nucleoplasmin from amphibian eggs. Dev Growth Differ 35:59–66

Janke A, Xu X, Arnason U (1997) The complete mitochondrial genome of the wallaroo (*Macropus robustus*) and the phylogenetic relationship among Monotremata, Marsupialia and Eutheria. Proc Nat Acad Sci USA 94:1276–1281

Johnston DS, Wright WW, Shaper JH, Hokke CH, Van den Eijnden DH, Joziasse DH (1998) Murine sperm–zona binding, a fucosyl residue is required for a high affinity sperm-binding ligand. A second site on sperm binds a nonfucosylated, beta-galactosyl-capped oligosaccharide. J Biol Chem 274:1888–1895

Jones R (1989) Membrane remodelling during sperm maturation in the epididymis. Oxf Rev Reprod Biol 11:285–337

Jones R (1998a) Plasma membrane structure and remodelling during sperm maturation in the epididymis. J Reprod Fertil (Suppl) 53:73–84

Jones RC (1998b) Evolution of the vertebrate epididymis. J Reprod Fertil (Suppl) 53:163–181

Jones R, Brown CR (1987) Identification of a zona-binding protein from boar spermatozoa as proacrosin. Exp Cell Res 171:503–508

Jungnickel MK, Harman AJ, Rodger JC (1999) Ultrastructural observations on in vivo fertilisation in the brushtail possum, *Trichosurus vulpecula*, following PMSG/LH superovulation and artificial insemination. Zygote 7:307–320

Kim N-H, Chung K-S, Day BN (1997) The distribution and requirements of microtubules and microfilaments during fertilization and parthenogenesis in pig oocytes. J Reprod Fertil 111:143–149

Kirchhoff C, Hale G (1996) Cell to cell transfer of glycosylphosphatidylinositol-anchored membrane proteins during sperm maturation. Mol Hum Reprod 2:177–184

Kirsch JAW, Mayer GC (1998) The platypus is not a rodent: DNA hybridization, amniote phylogeny and the palimpsest theory. Philos Trans R Soc Lond B 353:1221–1237

Kirsch JAW, Lapointe F-J, Springer MS (1997) DNA-hybridisation studies of marsupials and their implications for metatherian classification. Aust J Zool 45:211–280

Kline D, Zagray JA (1995) Absence of an intracellular pH change following fertilization of the mouse egg. Zygote 3:305–311

Krutzsch PH, Crichton EG (1991) Fertilization in bats. In: Dunbar BS, O'Rand MG (eds) A comparative overview of mammalian fertilization. Plenum Press, New York, pp 137–149

Le Guen PL, Crozet N, Huneau D, Gall L (1989) Distribution and role of microfilaments during early events of sheep fertilization. Gamete Res 22:411–425

Leno GH, Mills AD, Philpott A, Laskey RA (1996) Hyperphosphorylation of nucleoplasmin facilitates *Xenopus* sperm decondensation at fertilization. J Biol Chem 271:7253–7256

Leyton L, Saling P (1989) Evidence that aggregation of mouse sperm receptors by ZP3 triggers the acrosome reaction. J Cell Biol 108:2163–2168

Liang L-F, Dean J (1993) Oocyte development: molecular biology of the zona pellucida. Vitamins Hormones 47:115–159

Lin M, Rodger JC (1999) Acrosome formation during sperm transit through the epididymis in two marsupials, the tammar wallaby (*Macropus eugenii*) and the brushtail possum (*Trichosurus vulpecula*). J Anat 194(2):223–232

Lin M, Harman A, Rodger JC (1997) Spermiogenesis and spermiation in a marsupial, the tammar wallaby (*Macropus eugenii*). J Anat 190:377–395

Litscher ES, Wassarman PM (1996) Characterization of a mouse ZP3-derived glycopeptide, gp55, that exhibits sperm receptor and acrosome-reaction-inducing activity in vitro. Biochemistry 35:3980–3985

Long CR, Pinto-Correia C, Duby RT, Ponce de Leon FA, Boland MP, Roche JF, Robl JM (1993) Chromatin and microtubule morphology during the first cell cycle in bovine zygotes. Mol Reprod Dev 36:23–32

Mahi-Brown CA (1991) Fertilization in dogs. In: Dunbar BS, O'Rand MG (eds) A comparative overview of mammalian fertilization. Plenum Press, New York, pp 281–297

Martinez SP, Conesa D, Cuasnicu PS (1995) Potential contraceptive use of epididymal proteins: evidence for the participation of specific antibodies against rat epididymal protein DE in male and female fertility inhibition. J Reprod Immunol 29:31–45

Mate KE (1996) Cytoplasmic maturation of the marsupial oocyte during the periovulatory period. Reprod Fertil Dev 8:509–519

Mate KE (1998) A timetable of zona pellucida formation in the tammar wallaby *(Macropus eugenii)* and brushtail possum *(Trichosurus vulpecula)*. Anim Reprod Sci 53:237–252

Mate KE, McCartney CA (1998) Sequence and analysis of the zona pellucida 2 cDNA (*ZP2*), from a marsupial, the brushtail possum, *Trichosurus vulpecula*. Mol Reprod Dev 51:322–329

Mate KE, Rodger JC (1991) Stability of the acrosome of the brush-tailed possum (*Trichosurus vulpecula*) and tammar wallaby (*Macropus eugenii*) in vitro and after exposure to conditions and agents known to cause capacitation or acrosome reaction of eutherian spermatozoa. J Reprod Fertil 91:41–48

Mate KE, Rodger JC (1993) Role of diacylglycerols and calcium in the marsupial acrosome reaction. J Reprod Fertil 99:367–373

Mate KE, Rodger JC (1996) Capacitation and the acrosome reaction in marsupial spermatozoa. Reprod Fertil Dev 8:595–603

Mate KM, Giles I, Rodger JC (1992) Evidence that cortical granule formation is a periovulatory event in marsupials. J Reprod Fertil 95:719–728

Mate KE, Kosower NS, White IG, Rodger JC (1994) Fluorescent localisation of thiols and disulphides in marsupial spermatozoa by bromobimane labelling. Mol Reprod Dev 37:318–325

Mate KE, Sidhu KS, Molinia FC, Glazier AM, Rodger JC (2000) Sperm binding and penetration of the zona pellucida but not sperm–egg fusion in an Australian marsupial, the brushtail possum (*Trichosurus vulpecula*). Zygote (in press)

McCartney CA, Mate KE (1999) Cloning and characterisation of a zona pellucida 3 cDNA from a marsupial, the brushtail possum *Trichosurus vulpecula*. Zygote 7(1):1–9

McLay DW, Clarke HJ (1997) The ability to organise sperm DNA into functional chromatin is acquired during meiotic maturation in murine oocytes. Dev Biol 186(1):73–84

McLeskey SB, Dowds C, Carballada R, White RR, Saling PM (1998) Molecules involved in mammalian sperm–egg interaction. Int Rev Cytol 177:57–113

Meizel S (1984) The importance of hydrolytic enzymes to an exocytotic event, the mammalian sperm acrosome reaction. Biol Rev 59:125–157

Merry NE, Johnson MH, Gehring CA, Selwood L (1995) Cytoskeletal organization in the oocyte, zygote, and early cleaving embryo of the stripe-faced dunnart (*Sminthopsis macroura*). Mol Reprod Dev 41:212–214

Miller DJ, Gong X, Decker G, Shur BD (1993) Egg cortical granule N-acetylglucosaminidase is required for the mouse zona block to polyspermy. J Cell Biol 123:1431–1440

Miyazaki S (1995) Calcium signalling during mammalian fertilization. Ciba Found Symp 188:235–247

Miyazaki S-I, Yuzaki M, Nakada K, Shirakawa H, Nakanishi S, Nakade S, Mikoshiba K (1992) Block of Ca^{2+} wave and Ca^{2+} oscillation by antibody to the inositol 1,4,5-trisphosphate receptor in fertilized hamster eggs. Science 257:251–255

Molinia FC, Gibson RJ, Brown AM, Glazier AM, Rodger JC (1998) Successful fertilization after superovulation and laparoscopic intrauterine insemination of the brushtail possum, *Trichosurus vulpecula* and tammar wallaby, *Macropus eugenii*. J Reprod Fertil 113:9–17

Moore HDM (1995) Sperm maturation and transport in the excurrent ducts. In: Gruzinska J, Yovich J (eds) Gametes: the spermatozoon. Cambridge University Press, Cambridge, pp 140–156

Moore HDM (1996) The influence of the epididymis on human and animal sperm maturation and storage. Hum Reprod 11:103–110

Moore HDM, Bedford JM (1983) The interaction of mammalian gametes in the female. Sperm/egg interactions in vivo. In: Hartmann JF (ed) Mechanisms and control of mammalian fertilization. Academic Press, New York, pp 453–497

Moore HDM, Taggart DA (1993) In vitro fertilization and embryo culture in the grey short-tailed opossum, *Monodelphis domestica*. J Reprod Fertil 98:267–274

Moos J, Visconti PE, Moore GD, Schultz RM, Kopf GS (1995) Potential role of mitogen activated protein kinase in pronuclear envelope assembly and disassembly following fertilization of mouse eggs. Biol Reprod 53:692–699

Moos J, Xu Z, Schultz RM, Kopf GS (1996) Regulation of nuclear envelope assembly/disassembly by MAP kinase. Dev Biol 175:358–361

Motlik J, Kubelka M (1990) Cell-cycle aspects of growth and maturation of mammalian oocytes. Mol Reprod Dev 27:366–375

Myles DG, Primakoff P (1984) Localized surface antigens of guinea pig sperm migrate to new regions prior to fertilization. J Cell Biol 99:1634–1641

Myles DG, Kimmel LH, Blobel CP, White JM, Primakoff P (1994) Identification of a binding site in the disintegrin domain of fertilin required for sperm–egg fusion. Proc Natl Acad Sci USA 91:4195–4198

Oh YK, Brackett BG (1975) Ultrastructure of rabbit ova recovered from ovarian follicles and inseminated in vitro. Fertil Steril 26:665–685

Olson GE (1980) Changes in intramembranous particle distribution in the plasma membrane of *Didelphis virginiana* spermatozoa during maturation in the epididymis. Anat Rec 197:471–488

O'Rand MG, Nikolajczyk BS (1991) Fertilization in the rabbit. In: Dunbar BS, O'Rand MG (eds) A comparative overview of fertilization. Plenum Press, New York, pp 271–279

Palermo GD, Colombero LT, Rosenwaks Z (1997) The human sperm centrosome is responsible for normal syngamy and early embryonic development. Rev Reprod 2:19–27

Parkes JE, Hammerstedt RH (1985) Developmental changes occurring in the lipids of ram epididymal spermatozoa plasma membrane. Biol Reprod 32:653–668

Perry ACF, Wakayama T, Yanagimachi R (1999) A novel trans-complementation assay suggests full mammalian oocyte activation is coordinately initiated by multiple, submembrane sperm components. Biol Reprod 60:747–755

Phelps BM, Primakoff P, Koppel DE, Low MG, Myles DG (1988) Restricted lateral diffusion of PH-20, a PI-anchored sperm membrane protein. Science 240:1780–1782

Phelps BM, Koppel DE, Primakoff P, Myles DG (1990) Evidence that proteolysis of the surface is an initial step in the mechanism of formation of sperm cell surface domains. J Cell Biol 111:1839–1847

Phillips DM (1991) Structure and function of the zona pellucida. In: Familiari G, Makabe S, Motta PM (eds) Ultrastructure of the ovary. Kluwer Academic, The Hague, pp 63–72

Phillips DM, Jones R, Shalgi R (1991) Alterations in distribution of surface and intracellular antigens during epididymal maturation of rat spermatozoa. Mol Reprod Dev 29:347–356

Primakoff P, Hyatt H, Myles DG (1985) A role for the migrating sperm surface antigen PH-20 in guinea pig sperm binding to the egg zona pellucida. J Cell Biol 101:2239–2244

Primakoff P, Hyatt H, Tredick-Kline J (1987) Identification and purification of a sperm surface protein with a potential role in sperm–egg membrane fusion. J Cell Biol 104:1141–1149

Rankin TL, Tong ZB, Castle PE, Lee E, Gore-Langton R, Nelson LM, Dean J (1998) Human ZP3 restores fertility in Zp3 null mice without affecting order-specific sperm binding. Development 125:2415–2424

Raychoudhury SS, Suarez SS (1991) Porcine sperm binding to oviductal explants in culture. Theriogenology 36:1059–1070

Raz T, Ben-Yosef D, Shalgi R (1998a) Segregation of the pathways leading to cortical reaction and cell cycle activation in the rat egg. Biol Reprod 58:94–102

Raz T, Skutelsky E, Amihai D, Hammel I, Shalgi R (1998b) Mechanisms leading to cortical reaction in the mammalian egg. Mol Reprod Dev 51:295–303

Retief JD, Krajewski C, Westerman M, Dixon GH (1995) The evolution of protamine P1 genes in dasyurid marsupials. J Mol Evol 41:549–555

Richardson RT, Yamasaki N, O'Rand MG (1994) Sequence of a rabbit sperm zona pellucida binding protein and localization during the acrosome reaction. Dev Biol 165:688–701

Rochwerger L, Cuasnicu PS (1992) Redistribution of a rat sperm epididymal glycoprotein after in vitro and in vivo capacitation. Mol Reprod Dev 31:34–41

Rochwerger L, Cohen D, Cuasnicu PS (1992) Mammalian sperm–egg fusion: the rat egg has complementary sites for a sperm protein that mediates gamete fusion. Dev Biol 153:83–90

Rodger JC (1991) Fertilization of marsupials. In: Dunbar BS, O'Rand MG (eds) A comparative overview of mammalian fertilization. Plenum Press, New York, pp 117–135

Rodger JC, Bedford JM (1982) Separation of sperm pairs and sperm–egg interaction in the opossum, *Didelphis virginiana*. J Reprod Fertil 64:171–179

Rodger JC, Young RJ (1981) Glycosidase and cumulus dispersal activities of acrosomal extracts from opossum (marsupial) and rabbit (eutherian) spermatozoa. Gamete Res 4:507–514

Saling PM, Sowinski J, Storey BT (1979) An ultrastructural study of epididymal mouse spermatozoa binding to zonae pellucidae in vitro: sequential relationship to acrosome reaction. J Exp Zool 209:229–238

Salzmann GS, Greve JM, Roller RJ, Wassarman PM (1983) Biosynthesis of the sperm receptor during oogenesis in the mouse. EMBO J 2:1451–1456

Sathananthan AH, Trounson AO (1982) Ultrastructural observations on cortical granules in human follicular oocytes cultured in vitro. Gamete Res 5:191–198

Schatten G (1994) The centrosome and its mode of inheritance: the reduction of the centrosome during gametogenesis and its restoration during fertilization. Dev Biol 165:299–335

Schatten G, Schatten H (1987) Fertilization: motility, the cytoskeleton and the nuclear architecture. Oxford Rev Reprod Biol 9:322–378

Schatten G, Simerly C, Schatten H (1985) Microtubule configurations during fertilization, mitosis, and early development in the mouse and the requirement for egg microtubule-mediated motility during mammalian fertilization. Proc Natl Acad Sci USA 82:4152–4156

Selwood L (1982) A review of maturation and fertilization in marsupials with special reference to the dasyurid: *Antechinus stuartii*. In: Archer M (ed) Carnivorous marsupials. Royal Zoological Society New South Wales, Sydney, pp 65–76

Selwood L, McCallum F (1987) Relationship between longevity of spermatozoa after insemination and the percentage of normal embryos in brown marsupial mice *(Antechinus stuartii)*. J Reprod Fertil 79:495–503

Setiadi D, Lin M, Rodger JC (1997) Posttesticular development of spermatozoa of the tammar wallaby (*Macropus eugenii*). J Anat 190:275–288

Shalgi R, Phillips DM, Jones R (1989) Status of the rat acrosome during sperm–zona pellucida interactions. Gamete Res 22:1–13

Shur BD, Hall NG (1982) A role for mouse sperm surface galactosyltransferase in sperm binding to the egg zona pellucida. J Cell Biol 95:574–579

Si Y, Okuno M (1999) Role of tyrosine phosphorylation of flagellar proteins in hamster sperm hyperactivation. Biol Reprod 61:240–246

Sidhu KS, Guraya SS (1989) Cellular and molecular biology of capacitation and acrosome reaction in mammalian spermatozoa. Int Rev Cytol 118:231–280

Sidhu KS, Mate KE, Rodger JC (1998) Sperm–oviduct epithelial cell monoculture: an in vitro model of sperm–female tract interactions in a marsupial, the tammar wallaby (*Macropus eugenii*). J Reprod Fertil 114:55–61

Sidhu KS, Mate KE, Molinia FC, Rodger JC (2000) Induction of thumbtack sperm during co-culture with oviduct epithelial monolayers in a marsupial, the brushtail possum (*Trichosurus vulpecula*). Biol Reprod 61:1356–1361

Sistina Y, Rodger JC (1998) Arachidonic acid-induced acrosomal loss in the spermatozoa of a marsupial, the tammar wallaby (*Macropus eugenii*). Reprod Fertil Dev 9:803–809

Sistina Y, Lin M, Mate KE, Rodger JC (1993) Induction of the marsupial sperm acrosome reaction in vitro by treatment with diacylglycerols. J Reprod Fertil 99:335–341

Smith TT, Yanagimachi R (1990) The viability of hamster spermatozoa stored in the isthmus of the oviduct: the importance of sperm-epithelium contact for sperm survival. Biol Reprod 42:450–457

Smith TT, Yanagimachi R (1991) Attachment and release of spermatozoa from the caudal isthmus of the hamster oviduct. J Reprod Fertil 91:567–573

Stauss CR, Votta TJ, Suarez SS (1995) Sperm motility hyperactivation facilitates penetration of the hamster zona pellucida. Biol Reprod 53:1280–1285

Storey BT (1995) Interactions between gametes leading to fertilization: the sperm's eye view. Reprod Fertil Dev 7:927–942

Suarez SS (1996) Hyperactivated motility in sperm. J Androl 17:331–335

Suarez SS (1998) The oviductal sperm reservoir in mammals: mechanisms of formation. Biol Reprod 58:1105–1107

Sun QY, Rubinstein S, Breitbart H (1999) MAP kinase activity is downregulated by phorbol ester during mouse oocyte maturation and egg activation in vitro. Mol Reprod Dev 52:310–318

Sutovsky P, Navara CS, Schatten G (1996) Fate of the sperm mitochondria, and the incorporation, conversion, and disassembly of the sperm tail structures during bovine fertilization. Biol Reprod 55:1195–1205

Taggart DA (1994) A comparison of sperm and embryo transport in the female reproductive tract of marsupial and eutherian mammals. Reprod Fertil Dev 6:451–472

Taggart DA, Temple-Smith PD (1990) An unusual mode of progressive motility in spermatozoa from the dasyurid marsupial, *Antechinus stuartii*. Reprod Fertil Dev 2:107–114

Taggart DA, Temple-Smith PD (1991) Transport and storage of spermatozoa in the female reproductive tract of the brown marsupial mouse, *Antechinus stuartii* (Dasyuridae). J Reprod Fertil 93:97–110

Taggart DA, O'Brien HP, Moore HDM (1993) Ultrastructural characteristics of in vivo and in vitro fertilization in the grey short-tailed opossum, *Monodelphis domestica*. Anat Rec 237:21–37

Talbot P, DiCarlantonio G (1984) Ultrastructure opossum oocyte investing coats and their sensitivity to trypsin and hyaluronidase. Dev Biol 103:159–167

Temple-Smith PD (1987) Sperm structure and marsupial phylogeny. In: Archer M (ed) Possums and opossums: studies in evolution. Surrey Beatty and the Royal Zoological Society of New South Wales, Sydney, pp 171–193

Temple-Smith PD, Bedford JM (1976) The features of sperm maturation in the epididymis of a marsupial, the brush-tailed possum, *Trichosurus vulpecula*. Am J Anat 147:471–500

Temple-Smith PD, Bedford JM (1980) Sperm maturation and the formation of sperm pairs in the epididymis of the opossum, *Didelphis virginiana*. J Exp Zool 214:161–171

Tombes RM, Simerly C, Borisy GG, Schatten G (1992) Meiosis, egg activation, and nuclear envelope breakdown are differentially reliant on Ca^{2+}, whereas germinal vesicle breakdown is Ca^{2+} independent in the mouse oocyte. J Cell Biol 117:799–811

Tomes C, McMaster C, Saling P (1996) Activation of mouse sperm phosphatidylinositol-4,5-bisphosphate-phospholipase C by zona pellucida is modulated by tyrosine phosphorylation. Mol Reprod Dev 43:196–204

Topfer-Petersen E, Calvete JJ (1995) Molecular mechanisms of the interaction between sperm and the zona pellucida in mammals – studies on the pig. Int J Androl 18:20–26

Topfer-Petersen E, Henschen A (1988) Zona pellucida binding and fucose binding of boar sperm acrosin is not correlated with proteolytic activity. Hoppe-Seylers Z Biol Chem 369:69–76

Tyndale-Biscoe CH (1973) Life of marsupials. Arnold, London

Tyndale-Biscoe CH, Renfree MB (1987) Reproductive physiology of marsupials. Cambridge University Press, Cambridge

Usui N, Ogura A, Kimura Y, Yanagimachi R (1997) Sperm nuclear envelope: breakdown of intrinsic envelope and de novo formation in hamster oocytes or eggs. Zygote 5:35–46

Visconti PE, Kopf GS (1998) Regulation of protein phosphorylation during sperm capacitation. Biol Reprod 59:1–6

Visconti PE, Galantino-Homer H, Moore GD, Bailey JL, Ning X, Fornes M, Kopf GS (1998) The molecular basis of sperm capacitation. J Androl 19:242–248

Wang WH, Sun Q-Y, Hosoe M, Shioya Y, Day BN (1997) Quantified analysis of cortical granule distribution and exocytosis of porcine oocytes during meiotic maturation and activation. Biol Reprod 56:1376–1382

Ward CR, Kopf GS (1993) Molecular events mediating sperm activation. Dev Biol 158:9–34

Wassarman PM (1999) Mammalian fertilization: molecular aspects of gamete adhesion, exocytosis, and fusion. Cell 96:175–183

Wassarman PM, Florman HM, Greve JM (1985) Receptor-mediated sperm–egg interactions in mammals. In: Metz CB, Monroy A (eds) Biology of fertilization, vol 2: biology of the sperm. Academic Press, Florida, pp 341–360

Williams CJ, Mehlmann LM, Jaffe LA, Kopf GS, Schultz RM (1998) Evidence that Gq family G proteins do not function in mouse egg activation at fertilization. Dev Biol 198:116–127

Wolfsberg TG, Primakoff P, Myles DG, White JM (1995) ADAM, a novel family of membrane proteins containing a disintegrin and metalloprotease domain: multipotential functions in cell–cell and cell-matrix interactions. J Cell Biol 131:275–278

Wolny YM, Fissore RA, Wu H, Reis MM, Colombero LT, Ergun B, Rosenwaks Z, Palermo GD (1999) Human glucosamine-6-phosphate isomerase, a homologue of hamster oscillin, does not appear to be involved in Ca^{2+} release in mammalian oocytes. Mol Reprod Dev 52:277–287

Yanagimachi R (1994) Mammalian fertilization. In: Knobil E, Neill JD (eds) The physiology of reproduction, 2nd edn, vol 1. Raven Press, New York, pp 189–317

Yuan R, Primakoff P, Myles D (1997) A role for the disintegrin domain of cyritestin, a sperm surface protein belonging to the ADAM family, in mouse sperm–egg plasma membrane adhesion and fusion. J Cell Biol 137:105–112

Zamboni L (1970) Ultrastructure of mammalian oocytes and ova. Biol Reprod (Suppl) 2:44–63

CHAPTER 7

Fertilization in Protozoa and Metazoan Animals: A Comparative Overview

JUAN J. TARÍN[1]

1 Introduction

This review aims to summarize the information provided in this book and give an integrated view of the process of fertilization in two different and separate groups of eukaryotic organisms: protozoa and metazoan animals. As a summary/concluding chapter, this essay provides only a schematic synthesis of the topics treated in this book. For specific details, readers are referred to previous sections of this book as well as to the studies cited throughout this chapter. The rapid pace at which the field of fertilization advances, in particular the area of cell signalling, may render some of the concepts and conclusions laid down here obsolete by the time this review leaves the press. Being aware of this circumstance, I beg readers to be tolerant of any inconsistency in the state of the art of this field when reading this essay. I am open to any comment/suggestion readers wish to make for improving further editions of this synthesis.

2 Definition of Fertilization

A classical definition of fertilization is the fusion of one spermatozoon with an oocyte (syngamy), involving both activation of the oocyte to develop and fusion of the gamete nuclei (karyogamy) to produce a zygote nucleus (a synkaryon) (Greek terms: *syn*, with, together; *gamos*, marriage; *karyon*, nucleus). This definition, however, needs to be modified and extended to include some processes used by some protozoa such as (1) the fusion of gametes sharing the same morphology (isogamy); (2) the fusion of two gamete nuclei produced by the same parental cell (autogamy); and (3) the fusion of gamonts or individuals that produce gametes (gamontogamy). We should bear in mind also that in many species from some groups of animals such as

[1] Department of Animal Biology, Faculty of Biological Sciences, University of Valencia, Dr. Moliner 50, Burjassot, 46100 Valencia, Spain

Juan J. Tarín and Antonio Cano (Eds.)
Fertilization in Protozoa and Metazoan Animals

ctenophores, oligochaetes, arachnids, insects, elasmobranchs, urodeles, reptiles and birds more than one spermatozoon may enter the oocyte under normal conditions (physiological polyspermy). Although polyploid individuals may result from the entry of more than one spermatozoon into the oocyte, supernumerary sperm nuclei are either extruded from the oocyte or neutralized/disintegrate within the oocyte cytoplasm. Furthermore, karyogamy or fusion of male and female pronuclei is not a universal event in the animal kingdom. For instance, in molluscs, teleost fish, newts and eutherian mammals the male and female pronuclei never fuse to form a single zygote nucleus. The male and female pronuclei just become closely apposed, the pronuclear envelopes disintegrate, and the maternal and paternal chromosomes mingle to form the metaphase spindle of the first mitotic division. Another remarkable exception to the usual pattern of oocyte fertilization that takes place in metazoan animals is seen in sponges and chaetognaths (the marine arrow worms) which have accessory cells that mediate the passage of sperm to the oocyte. However, the most unusual type of fertilization in the animal kingdom is offered perhaps by dicyemid mesozoans (parasitic invertebrates with no digestive tract or other internal cavity), since in these forms the oocyte fuses with the sperm while both are actually within another cell (S. Stricker, pers. comm.). For these reasons, we should look for an alternative and wider definition of fertilization inclusive of the diversity of the process among all species. This definition could be as follows: fertilization is the series of cellular/molecular processes that results in the union of two germ cells or germ nuclei to give rise to a new individual with a new and unique genetic makeup.

3
Somatic and Germ Cell Lines in Protozoa and Metazoa

In Metazoa, as a rule, the somatic and germ cell lines are present within the same organism. Haploid gametes form a diploid individual after fertilization. In contrast, cells from unicellular protozoa pass alternatively through a vegetative and a germinal phase. Protozoan zygotes are diploid and after undergoing two meiotic divisions give rise to haploid cells that, depending on the particular organism analysed, are called vegetative cells, gamonts or asexual cells. These haploid cells thereafter transform into gametes that fuse in pairs to form a diploid zygote. For instance, the so-called vegetative cells from the flagellate *Chlamydomonas reinhardtii* are haploid. Each individual cell differentiates into a haploid gamete that fuses with another gamete of opposing mating type. As a result of this fusion, a diploid zygote is formed. This zygote undergoes a pre-meiotic synthesis of DNA and two meiotic divisions with the resulting production of four haploid vegetative cells, which have the capability of reproducing asexually by repeated cell divisions. In the human malaria parasite *Plasmodium falciparum*, a diploid zygote is formed after the fusion of a male microgamete with a female macrogamete. Then, the zygote enters into meiosis to produce haploid cells, which divide many times and change their morphology and physical/biochemical properties as they pass through different asexual phases within the outer layer of the mosquito mid-gut wall, human liver and human blood. In Foraminifera, the haploid nucleus of a gamont divides several times and each one of the resulting nuclei together with a fraction of protoplasm organizes as a gamete. Subsequently, two gametes coming from the same gamont (*Rotaliella heterocaryotica*) or from different gamonts (*Patellina corrugata*) fuse and form a diploid zygote. The zygote undergoes two mitotic divisions, which result in the generation of four nuclei. In the heterokaryotic species *Rotaliella heterocaryotica*, three of these nuclei (generative nuclei) undergo two meiotic divisions, creating a total of twelve haploid gamonts. The fourth nucleus differentiates into a somatic nucleus that degenerates before the formation of the twelve haploid gamonts (for review, see Mikami, this Vol.).

By analogy with the metazoan counterpart, haploid cells from Protozoa can be considered germ cells despite the fact that many authors call them vegetative cells, gamonts or asexual cells. Like metazoan primordial germ cells and gonial cells, protozoan germ cells can multiply asexually and pass through one or more developmental phases (each one of them exhibiting different morphological and physical/biochemical properties) before differentiating into mature gametes. However, there are some differences between Protozoa and Metazoa in the ploidy status and properties of germ cells. Firstly, metazoan cells from the germline can be diploid (primordial germ cells, gonial cells and secondary gametes), tetraploid (primary gametes) and haploid (true gametes that have completed the second meiotic division) instead of being confined to a single haploid state as seen in protozoan germ cells. Secondly, protozoan germ cells exhibit some reproductive traits that are absent in Metazoa, such as the potential for fusing two or more non-differentiated gametes (gamonts) displayed by *Patellina corrugata*.

Species belonging to the phylum Ciliophora present clear differences in their reproductive traits when compared to the remaining protozoan phyla. The process of conjugation showed by *Paramecium caudatum* starts with the adhesion of two gamonts, each one containing a diploid micronucleus and a polygenomic macronucleus. For this reason, these gamonts cannot be regarded as non-differentiated haploid gametes as suggested above for foraminifera. They are polyploid cells carrying both the vegetative cell line, represented by a polygenomic macronucleus, and the germ cell line, characterized by a diploid micronucleus. The micronucleus of a conjugated gamont from *Paramecium caudatum* undergoes a pre-meiotic DNA synthesis followed by two meiotic divisions. Three of the resulting haploid nuclei degenerate, whereas the remaining nucleus divides, forming a micronucleus that undergoes reciprocal exchange with the micronucleus from the other gamont. The resulting cell multiplies by binary fission after conjugation is complete. Segregation of DNA in vegetative and germ nuclei within a single cell is also observed in foraminifera in real vegetative cells (agamonts) before the onset of the first meiotic division (for review, see Mikami, this Vol.).

4 Comparative Analysis of Fertilization in Metazoa

Because of the remarkable phylogenetic distances among the different groups of Metazoa analysed in the present monograph (see Table 1), it is not surprising to observe considerable diversity among the groups as far as the process of fertilization is concerned (see Tables 2 and 3). However, there are also common themes in some cellular and molecular events, including receptor–ligand interactions, signalling cascades, specific proteolysis and nuclear transformations.

4.1 Sperm Capacitation and the Environment of Fertilization

Capacitation is the pool of physiological (functional) changes undergone after release from the male tract that render the spermatozoa competent to fertilize (Yanagimachi 1994). It appears that the spermatozoa from all eutherian mammals and perhaps all marsupials as well as some platyhelminths, nematoda and viviparous fishes must undergo the process of capacitation as a prerequisite for fertilization. In these groups of animals, fertilization takes place inside the female body, i.e. they exhibit internal fertilization. This association, however, does not imply a cause–effect relationship. In mammals, the process of capacitation is thought to have evolved to remove or reverse the stabilization imposed on the surface of the spermatozoon during regulated storage in the epididymis, rather than as a response to adapt to internal fertilization per se. This concept, first proposed by Bedford (1991), is supported by the fact that avian spermatozoa, which fertilize oocytes in the infundibulum, have a rapid migration through the epididymis, show few maturational changes and fertilize oocytes in vitro immediately, apparently without undergoing a process of capacitation (for review, see Wishart and Horocks, this Vol.). However, the quick reactivation step at ovulation undergone by avian spermatozoa after prolonged storage in the female tract may be equivalent to a capacitation process. Monotreme spermatozoa may follow a similar pattern to that exhibited by avian spermatozoa as far as the reactivation step is concerned. It is worthy of note that the change in the pattern of sperm motility associated with sperm capacitation, termed sperm hyperactivation, has been reported to occur only in eutherian mammals (for review, see Mate et al., this Vol.).

4.2 Composition of Oocyte Envelope(s)

The zona pellucida of the few eutherian mammals studied so far has been found to consist of three sulfated glycoproteins termed ZP1, ZP2 and ZP3. ZP1 is a structural protein that cross-links filaments made of ZP2 and ZP3

Table 1. Groups of protozoa and metazoan animals analysed in the present monograph

- Kingdom Protista
  - Unicellular microorganisms: the Protozoa
    - Phylum Sarcomastigophora
      - Subphylum Mastigophora
        - Class Phytomastigophorea (photosynthetic flagelates; genus *Chlamydomonas*)
      - Subphylum Sarcodina
        - Superclass Rhizopodans (foraminiferans; genera *Rotaliella* and *Patellina*)
    - Phylum Apicomplexa
      - Class Sporozoea (coccidians; genus *Plasmodium*)
    - Phylum Ciliophora
      - Class Oligohymenophorea (paramecia; genus *Paramecium*)
- Kingdom Animalia
  - Multicellular organisms: Metazoa
    - 1. Without true tissues
      - Phylum Porifera (sponges)
    - 2. With true tissues
      - 2.1. Diploblastic eumetazoa (lacking true mesoderm)
        - Phylum Cnidaria (sea anemones, medusae)
        - Phylum Ctenophora (comb jellies)
      - 2.2. Triploblastic eumetazoa (with true mesoderm)
        - 2.2.1. Acoelomates (without a body space other than digestive tract)
          - Phylum Platyhelminthes (flatworms)
          - Phylum Rhynchocoela (ribbon worms)
        - 2.2.2. Pseudocoelomates (with a persistent pseudocoel between gut and body wall)
          - Phylum Nematoda (roundworms)
        - 2.2.3. Coelomates or Eucoelomates (with a true coelom or mesodermal cavity)
          - Phylum Mollusca
          - Phylum Annelida
          - Phylum Arthropoda
          - Phylum Echinodermata
          - Phylum Chordata
            - Subphylum Protochordata (Urochordata and Cephalochordata)
            - Subphylum Vertebrata
              - Class Osteichthyes
                - Infraclass Teleostei (bony fishes)
              - Class Amphibia
                - Order Caudata or Urodela (salamanders and newts)
                - Order Anura (frogs and toads)
              - Class Aves (non-passerine birds)
              - Class Mammalia
                - Subclass Prototheria (monotremes)
                - Subclass Theria (metatherian and eutherian mammals)

heterodimers (for review, see Mate et al., this Vol.). During fertilization, the plasma membrane over the acrosomal cap of acrosome-intact mouse spermatozoa initially binds to ZP3, which induces the sperm acrosome reaction. Thereafter, sperm penetrates through the zona pellucida probably by physical (cutting thrust) means (Bedford 1998). This fact is in clear contrast with that of subtherian groups in which a lytic function for acrosomal enzymes seems

Table 2. Several fertilization traits displayed by all groups of animals analysed in this book

Group	Fertilization environment	Pre-fertilization meiotic arrest and/or normal maturation state at fertilization	Site of sperm entry into oocyte	Mechanisms for prevention of polyploidy
Porifera[a]	Internal in most species	Prophase I	A modified choanocyte engulfs a spermatozoon and transfers the spermiocyst (vesicle containing modified spermatozoon) to oocyte	Some mechanisms preventing polyspermy may exist
Cnidaria[a]	External in most marine species	Interphase after having completed meiosis in *Campanularia, Metridium, Orthopyxis, Phialidium* and *Tubularia*	Cortical area where first polar bodies are extruded or female pronucleus is located	Formation of fertilization envelope does not occur in majority of species and cortical granules are not discharged. Some species, such as *Bunodosoma cavernata*, release cortical granules at fertilization which forms a layer around oocyte
	Internal in most freshwater species			
Ctenophora[a]	External in most species	Optimal fertilization rates between germinal vesicle breakdown and extrusion of first polar body in *Beroe ovata*. Oocytes are shed when undergoing first meiotic division and complete meiosis whether they are fertilized or not[b]	At any site	Block to self-fertilization at levels of vitelline envelope and jelly coat

	Internal in *Coeloplana* and *Tjalfiella*		–	Intracytoplasmic degeneration of supernumerary spermatozoa
Platyhelminthes[a]	Internal	Prophase I or early metaphase I	At any site	Likely formation of fertilization envelope in *Turbellaria*
	Hypodermic insemination in *Turbellaria*			Likely intracytoplasmic degeneration of supernumerary spermatozoa in *Turbellaria*
Rhynchocoela[a]	External in most species	Metaphase I	At any site	Electrical fast-block at plasma membrane in *Cerebratulus*
Nematoda[a]	Internal	Prophase I or metaphase I in most species	?	?
Mollusca[a]	Internal in most gastropods and cephalopods	Metaphase I in most species including *Dentalium* (Scaphopoda), *Crassostrea*, *Cumingia*, *Hiatella*, *Limaria*, *Mytilus* and *Ruditapes* (Bivalva) and *Helix*, *Lymmea*, *Patella* and *Phyllaplysia* (Gastropoda)	Micropyle	Depolarization of oocyte plasma membrane in *Mytilus* and *Spisula*
	External in certain marine bivalves and primitive gastropods	Prophase I in *Barnea*, *Mactra*, *Pholas* and *Spisula* (Bivalva)	Crater region in vegetal pole in *Unio*	Formation of fertilization envelope in *Mytilus galloprovincialis*
	Hypodermic insemination in opisthobranch molluscs			Inhibition of acrosome reaction in *Mytilus edulis*
				Likely intracytoplasmic degeneration of supernumerary spermatozoa in *Spisula*

Table 2. *Continued*

Group	Fertilization environment	Pre-fertilization meiotic arrest and/or normal maturation state at fertilization	Site of sperm entry into oocyte	Mechanisms for prevention of polyploidy
Annelida[a]	Internal in most polychaetes and oligochaetes	Metaphase I in most species including *Chaetopterus*, *Pectinaria* and *Sabellaria* (Polychaeta)	Likely at any site	Retraction of oocyte-surface microvilli that have not fused with sperm in most species
	Hypodermic insemination in Hirudinea	Prophase I in some species		Intracytoplasmic degeneration of supernumerary spermatozoa in oligochaetes
Arthropoda[a]	Internal in most species	Fertilization occurs at any stage between prophase I and completion of meiosis	Micropyle	Hyperpolarization of oocyte plasma membrane in decapod crustaceans
		Prophase I in *Cyclops* (Crustacea)		Formation of fertilization envelope
		Metaphase I in *Artemia* (Crustacea), *Drosophila* (Insecta, Diptera) and *Melanoplus* (Insecta, Orthoptera)		
Echinodermata[a]	External	Interphase after having completed meiosis in Echinoidea and most Crinoidea (only one crinoid species is supposed to fertilize prophase I)[c]	At any site membrane	Depolarization of oocyte plasma

		Optimal fertilization rates between germinal vesicle breakdown and extrusion of first polar body in Asteroidea and Holothurioidea[c]		Formation of fertilization envelope
		Between metaphase I and interphase after having completed meiosis in Ophiuroidea[c]		Formation of hyaline layer over plasma membrane by cortical granules
Protochordata[a]	External in most species	Metaphase I in most urochordates	At any site, although preferentially in animal hemisphere	Modification of vitelline-coat glycoproteins by follicle cells surrounding vitelline envelope in *Ascidia* and *Phallusia*
	Internal in most colonial species	Metaphase II in most cephalochordates		Depolarisation of the oocyte plasma membrane in *Ascidia* and *Phallusia*
Teleost fishes[d]	External	Metaphase II	Micropyle in animal hemisphere	Narrow micropylar canal
	Internal			Outgrowth of early fertilization cone Outflow of perivitelline fluid through micropylar canal
				Modification of chrorion glycoproteins
				Closure of micropylar canal
Urodeles[e]	Internal in most species	Metaphase II	At any site, although preferentially in animal hemisphere	Depolarization of oocyte plasma membrane in monospermic species

Table 2. *Continued*

Group	Fertilization environment	Pre-fertilization meiotic arrest and/or normal maturation state at fertilization	Site of sperm entry into oocyte	Mechanisms for prevention of polyploidy
	External in Sirenidae, Cryptobranchoidae and Hynobiidae			Modification of outer jelly layers
	Internal in most species External in Sirenidae, Cryptobranchoidae and Hynobiidae			Intracytoplasmic degeneration of supernumerary spermatozoa
Caecilians[e]	Internal	Metaphase II	?	?
Anurans[e]	External in most species	Metaphase II	Animal hemisphere or vegetal hemisphere near equator	Depolarization of oocyte plasma membrane
			Animal dimple in frog *Discoglossus*	Formation of fertilization envelope
			Animal hemisphere or vegetal hemisphere near equator	Formation of smooth layer on tips of plasma membrane microvilli facing perivitelline space in *Xenopus laevis*
			Animal dimple in frog *Discoglossus*	
			Animal hemisphere or vegetal hemisphere near equator	Modification of outer jelly layers
			Animal dimple in frog *Discoglossus*	

Birds (non-passerine)[f]	Internal	Metaphase II	Animal pole	Formation of outer vitelline layer
				Intracytoplasmic degeneration of supernumerary spermatozoa
Monotremes[g]	Internal	Metaphase II	?	Likely formation of oviduct-deposited mucoid coat over zona pellucida
Marsupials[g]	Internal	Metaphase II	?	Likely modification of zona pellucida in species whose oocytes have cortical granules
				Likely formation of cortical granule envelope within the perivitelline space in some species including *Monodelphis* and *Sminthopsis*
				Formation of oviduct-deposited mucoid layer over zona pellucida
Eutherian mammals[g]	Internal	Metaphase II	At any site	Modification of zona pellucida in many species, although relative contribution varies in different species
				Modification of plasma membrane in many species, although relative contribution varies in different species

[a] Masui (1985); Whitaker (1996); Colas and Dubé (1998); Kishimoto (1998); Stricker (1999); Schatten and Chakrabarti (this Vol.)
[b] Carré et al. (1991)
[c] S. Stricker, (pers. comm.)
[d] Iwamatsu (this Vol.)
[e] Iwao (this Vol.)
[f] Wishart and Horrocks (this Vol.)
[g] Masui (1985); Yanagimachi (1994); Mate et al. (this Vol.)

Table 3. Several fertilization traits displayed by some groups of animals analysed in this book[a]

	Echinoderms[b]	Ascidians[b]	Teleost fishes[c]	Amphibians[d]	Birds (non-passerine)[e]	Mammals[f]
Site where acrosome reaction takes place	Oocyte jelly coat	Chorion, a.k.a. vitelline coat in species whose spermatozoa contain an acrosome	No acrosome in teleost spermatozoa	Pars recta secretions, components of jelly layers and outer surface of vitelline envelope in toad *Bufo japonicus*	Inner vitelline envelope molecules	Zona pellucida in higher eutherians and likely in most marsupials
				Outermost jelly layer in frog *Discoglossus pictus*		Likely cumulus oophorus in ancient eutherians such as insectivore shrews from subfamilies Soricinae and Crocidurinae
				Innermost jelly layers in newt *Pleurodeles waltl*		
Area of spermatozoon that fuses with the oocyte plasma membrane	Inner acrosomal membrane covering tip of acrosomal process	Inner acrosomal membrane covering tip of acrosomal process	Plasma membrane of apical or equatorial region of sperm head	Inner acrosomal membrane at top of sperm head in anuran *Discoglossus pictus* and urodele *Pleurodeles waltl*	Inner acrosomal region overlaying perforatorium *or*, according to alternative evidence, plasma membrane over post-acrosomal region in acrosome-reacted spermatozoa	Plasma membrane over equatorial segment after spermatozoa have undergone acrosome reaction in eutherians
						Likely, plasma membrane overlying a narrow peripheral region of acrosome equivalent to eutherian equatorial segment that remains intact following acrosome reaction in *Monodelphis* and *Sminthopsis* (marsupials)

						Inner acrosomal membrane in *Didelphis*, *Trichosurus* and *Macropus* (marsupials)
Changes in membrane potential of oocytes	Depolarization of plasma membrane	Depolarization of plasma membrane	Short depolarization followed by a longer hyperpolarization phase ofplasma membrane	Depolarization of plasma membrane in anurans No change in membrane potential or even hyperpolarization of plasma membrane in urodels	?	In rabbits, slow depolarization of plasma membrane followed by slow repolarization In hamsters and mice, a series of repetitive membrane hyperpolarizations which are concomitant with transient rises in in $[Ca^{2+}]_i$ and superimposed on a slow membrane hyperpolarization In humans, a gradual hyperpolarization of the plasma membrane, 60–120 min after insemination, followed by a repolarization
Pattern of sperm-induced $[Ca^{2+}]_i$ rise in oocytes	A single rise	A large initial increase in $[Ca^{2+}]_i$ followed by a series of $[Ca^{2+}]_i$ oscillations	A single rise	A single rise	?	A large initial increase in $[Ca^{2+}]_i$ followed by a series of $[Ca^{2+}]_i$ oscillations in eutherians, which last until pronuclear stage
Inducing factor of cortical vesicle/alveolus/granule exocytosis of oocytes	Increase in $[Ca^{2+}]_i$	Oocytes lack cortical granules	Increase in $[Ca^{2+}]_i$	Increase in $[Ca^{2+}]_i$ in anurans	Oocytes lack cortical granules	Initial rise in $[Ca^{2+}]_i$ and activation of phospholipase C (PLC) in eutherians

Table 3. *Continued*

	Echinoderms[b]	Ascidians[b]	Teleost fishes[c]	Amphibians[d]	Birds (non-passerine)[e]	Mammals[f]
				Urodele oocytes lack cortical granules		Oocytes lack cortical granules in monotremes and some marsupials including *Monodelphis* and *Trichosurus*
Changes in pH_i of oocytes	Rise in pH_i	No pH_i change in *Ciona*	Rise in pH_i	Rise in pH_i	?	No pH_i change in rats and mice
Ooplasmic segregation	?	Yes	Yes	Yes	?	Yes

[a] Most invertebrate phyla have not been included because of scarce or non-existent information gathered so far in these particular groups of animals.
[b] Schatten and Chakrabarti (this Vol.)
[c] Iwamatsu (this Vol.)
[d] Iwao (this Vol.)
[e] Wishart and Horrocks (this Vol.)
[f] Yanagimachi (1994); Mate et al. (this Vol.)

to be the rule (for review, see Bedford 1998). In addition, the rapid lateral oscillations used by at least human and hamster spermatozoa to penetrate the zona pellucida would not support the notion that ZP2 binds to the inner acrosomal membrane, in this way keeping the spermatozoon within the zona pellucida (J.M. Bedford, pers. comm.).

The oocyte envelope glycoproteins from marsupials, birds, amphibians and teleost fishes share a high homology with the glycoproteins of the eutherian zona pellucida. For instance, the zona pellucida of the marsupial brushtail possum is composed of a matrix of three weakly acidic glycoproteins which have approximately 50% homology to the eutherian ZP1, ZP2 and ZP3 (for review, see Mate et al., this Vol.). In the chicken, one of the inner perivitelline layer glycoproteins termed IPVL1, GP1, gp42 or chZPC shows 41% of the amino acid sequence identical to the eutherian ZP3 (Waclawek et al. 1998). In the anuran *Xenopus laevis*, the glycoproteins gp41 and gp69/64 of the vitelline envelope are homologous to the eutherian ZP3 (35–39% identity) and ZP2 (34–37% identity), respectively (for review, see Iwao, this Vol.). In the teleost fish *Pseudopleuronectes americanus*, the glycoprotein wf♀; of the oocyte chorion has a region of 335 amino acids homologous with the major zona pellucida glycoprotein of the rabbit, rc55 (37% identity), and with the mouse zona pellucida protein ZP2 (28% identity) (Lyons et al. 1993). It is interesting to note that mammalian and amphibian zona pellucida glycoproteins are synthesized by the oocytes whereas fish and bird glycoproteins are synthesized by the liver and granulosa cells, respectively (cited by Waclawek et al. 1998). In some eutherian mammals such as humans, rhesus monkey, cow and rabbit, however, ZP genes are expressed in both oocytes and granulosa cells (for reviews, see Bedford 1998; and Mate et al., this Vol.).

During amphibian and echinoderm fertilization, the oocyte jelly coat triggers the sperm acrosome reaction, which is required for sperm binding and fusion with the oocyte plasma membrane. In amphibians, the jelly coat contains three to six layers of highly glycosylated proteins with species-specific patterns of oligosaccharide chains (for review, see Iwao, this Vol.). In the sea urchin, the jelly coat consists of a major acrosome reaction-inducing fucosylated glycoprotein and a group of sperm-activating peptides (SAPs) which activate the metabolism of endogenous phosphatidylcholine and promote the acrosome reaction as a specific co-factor of the acrosome-inducing glycoprotein (Suzuki 1995; for review, see Foltz 1995). In the sea urchin *Strongylocentrotus purpuratus*, the extremely thin vitelline envelope that surrounds the oocyte plasma membrane is composed of eight major glycoproteins that are heavily mannosylated and contain fucose and N-acetylglucosamine moieties (Correa and Carroll 1997). Among these molecules, a 350-kDa glycoprotein is the oocyte receptor for sperm (for review, see Ohlendieck and Lennarz 1995).

The jelly coat of starfish oocytes consists of a highly sulfated glycoprotein named acrosome reaction-inducing substance (ARIS), a group of sulfated steroidal saponins named Co-ARIS, and a group of glutamine-rich tetratriacontapeptides SAPs denominated asterosaps. Starfish asterosaps are much larger than sea urchin SAPs and do not show any significant sequence similarities to known proteins (Nishigaki et al. 1996). It appears that the three components of the starfish jelly coat are responsible for the induction of the acrosome reaction, although ARIS is the most important among them (Koyota et al. 1997).

4.3 Sperm Binding to Oocyte Envelope(s)

In general, specific glycoproteins of oocyte extracellular envelope(s) are involved in sperm recognition and binding as well as in the induction of the sperm acrosome reaction. At the present time, most data suggest that the binding of spermatozoa to the zona pellucida is a carbohydrate-mediated event in eutherian mammals (for review, see Wassarman 1999). In particular, it appears that ZP3 O-linked oligosaccharides play a direct role in sperm receptor function. It seems likely, however, that at least in ancient eutherians such as the crocidurine shrew *Suncus*, a barbed perforation that is exposed by the cumulus–oophorus-induced acrosome reaction mediates the binding of already acrosome-less reacted spermatozoa to the zona pellucida (for review, see Bedford 1999). Carbohydrates also play an important function in the binding of bird, amphibian, ascidian, echinoderm, arthropod, mollusc and hydrozoan spermatozoa to their respective oocyte envelopes (see Table 4).

In mammals, a large list of ZP3-binding proteins expressed on the sperm plasma membrane [among others, the 95-kDa tyrosine kinase receptor, β-1,4-galactosyltransferase and sperm protein-56 in the mouse; zonadhesin and spermadhesins in the pig; and phospholipase A_2 and the zona receptor kinase (ZRK) in humans] has been proposed (for reviews, see Tulsiani et al. 1997; McLesky et al. 1998; Brewis and Wong 1999; Wassarman 1999; Mate et al. , this Vol.). Due to the large number and diverse nature of candidate ZP3-binding proteins, it is likely that multiple sperm receptors interact in concert with the zona pellucida. It has been proposed that sperm–zona interactions may occur as either protein–protein, carbohydrate–protein, carbohydrate–carbohydrate or a combination of these binding phenomena (Brewis and Wong 1999).

Spermatozoa from the urodele *Cynops pyrrhogaster* bind to oocyte receptors on the vitelline membrane via heparin-like molecules on the sperm plasma membrane. Ascidians employ sperm glycosidases (alpha-L-fucosidase in *Halocynthia* and *Ciona*, and N-acetylglucosaminidase in *Phallusia*) as binding proteins for vitelline envelope proteins. In the sea urchin, the sperm molecule that recognizes the fucosylated glycoproteins of the jelly coat is a 210-kDa sperm plasma membrane glycoprotein with extensive homology to the human polycystic kidney disease protein, PKD1 (Hughes et al. 1999). Subsequent to the acrosome reaction, the major protein of the acrosomal

Table 4. Carbohydrates from oocyte extracellular envelope(s) that may play an important function in sperm recognition and binding

Group	Carbohydrates
Mammals[a]	O-linked oligosaccharides of zona pellucida glycoprotein ZP3
Birds[b]	N-linked oligosaccharides of glycoproteins of inner vitelline envelope (likely the glycoprotein IPVL1) in chicken
Amphibians[c]	Oligosaccharide chains of vitelline envelope glycoprotein gp69/64 in anuran *Xenopus laevis*
	Carbohydrate moieties of major 36- to 39-kDa glycoproteins of vitelline envelope in anuran *Bufo japonicus*
Ascidians[d]	Sulfated O-linked glycans of vitelline envelope in *Halocynthia roretzi*
	Fucosyl residues of vitelline envelope in *Ciona intestinalis*
	N-acetylglucosamine residues of vitelline envelope in *Phallusia mammillata*
Echinoderms[e]	A saccharide polymer, rich in fucose residues, of ARIS (one of the three components of jelly coat) in starfish *Asterias amurensis*
	Fucosylated glycoproteins of jelly coat in sea urchins. In *Strongylocentrotus*, subsequent to acrosome reaction, two sulfated O-linked oligosaccharide chains from a 350-kDa vitelline envelope glycoprotein act as a secondary oocyte receptor for acrosome-reacted sperm
Arthropods[f]	Alpha-mannose and beta-N-acetylglucosamine residues present on oocyte chorion at site of sperm entry in *Drosophila melanogaster*
	In the penaeoidean shrimp *Sicyonia ingentis*, primary binding of sperm to vitelline envelope appears also to be mediated by a carbohydrate-dependent process
Molluscs[g]	Carbohydrate moieties of a glycoprotein (VERL) of vitelline envelope in abalones (*Haliotis*)
Cnidarians[h]	Carbohydrate moieties sensitive to mannosidase and N-acetylglucosaminidase of jelly coat in hydrozoan *Aequorea victoria*

[a] Wassarman (1999)
[b] Wishart and Horrocks (this Vol.)
[c] Iwao (this Vol.)
[d] Folz (1995); Schatten and Chakrabarti (this Vol.)
[e] Folz (1995); Ohlendieck and Lennarz (1995); Koyota et al. (1997); Schatten and Chakrabarti (this Vol.)
[f] Folz (1995); Schatten and Chakrabarti (this Vol.)
[g] Swanson and Vacquier (1997)
[h] Freeman (1996)

vesicle, bindin, becomes exposed and remains associated on the acrosomal process. Bindin is the protein on the sperm that mediates binding to the oocyte receptor on the vitelline envelope. In *Drosophila melanogaster*, two glycosidases, beta-N-acetylglucosaminidase (GlcNAc'ase) and alpha-mannosidase (Man'ase), have been detected in spermatozoa and might play a role in the recognition of the carbohydrate residues of the oocyte shell. Finally, in abalones, lysin (which is exocytosized during the acrosome reaction) is the sperm recognition protein that binds to the vitelline envelope receptor for lysin

(VERL) (Swanson and Vacquier 1997; for reviews, see Folz 1995; Ohlendieck and Lennarz 1995; Schatten and Chakrabarti, this Vol.; Iwao, this Vol.; Wishart and Horrocks, this Vol.).

4.4 Acrosome Reaction

Studies in echinoderms and eutherian mammals suggest that, although there are some differences between species, the basic cellular/molecular mechanism of the acrosome reaction is conserved through the different invertebrate and vertebrate phyla. In the sea urchin (*Lytechinus pictus*), for instance, molecules from the oocyte jelly coat induce changes in sperm including Na^+ and Ca^{2+} influx, H^+ and K^+ efflux, a transient intracellular pH (pH_i) alkalinization, a transient increase in intracellular levels of cAMP and cGMP, membrane depolarization (positive shift in plasma membrane potential), increase in activity of adenylate cyclase and cAMP-dependent protein kinase, inactivation of guanylate cyclase, histone phosphorylation, increase in fatty acid oxidation and polymerization of actin into the acrosomal rod (for review, see Schatten and Chakrabarti, this Vol.). Likewise, in eutherian mammals, sperm passage through cumulus cells or sperm binding to zona pellucida ZP3 molecules activates G proteins and voltage-sensitive T type Ca^{2+} channels, Na^+/H^+ exchanger or Na^+-dependent Cl^-/HCO_3^- exchanger with the resulting Na^+ and Ca^{2+} influx, H^+ efflux, pH_i alkalinization and depolarization of the plasma membrane (for review, see Mate et al., this Vol.). The increase in pH_i may also activate Ca^{2+}/calmodulin-dependent adenylyl cyclase, protein phosphatase, tyrosine kinase, protein kinases (A and C) and various phospholipases (e.g., A_2, C and D; for review, see Wassarman 1999). However, closely related species such as the sea urchin and the starfish may show differences in the sequence of events leading to the acrosome reaction. In sea urchin sperm, for instance, both an increase in $[Ca^{2+}]$ and a rise in pH_i are necessary for inducing an acrosome reaction. In contrast, an increase in pH_i may not be absolutely obligatory for triggering the acrosome reaction in starfish sperm. In this species (*Asterina pectinifera*), an increase in $[Ca^{2+}]$ induces exocytosis of acrosomal contents, whereas a rise in pH_i, observed immediately after exocytosis, is coincident with the formation of an acrosomal rod and the initiation of vigorous flagellar motility (Sase et al. 1995).

4.5 Binding and Fusion of Spermatozoon to Oocyte Plasma Membrane

At fertilization, the spermatozoon initially binds to and then fuses with the oocyte plasma membrane (Fig. 1). It is likely that one or several disintegrin-like proteins participate in sperm–oocyte binding and/or fusion (for review, see Fénichel and Durand-Clément 1998). Among several eutherian sperm-surface disintegrin-like proteins reported to be involved in sperm–oocyte binding and/or fusion, fertilin (PH-30) is the most extensively studied. Fertilin is a surface heterodimer of α and β N-glycosylated subunits, both of which are members of a widely distributed family of membrane proteins containing a metalloprotease-like, a disintegrin-like and a cysteine-rich domain (MDC). The subunit α, in addition, possesses a moderately hydrophobic sequence in its cysteine-rich domain that may function as a "fusion peptide" (Wassarman 1999). Several members of this family include the proteins called cyristestin and xMDC16 localized on the plasma membrane of mouse and *Xenopus laevis* spermatozoa, respectively (for reviews, see Fénichel and Durand-Clément 1998; Iwao, this Vol.). Integrin expression on oocytes has been reported in several species including sea urchin, *Discoglossus pictus* (anuran), mouse, hamster and humans. In addition, experimental data suggest that integrins, by clustering in a multimolecular complex, may be able to transduce signals through the oocyte membrane after ligand binding, including the regulation of pH_i or intracytoplasmic Ca^{2+} concentration ($[Ca^{2+}]_i$), phosphorylation of proteins on tyrosine or serine/threonine residues and inositol lipid turnover. In addition, other sperm-surface components different from the integrin/disintegrin family may participate in sperm–oocyte binding and/or fusion. For instance, the epididymal maturation antigen DE/AEG, the membrane cofactor protein (MCP or CD46), the complement component C1q receptor, protectin (CD59) and sulfogalactosylglycerolipid (SGG) have been involved in mammalian sperm–oocyte binding and/or fusion (Ahnonkitpanit et al. 1999; for review, see Fénichel and Durand-Clément 1998). In the sea urchin, the binding of bindin (which bears no sequence similarity to the intregrin family) to the oocyte receptor on the vitelline envelope does not induce oocyte activation. This fact suggests that a second sperm component may be involved in the overall process of sperm–oocyte binding/fusion. We should bear in mind at this point that the vitelline envelope of sea urchins lies in close juxtaposition to the plasma membrane and, therefore, the plasma membrane and the vitelline envelope may be considered as a structural/functional unit as far as the process of sperm binding/fusion to the plasma membrane is concerned (for review, see Ohlendieck and Lennarz 1995). In the echiuran worm *Urechis caupo*, a peptide (Val-Ala-Lys-Lys-Pro-Lys) derived from a highly sperm-basic protein isolated from acrosome-reacted sperm likely binds to a trypsin-sensitive receptor on the oocyte plasma membrane. Such a receptor–ligand binding triggers oocyte activation as defined by changes in membrane potential, completion of meiosis and DNA synthesis (for review, see Foltz 1995).

4.6
Oocyte Activation

After sperm fusion with the oocyte plasma membrane, activation of a series of biochemical and molecular pathways within the oocyte leads to the formation of a new individual. In some species such as the marine shrimp *Sicyonia ingentis*, it appears that oocyte activation is not triggered by the fertilizing sperm. Instead, oocytes are activated by seawater Mg^{2+} during spawning (for

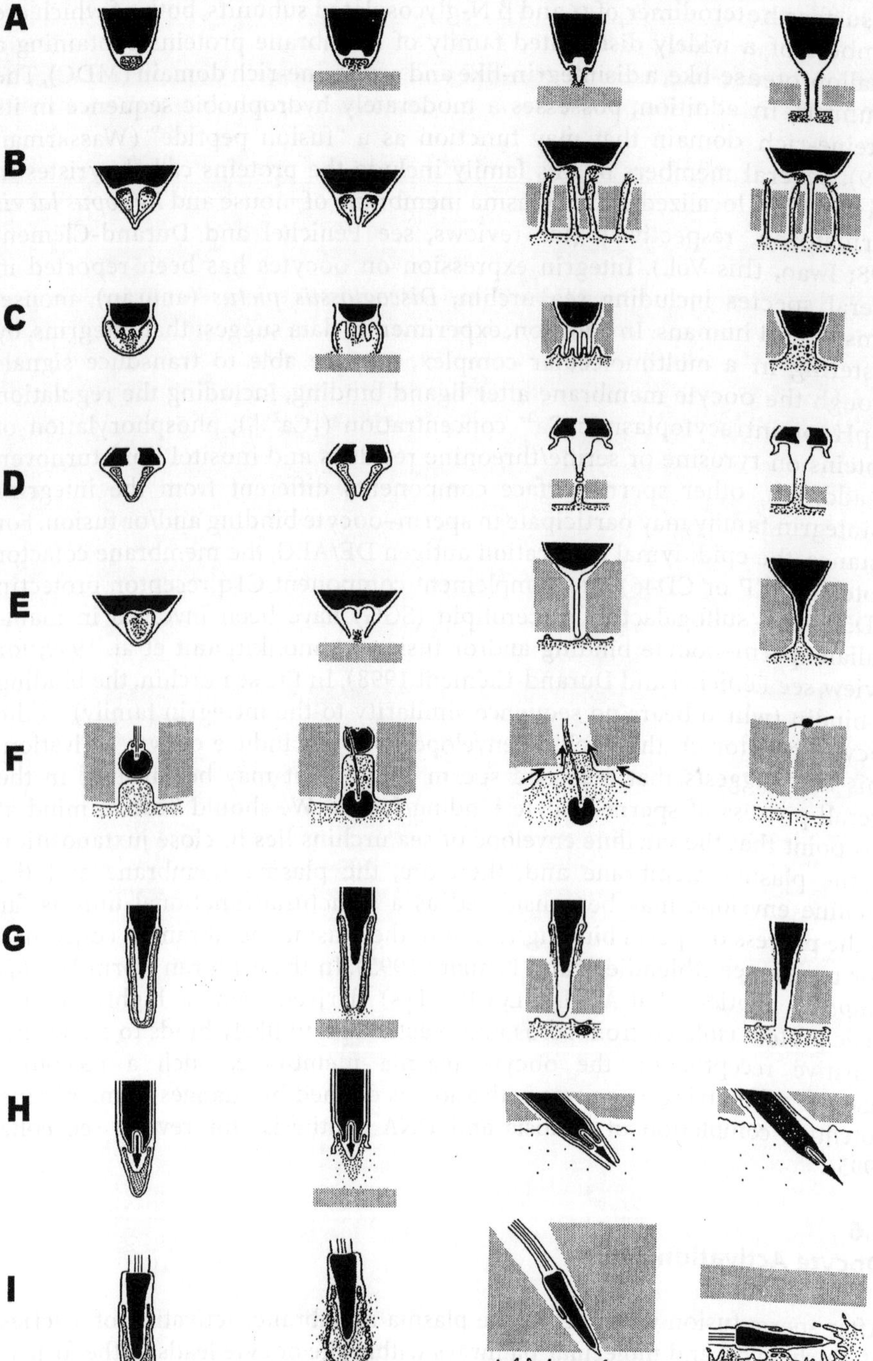
A
B
C
D
E
F
G
H
I

review, see Schatten and Chakrabarti, this Vol.). Although there are again exceptions (see Table 3), an increase in $[Ca^{2+}]_i$ shortly preceded (echinoderms, ascidians and teleost fishes) or followed (anurans) by depolarization of the oocyte plasma membrane are the first signals associated with oocyte activation (for reviews, see Schatten and Chakrabarti, this Vol.; Iwamatsu, this Vol.; Iwao, this Vol.). These events are followed by cortical alveolus/granule exocytosis, reorganization of the cortical cytoplasm, a rise in pH_i, resumption of meiosis, formation of female and male pronuclei, pronuclear migration to the centre of the oocyte, protein and DNA synthesis, pronuclear fusion or pronuclear apposition, nuclear envelope breakdown and formation of the chromosomal plate of the first mitotic division. The intermingling of chromosomes and formation of the first mitotic chromosomal plate can be considered as the end of fertilization and the beginning of embryonic development (Yanagimachi 1994).

4.6.1 Changes in Plasma Membrane Potential

Immediately after sperm fusion with the oocyte plasma membrane, oocytes from many groups including nemerteans, molluscs, annelids, echinoderms, ascidians, lampreys and anurans undergo a depolarization of the plasma membrane potential, called the fertilization potential, which prevents further spermatozoa from fusing with the oocyte (for review, see Jaffe and Gould 1985; Tables 2 and 3; Fig. 2). Decapod crustaceans, however, appear to be an exception. In these species, a hyperpolarization of the oocyte membrane (a negative shift in the membrane potential) in response to fertilization seems to constitute a long-lasting electrical block to polyspermy (Goudeau and Goudeau 1989). Invertebrate oocytes use non-specific ion channels, which are not gated by Ca^{2+}, to depolarize the plasma membrane (for review, see Dale 1994). After the first sperm-induced inward current (the sperm current), depolarization of the plasma membrane is sustained by activation of a voltage-dependent inward current (the activation current) induced by the increase in $[Ca^{2+}]_i$ (David et al.

◄ **Fig. 1.** Acrosome reaction and sperm–oocyte fusion in different species. **A** Sea urchin, *Tripneustes gratilla*; **B** mollusc, *Chama macerophylla*; **C** annelid, *Hydroides hexagonous*; **D** annelid, *Neanthes japonica*; **E** hemicordate, *Saccoglossus kowalevskii*; **F** fish, *Cyprinus carpio*. Note that fish spermatozoa have no acrosome; **G** amphibian, *Bufo japonica*; **H** bird, *Gallus gallus*. Note arrow-like subacrosomal rod at tip of sperm head; **I** mammal, *Mesocricetus auratus*. In most animals, inner acrosomal membrane fuses first with oocyte plasma membrane. In fishes and eutherian mammals, it is the sperm plasma membrane covering the sperm head (fishes) or equatorial segment (eutherian mammals) that fuses first with the oocyte plasma membrane. In many invertebrate species, the acrosomal membrane elongates during acrosome reaction to form one or more acrosomal process(es), each of which fuses with the oocyte plasma membrane. *Grey areas* represent vitelline envelope (commonly called chorion in fishes, inner perivitelline layer in birds and zona pellucida in mammals). (Modified from Yanagimachi 1988)

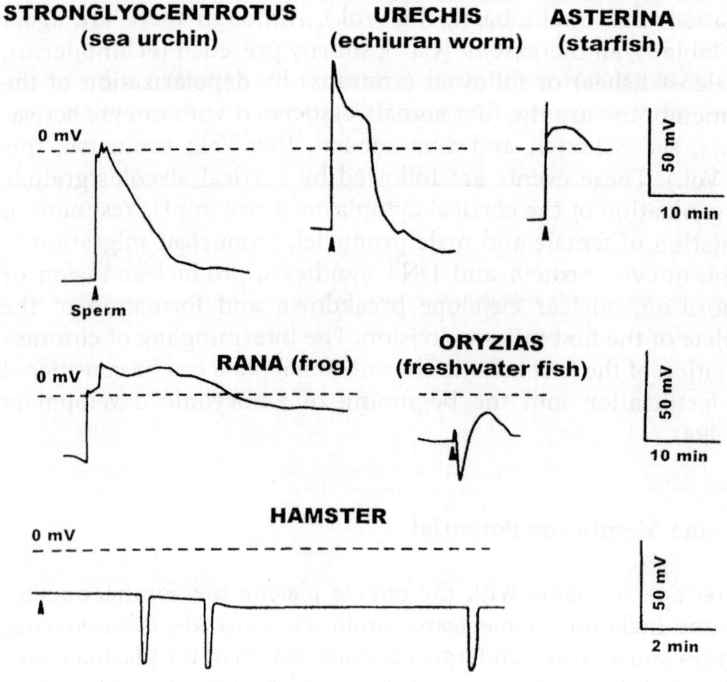

Fig. 2. Fertilization potentials at fertilization in oocytes from different species. *Arrowheads* indicate time of insemination. (Modified from Jaffe and Gould 1985)

1988; for review, see Chambers and McCulloh 1990). Anuran oocytes utilize K^+, voltage-activated Na^+ and Ca^{2+}-activated Cl^- channels to depolarize the plasma membrane. Like the lamprey *Lampreta japonica* (Agnatha) (Kobayashi et al. 1994), the increase in $[Ca^{2+}]_i$ at fertilization directly activates Cl^- channels, and the resultant Cl^- efflux forms the ionic basis for the fertilization potential in anuran oocytes (Young et al. 1984). Although the depolarization of the oocyte plasma membrane represents a barrier for polyspermic fertilization, it appears that the voltage-sensitive component that prevents sperm fusion with the oocyte plasma membrane is contributed by the spermatozoon rather than by the oocyte plasma membrane (for review, see Iwao, this Vol.).

In some groups of animals, including teleost fishes (Nuccitelli 1980) and mammals (for review, see Yanagimachi 1994), an electrical polyspermy block at the level of the plasma membrane is not observed. The freshwater teleost fish *Oryzias* exhibits a short depolarization of the plasma membrane followed by a longer hyperpolarization phase. The transient depolarization is due to a non-specific leak of ions, whereas the hyperpolarization phase is caused by an increase in K^+ permeability. Such an increase may be due to the addition of K^+ channels from the envelopes of the cortical alveoli to the plasma membrane after cortical granule discharge, although an opening of K^+ channels induced by a rise in $[Ca^{2+}]_i$ may also contribute to the increased permeability to K^+ (for review, see Iwamatsu, this Vol.). The periodic plasma membrane hyperpolarizations concomitant with transient rises in $[Ca^{2+}]_i$ displayed by golden-hamster and mouse oocytes, as well as the gradual hyperpolarization of the plasma membrane exhibited by human oocytes, are caused by a Ca^{2+}-activated K^+ conductance (for reviews, see Dale 1994; Yanagimachi 1994). In the mouse (Jaffe et al. 1983) and rabbit (McCulloh et al. 1983), the shift in plasma membrane potential at fertilization appears insufficient to block polyspermy by a mechanism dependent upon the magnitude of oocyte membrane potential. It has been shown that, at least in the mouse, the incorporation of sperm plasma membrane components into the oocyte plasma membrane contributes to the plasma membrane block to polyspermy (Maleszewski et al. 1996).

4.6.2
Increase in $[Ca^{2+}]_i$

As mentioned above, a transient rise in $[Ca^{2+}]_i$ is one of the first signals indicating that the oocyte has been activated by a fertilizing spermatozoon. The sperm activates the oocyte by causing either a single or a series of $[Ca^{2+}]_i$ oscillations. The pattern of $[Ca^{2+}]_i$ signals in many kinds of oocytes are pulse-like, although some species such as the newt show a long-lasting $[Ca^{2+}]_i$ signal (see Fig. 3). The peak height of $[Ca^{2+}]_i$ signals varies widely from one species to another mainly due to differences in oocyte size (for review, see Yoshimoto and Hiramoto 1991). It appears that a combination of store-operated, voltage-independent Ca^{2+} influx from the extracellular environment (emptying intracellular Ca^{2+} stores automatically induces Ca^{2+} influx; for review, see Parekh and Penner 1997) and internal Ca^{2+} release are utilized during fertilization in most animal species irrespectively of whether they are protostomes (animals whose embryonic blastopore develops into the adult mouth, e.g. nemertean worms, molluscs, and annelids) or deuterostomes (animals that do not derive their mouth from the blastopore, e.g. echinoderms, ascidians, fishes and mammals). The frog would be the only exception to this rule since its oocytes apparently exhibit only internal calcium release at fertilization (for review, see Stricker 1999).

Pattern of $[Ca^{2+}]_i$ Pulses. In most animals species, including eutherian mammals, ascidians, bivalves, nemertean worms and annelids, fertilization results in multiple $[Ca^{2+}]_i$ transients or pulses (see Fig. 4). In eutherian oocytes, most periodic $[Ca^{2+}]_i$ transients appear synchronously throughout the oocyte cytoplasm. In contrast, most $[Ca^{2+}]_i$ oscillations of ascidians, bivalves, annelids and the nemertean worm *Cerebratulus lacteus* occur in the form of waves (McDougall and Sardet 1995; Stricker 1997; for reviews, see Yoshimoto and Hiramoto 1991; Stricker 1999). In the polychaete annelid *Chaetopterus pergamentaceus* the first Ca^{2+} transient, which does not propagate in a wave-

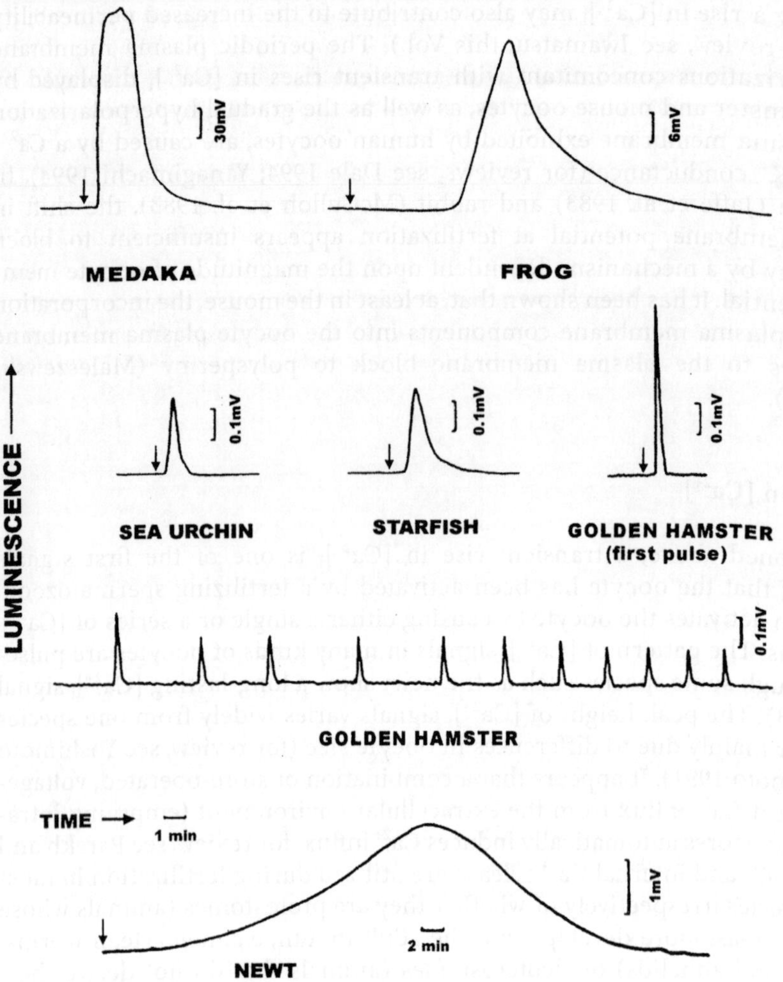

Fig. 3. Ca^{2+} signals at fertilization in oocytes from different species. *Arrows* indicate time of insemination. Time scale (1 min) is common to all signals except those for newt oocyte. (Modified from Yoshimoto and Hiramoto 1991)

like fashion, is followed by a series of globally propagated Ca^{2+} waves interspersed with additional non-propagated pulses (Eckberg and Miller 1995). In amphibians, teleost fishes, echinoderms, echiuran worms, some bivalves and cnidarians, fertilization results in a single $[Ca^{2+}]_i$ transient that may (amphibians, teleost fishes and echinoderms) or may not (the echiuran worm *Urechis caupo*) propagate from the point of sperm entry towards the opposite pole of the oocyte in a wave-like fashion.

It appears that multiple $[Ca^{2+}]_i$ transients, in addition to carrying "frequency-encoded" information and preventing potential toxic effects

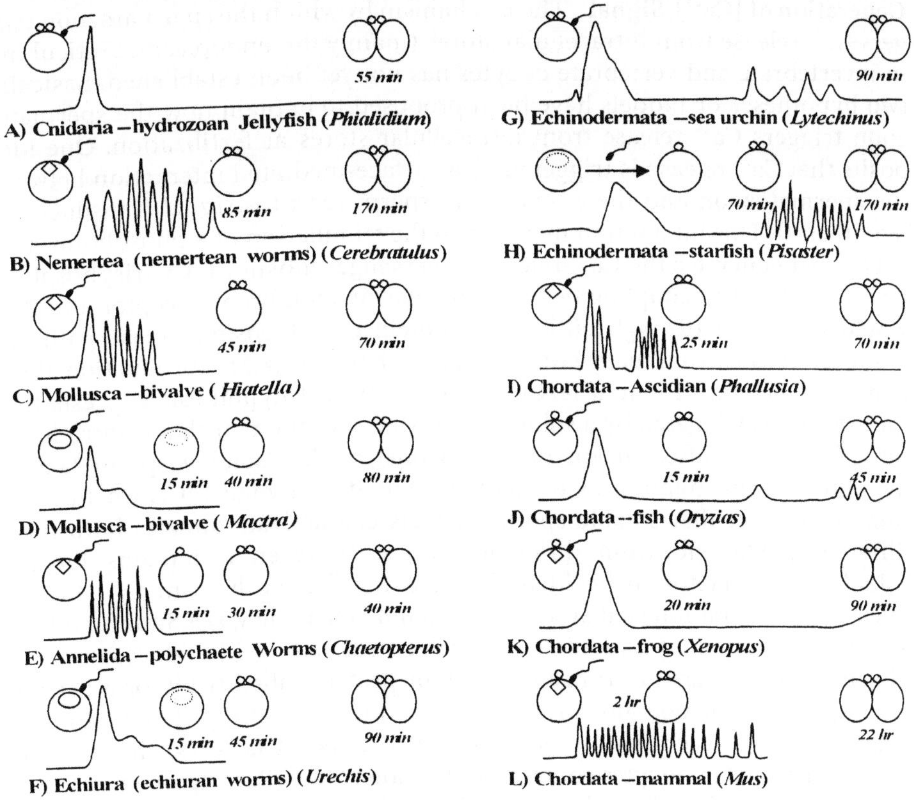

Fig. 4. Temporal patterns of Ca^{2+} signals at fertilization in oocytes from different species. Normal stage of oocyte maturation at fertilization: prophase-I-arrested (D, F); between germinal vesicle breakdown and polar body formation (H); metaphase-I-arrested (B, C, E, I); metaphase-II-arrested (J–L); and pronuclear oocyte after having completed meiosis (A, G). Amplitudes, durations and frequencies of calcium transients are not drawn to scale, and timings of first cell cycle events are only approximate. (Stricker 1999)

induced by a continuous elevation of $[Ca^{2+}]_i$, are necessary for releasing oocytes from meiotic arrest at the metaphase of the first (MI) or the second (MII) meiotic division (for review, see Stricker 1999). In fact, oocytes fertilized at prophase of the first meiotic division (prophase I) (starfishes and some bivalves such as *Mactra*) or at interphase after having completed meiosis (cnidarian and sea urchin), generate a solitary $[Ca^{2+}]_i$ transient (see Fig. 4). An apparent exception to this rule would be amphibian and fish oocytes which display a single $[Ca^{2+}]_i$ transient despite being fertilized at the MII stage. This paradox could be explained by the short transition from MII to pronucleus formation in these vertebrate groups when compared to eutherian mammals. Such a short post-fertilization phase may obviate the need for a repeated series of $[Ca^{2+}]_i$ transients (for review, see Stricker 1999).

Generation of $[Ca^{2+}]_i$ Signal. The mechanism by which the spermatozoon triggers Ca^{2+} release from intracellular stores (mainly the endoplasmic reticulum) in invertebrate and vertebrate oocytes has not yet been established. Basically, two hypotheses or models have been proposed to explain how the spermatozoon triggers Ca^{2+} release from intracellular stores at fertilization. One idea posits that Ca^{2+} release is triggered by a surface-mediated interaction between the spermatozoon and the oocyte (the "sperm-receptor signal transduction" hypothesis). This interaction may lead to the stimulation of a phospholipase C (PLC) that generates the Ca^{2+}-releasing messenger inositol 1,4,5-trisphosphate [Ins(1,4,5)P_3]. The coupling between the plasma membrane receptor and PLC would be linked through either a G-protein or a tyrosine kinase pathway. However, results from many studies suggest that Ins(1,4,5)P_3 production alone is not the primary event at fertilization. A different hypothesis suggests that Ca^{2+} release is triggered by a soluble activating sperm factor that is introduced into the oocyte after fusion of both gametes (the "soluble sperm factor" hypothesis). This hypothesis is nowadays the idea in vogue. It is worth mentioning the fact that soluble sperm extracts can activate oocytes both from different phyla and from different species. However, it remains for the soluble sperm factor to be identified and for the mechanism by which it releases Ca^{2+} from internal stores to be found (for review, see Swann and Lai 1997). Although many candidate molecules such as a 33-kDa glucosamine-6-phosphate deaminase, a truncated receptor protein called tr-kit, or a perinuclear sperm-borne oocyte-activating factor(s) (SOAF) (Perry et al. 1999) have been proposed, recent studies suggest that the sperm factor(s) that triggers Ca^{2+} oscillations in oocytes contains a PLC isoenzyme (Parrington et al. 1999). Glucosamine-6-phosphate deaminase is no longer believed to be the sperm-activating factor that triggers Ca^{2+} release from intracellular stores at fertilization; however, it remains to be demonstrated whether tr-kit and SOAF induce Ca^{2+} oscillations when injected into oocytes (for review, see Swann et al. 1999).

Types of Receptor/Ca^{2+} Release Channels. The release of Ca^{2+} from intracellular stores is controlled in most groups of animals by Ins(1,4,5)P_3-mediated signalling pathway (for review, see Stricker 1999). However, other types of receptor/Ca^{2+} release channels, including one sensitive to the alkaloid ryanodine [cyclic adenosine 5′-diphosphate-ribose (cADPR) is an endogenous activator of this receptor] and another one to the pyridine nucleotide [nicotinic acid adenine dinucleotide phosphate (NAADP)], can also be used by some species. For instance, all three Ca^{2+}-signalling mechanisms are present at fertilization in sea urchin oocytes (Perez-Terzic et al. 1995) and in the hermaphroditic ascidian *Phallausia mammillata* (Albrieux et al. 1998). It appears also that eutherian oocytes may utilize both Ins(1,4,5)P_3- and ryanodine-sensitive stores, although this issue is controversial (for review, see Dale et al. 1999).

4.6.3 Cortical Alveolus/Granule Exocytosis

Cortical alveoli/granules are derived from the Golgi apparatus and show a great diversity in their size, number and ultrastructure among the different animal species, sometimes even in different regions of the oocyte (for review, see Guraya 1982). They contain carbohydrates and proteins, many of them with protease, glycosidase or peroxidase activity, among other activities (for reviews, see Guraya 1982; Hoodbhoy and Talbot 1994). The increase in $[Ca^{2+}]_i$ after sperm–oocyte fusion causes cortical alveolus/granule exocytosis leading to morphological and biochemical changes of the oocyte envelope(s) as well as the formation of new layers on the oocyte plasma membrane, such as the echinoderm hyaline layer, the anuran smooth layer and the mammalian cortical granule envelope. Such modifications of the pre-existing oocyte envelope(s) and formation of a new layer on the plasma membrane are two of the multiple mechanisms for prevention of polyspermy that oocytes may display at fertilization (see Table 2). We should emphasize the fact that although the prime function of cortical granule exocytosis is to prevent polyspermic fertilization, embryo development can be initiated and proceeded without cortical granule exocytosis. It is known, for instance, that in certain molluscan species from the genera *Nopalia, Mytilus, Spisula* and *Acmaea*, the brachipod *Terebratalia transversa* and echiuran worm *Urechis caupo*, cortical granules are not exocytosized at fertilization, persisting as such at least to gastrulation (for review, see Guraya 1982). Likewise, the cortical reaction is not needed for initiating embryonic development in the sea urchin and fish (for review, see Yanagimachi 1994).

In sea-urchin, fish and amphibian oocytes, cortical alveolus/granule exocytosis takes place in a wave-like fashion following the propagation of the Ca^{2+} wave. This wave-like propagation of cortical granule exocytosis is not, however, evidenced in mammal oocytes, in which most of the cortical granules are exocytosized after the first Ca^{2+} release (for review, see Yanagimachi 1994).

Little is known about the molecular mechanism of cortical alveolus/granule exocytosis in the different animal phyla. Many studies in invertebrate and vertebrate species show that Ca^{2+} is the most important messenger for cortical granule release. Activation of protein kinase C (PKC) also has been considered to be involved in the mechanism of cortical granule release (for review, see Ducibella 1996). However, although there may be a pathway in which PKC activation causes cortical granule release, it appears that this pathway is not used by the spermatozoon at fertilization, at least in mammals (Jones 1998). Several lines of evidence show that cortical alveolus/granule release is energy dependent (for review, see Yanagimachi 1994) and requires a modification, shortening or depolymerization of the cortical layer of polymerized actin through the activation of Ca^{2+}-dependent actin-regulatory proteins (DiMaggio et al. 1997; for reviews, see Ducibella 1996; Iwamatsu, this Vol.). In this context, we should note that whereas antibodies to calmodulin or calmodulin antagonists inhibit cortical granule fusion with the plasma membrane in sea-urchin oocytes (for review, see Ducibella 1996), calmodulin antibodies and antagonists fail to block cortical alveolus/granule exocytosis in fish (Gilkey 1983; Iwamatsu 1992) and mouse (Xu et al. 1996) oocytes. A rise in $[Ca^{2+}]_i$ may also activate membrane-bound phospholipases A_2 and C which produce fusogenic lysophospholipids and diacylglycerol (for review, see Yanagimachi 1994).

4.6.4 Ooplasmic Segregation

After sperm–oocyte fusion, oocytes from many groups of animals (see Table 3) exhibit spectacular cytoplasmic movements that translocate oocyte components, including mitochondria, endoplasmic reticulum, pigment granules, RNA, cytoskeletal elements and plasma membrane molecules, from one region of the oocyte to another. This ooplasmic segregation is thought to be responsible for the process of cell diversification and embryonic axis formation in early development [for reviews, see Jeffery and Bates (1989); Edwards and Beard (1997)]. In most species, a wave of cortical contraction takes place after cortical alveolus/granule exocytosis. In ascidians, which lack cortical granules, cortical contraction occurs after the release of a cell-surface N-acetylglycosaminidase which blocks sperm binding to vitelline envelope receptors (McDougall et al. 1995). It appears that the increase in $[Ca^{2+}]_i$ after sperm–oocyte fusion induces a substantial reorganization of cytoskeleton elements including actin microfilaments, microtubules and/or mitotic asters. The resulting changes in the oocyte cytoskeleton provide the driving force for ooplasmic segregation (for reviews, see Jeffery and Bates 1989; Yoshimoto and Hiramoto 1991; Shimizu 1999). The wave of surface contraction that occurs in sea-urchin oocytes at fertilization seems not to be homologous to the cortical contraction exhibited by oocytes from other animal species. In fact, sea-urchin surface contraction takes place before the propagated increase in $[Ca^{2+}]_i$. It occurs coincidently with the membrane depolarization after adherence of the fertilizing spermatozoon and sperm–oocyte fusion (Eisen et al. 1984).

Figure 5, modified from Jeffery and Bates (1989), shows some examples of ooplasmic segregation in oocytes from different animal phyla. In the nematode *Caenorhabditis elegans* (Fig. 5A), after sperm entry in the posterior pole the oocyte undergoes contractions of the anterior cortex and pseudocleavage, involving formation and then regression of a cleavage furrow. At the time of fertilization, the germ line or P-lineage granules are small, numerous and dispersed randomly throughout the cytoplasm. However, thereafter they coalesce and become localized in the posterior pole (Strome and Wood 1983). In the oligochaete annelid *Tubifex rivulorum* (Fig. 5B), oocytes deform, appearing at the periphery of the oocyte, a yolk-free cytoplasm rich in membranous organelles such as mitochondria and endoplasmic reticulum. This peripheral cytoplasm subsequently separates into animal and vegetal portions that are

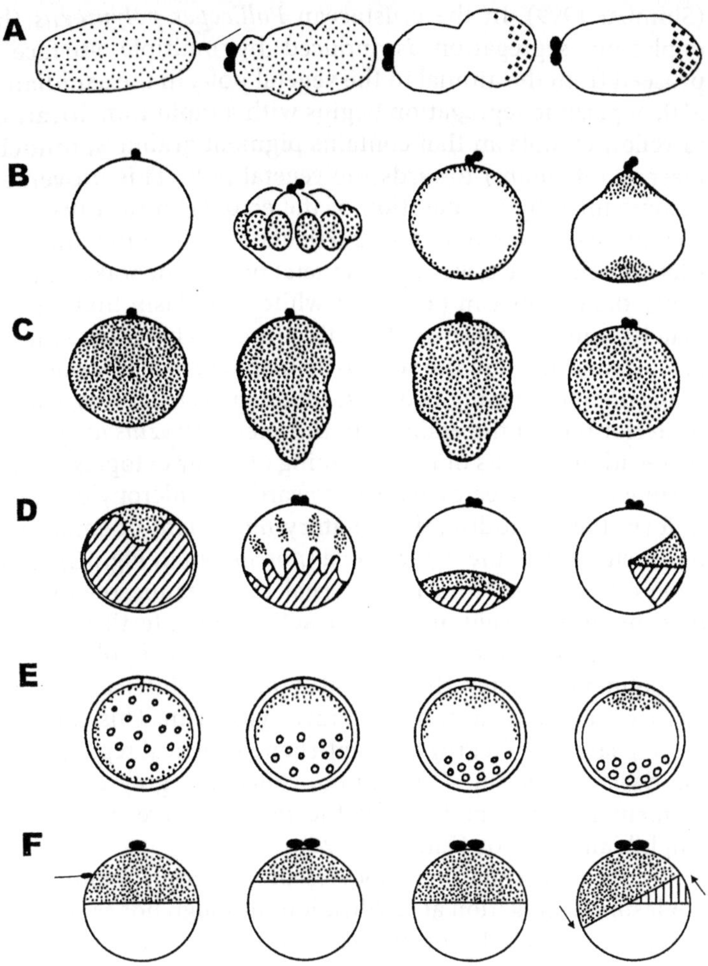

Fig. 5. Ooplasmic segregation in oocytes from different animal phyla. **A** The nematode *Caenorhabditis elegans*. *Dots* represent the germ line or P-lineage granules. **B** The oligochaete annelid *Tubifex rivulorum*. *Dots* represent the peripheral yolk-free cytoplasm rich in membranous organelles. **C** The crustacean *Pollicepes polymerus* (barnacle). **D** The ascidian *Styela plicata*. *Stripes* represent the myoplasm, *dots* the ectoplasm and the *unmarked area* the endoplasm. **E** The teleost *Oryzias latipes*. *Dots* represent the peripheral yolk-free cytoplasm and *spheres* large oil droplets. **F** In anurans, *dots* represent the pigmented cortex and *stripes* the grey crescent. (Modified from Jeffery and Bates 1989)

translocated to the poles of the oocyte. Zygotes of this type are seen in clitellates (i.e., oligochaetes and hirudinidans) but not in polychaetes. Zygotes from polychaetes show a unipolar yolk-free cytoplasm domain in the animal (*Nereis*, *Platynereis*, *Chaetopterus* and *Diopatra*) or vegetal (*Pseudopolydora*) pole. In addition, it appears that some oligochaete and branchiobdellid species undergo development without forming any domains of clear cytoplasm

(Shimizu 1999). In the crustacean *Pollicepes polymerus* (barnacle) (Fig. 5C), ooplasmic segregation is induced by multiple surface contractions that proceed from the animal to the vegetal pole. In the ascidian *Styela plicata* (Fig. 5D), ooplasmic segregation begins with a rapid translocation of the myoplasm (a yellow cytoplasm that contains pigment granules, mitochondria and endoplasmic reticulum) towards the vegetal pole. This movement of myoplasm is accompanied by a separation of the ectoplasm (a transparent cytoplasm that contains vesicular and fine particulate inclusions) into many protoplasmic islets which move towards the vegetal pole in the wake of the myoplasm. Meanwhile, the endoplasm (a grey or white cytoplasm that consists mainly of yolk platelets) moves towards the animal pole. Thereafter, the myoplasm and the ectoplasm shift from the vegetal hemisphere to a new position near the equator of the oocyte, forming two crescent-shaped structures, which define the future posterior pole of the embryo. In the teleost *Oryzias latipes* (Fig. 5E), ooplasmic segregation consists of the streaming of many cytoplasmic inclusions from the peripheral yolk-free cytoplasm toward the micropyle (point of sperm entry) to form the blastodisc; the saltatory movement of small peripheral cytoplasmic inclusions to the vegetal pole (not shown in Fig. 5E); and the movement of large oil droplets from the internal cytoplasmic region towards the vegetal pole. In anurans (Fig. 5F), oocyte activation at fertilization induces a contraction of the pigmented cortex towards the animal pole. After a re-expansion of the pigmented cortex, the cortical cytoplasm shifts about 30° towards the point of sperm entry, relative to the inner cytoplasm. This cortical rotation leads to the appearance of a grey crescent-shaped cortical region previously covered by the dark cytoplasm. This region, called the grey crescent, contains diffuse pigment granules and marks the region where gastrulation is initiated in amphibian embryos (Elinson 1975).

Recent data show that oocytes of eutherian mammals may also exhibit ooplasmic segregation at fertilization, although not as spectacularly as in other groups of animals. In the mouse, for instance, after sperm–oocyte fusion there is a microfilament-dependent transport of non-mitochondrial intracellular membranes to the cortical area between the plasma membrane and the male nucleus. The accumulation of intracellular membranes then vanishes during migration of the pronuclei to the centre of the oocyte (Tarín and Cano 1998). This mild cortical reorganization bears a resemblance to that displayed by physiologically polyspermic oocytes from the ctenophore *Beroe ovata* which experience a cortical reorganization around each sperm nucleus including a local tuft of microvilli, local exocytosis and accumulation of mitochondria and autofluorescence vesicles recruited from the surrounding cortical region. Such cortical differentiations disappear when the female pronucleus fuses with one of the sperm pronuclei (Carré et al. 1990). In human oocytes, after the extrusion of the second polar body there is a brief contraction of organelles away from the cortex of the oocyte, accompanied by a glassy cytoplasmic appearance which radiates out from the centre of the oocyte (Payne et al. 1997). However, it remains to be ascertained whether this organelle contraction is directly comparable to the cortical contraction and ooplasmic segregation exhibited by oocytes from other animal species.

4.6.5 Increase in pH_i

Oocytes from echinoderms, teleost fish and amphibians (see Table 3), as well as molluscan species that are fertilized at the prophase I stage and the echiuran worm *Urechis caupo*, all exhibit a rise in pH_i at fertilization. In contrast, no changes in pH_i are observed in the rat, mouse, and in the molluscan species that are fertilized at the MI stage (for reviews, see Ben-Yosef and Shalgi 1998; Colas and Dubé 1998). It appears that in the sea urchin, the CO_2 released by oocytes at fertilization together with the stimulation of Na^+-H^+ exchanger by a Ca^{2+}-calmodulin-dependent mechanism are responsible for this pH_i increase. As the pH_i increase is caused by the activation of the plasma membrane Na^+-H^+ antiporter, an influx of Na^+ and efflux of H^+ are observed shortly after sperm–oocyte fusion. *Spisula solidissima* and *Xenopus laevis* oocytes express also the Na^+-H^+ exchanger, although in *Spisula solidissima* it is regulated by PKC and in *Xenopus laevis* the alkalinization of the oocyte cytoplasm is independent of external Na^+. In this context, it is worth mentioning that mouse oocytes and embryos do not show Na^+-H^+ antiport activity, apparently. This may explain, at least in part, the unaltered pH_i exhibited by mouse and rat oocytes at fertilization (for review, see Ben-Yosef and Shalgi 1998).

4.6.6 Resumption of the Cell Cycle

Oocytes can exhibit a pre-fertilization arrest at prophase I, MI, MII or after completing meiosis at the pronucleus (G_1 or G_2) stage (see Table 2). In most species, fertilization provides the signal for the oocyte to resume the cell cycle and develop a new individual. However, in some species, including the starfish and many arthropods, resumption of the cell cycle is independent of sperm entry. In these, fertilization takes place at any stage between prophase I and the completion of meiosis. Table 2 shows that most vertebrate species exhibit a pre-fertilization arrest at the MII stage, whereas the rule for invertebrates is the prophase I or the MI stage.

Cell cycle resumption can be explained in terms of activation and inactivation of cell cycle control proteins, particularly the M-phase-promoting factor [MPF, which is a protein kinase consisting of two subunits identified as the cdc2 gene product catalytic subunit ($p34^{cdc2}$) and the cyclin B regulatory subunit], mitogen-activated protein kinase (MAPK) and the product of the oncogene *mos* [Mos; cytostatic factor (CSF)], which can directly phosphorylate and activate an upstream activator of MAPK (MAPK kinase) leading to the activation of the MAPK pathway. In oocytes arrested at MI or MII, the fertilization $[Ca^{2+}]_i$ transient activates calmodulin, which activates CaMKinase II,

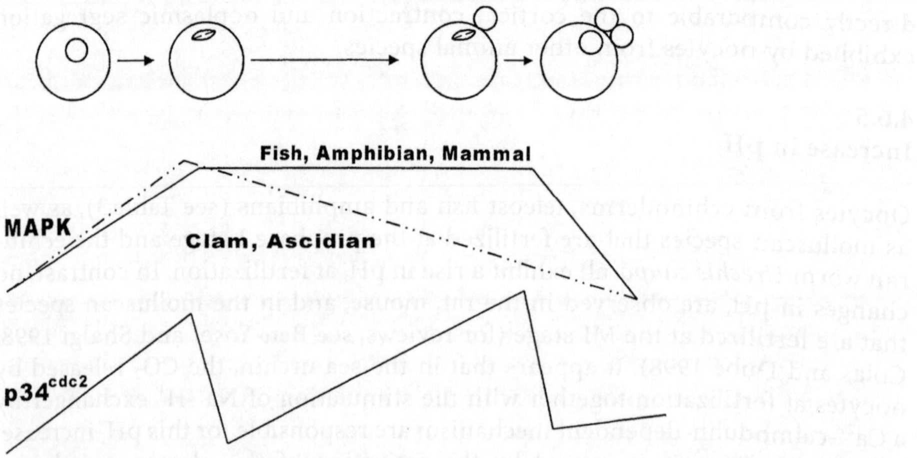

Fig. 6. Mitogen-activated protein kinase (MAPK) and p34^{cdc2} during oocyte maturation and fertilization. MAPK activity stays high throughout maturation in oocytes that exhibit a pre-fertilization arrest at MII (fish, amphibian and therian mammal) but falls before MII in those oocytes displaying a pre-fertilization arrest at prophase I (clam) or MI (ascidian). p34^{cdc2} activity shows a peak at metaphase of first and the second meiotic division. (Modified from Whitaker 1996)

which in turn activates degradation of Mos and cyclin, and hence causes MAPK and MPF inactivation. MAPK activity remains high throughout maturation in oocytes that exhibit a pre-fertilization arrest at MII (fishes, amphibians and therian mammals) but falls before MII in those oocytes displaying a pre-fertilization arrest at prophase I (the clam *Spisula*) or MI (the ascidian *Ascidiella*; Shibuya et al. 1992; McDougall and Levasseur 1998; for review, see Whitaker 1996; Fig. 6). The decrease in MAPK activity correlates temporally with the formation of the pronuclear envelopes. Sea-urchin oocytes that are fertilized at interphase of the first mitotic cell cycle show no MPF activity (p34^{cdc2} is dephosphorylated and there is no cyclin subunit with which to associate) and have little active MAPK. However, immediately after fertilization oocytes exhibit a transient increase in MAPK activity, which is likely triggered by the fertilization $[Ca^{2+}]_i$ increase, and p34^{cdc2} is phosphorylated. Thereafter, MAPK activity falls, only to rise again during mitosis in at least the first two cleavage cycles (Philipova and Whitaker 1998). In contrast, starfish oocytes, which arrest either at the G_1 (*Asterina pectinifera* and *Asterina miniata*) or G_2 (*Marsthasterias glacialis* and *Astropecten aranciacus*) pronuclear stage, if not fertilized at late meiosis I (natural situation), have elevated levels of MAPK activity after germinal vesicle breakdown. At fertilization of pronucleated oocytes, a fall is observed in MAPK activity, which is necessary and sufficient for triggering the G_1/S-phase (*Asterina pectinifera* and *Asterina miniata*) or G_2/M-phase transition (*Marsthasterias glacialis* and *Astropecten aranciacus*) (Kishimoto 1998).

4.6.7
Development of the Male Pronucleus

With the exception of some decapod crustaceans, sperm chromatin is tightly packed within the sperm nucleus. This is achieved by replacement of somatic histones by sperm-nuclear basic proteins (SNBPs). In contrast to the evolutionarily conservative somatic histones, SNBPs display a high degree of diversity among species, ranging from the sperm-specific histones in poriferans to the highly specialized mammalian protamines (Ausio et al. 1997). Although particular details involving the development of the male pronucleus at fertilization can differ from one species to another, there are some traits common to all the species analysed so far. In brief, (1) the poreless sperm nuclear membrane, if present, breaks down; (2) disassembly of the sperm nuclear lamina (a polymeric meshwork underlying the inner nuclear membrane); (3) chromatin decondensation with the concomitant replacement of all or most SNBPs by oocyte histones; (4) reformation of a nuclear membrane including pores and a lamina; and (5) swelling of the nucleus (for review, see Poccia and Collas 1997). It appears that in oocytes fertilized at the pronuclear stage, such as occurs in the sea urchin, the rate of nuclear expansion is uniform, whereas in oocytes inseminated at an arrested stage of meiosis, nuclear dispersion does not proceed at a uniform rate but consists of phases coordinate with stages of meiotic maturation of the maternal chromatin (Longo and Scarpa 1991).

A different mechanism(s) of sperm nuclear decondensation appears to exist among species. This variability depends basically on the extent to which SNBP phosphorylation, reduction of disulfide (-S-S-) bonds between SNBPs, and/or nucleoplasmin-mediated disassembly of SNBPs are required for sperm nucleus decondensation. The extremes would be represented by species relying totally (or mainly) on either SNBP phosphorylation (sea-urchin type) or nucleoplasmin-mediated disassembly of SNBP (*Xenopus* type). Intermediate forms would be represented by species with varying degrees of SNBP phosphorylation, -S-S- bond reduction and/or nucleoplasmin-mediated disassembly of SNBPs. In this context, we should bear in mind that teleost fish, birds, monotremes and all but one genus (the dasyurid *Planigale*) of marsupials lack cysteine residues in their SNBPs (Retief et al. 1995). Mammalian and teleost protamines and the high-molecular-weight SNBPs of flounder (fish) have abundant potential phosphorylation sites that become dephosphorylated during spermatogenesis. For this reason, SNBP phosphorylation may have a role in dispersion of the decondensed sperm chromatin in these groups of animals (for reviews, see Longo 1985; Poccia 1989).

Activation of the oocyte is not required for decondensation of the sperm nucleus but is a *sine qua non* for transformation of the decondensed sperm nucleus into a pronucleus. It is important to note also that the new nuclear membrane is required for DNA synthesis, regulated protein import and RNA export in those species such as the sea urchin (Longo and Kunkle 1977) whose male pronuclei are capable of RNA synthesis. Although most of the new nuclear envelope is derived from the endoplasmic reticulum, other oocyte cytoplasmic membranes distinct from endoplasmic reticulum as well as remnants of the sperm nuclear envelope contribute to the formation of the male pronuclear envelope (for review, see Poccia and Collas 1997).

4.6.8
Pronuclear Apposition and Karyogamy

While both the sperm and the oocyte contribute equal haploid genomes during fertilization, it appears that nearly all animals (rodent species are exceptions) rely on the paternal contribution of the centrosome. At fertilization, this centrosome directs the organization of microtubules which together with the oocyte microfilament system are involved in pronuclear migration to the centre of the oocyte. The sperm centrosome also uses oocyte constituents to form the zygote centrosome which serves as the poles of the spindle of the first mitotic division (for review, see Schatten 1994).

During migration, pronuclei undergo DNA synthesis, although in the sea urchin *Arbacia* DNA synthesis occurs after fusion of the male and female pronuclei (for review, see Longo 1985). When pronuclei are adjacent to one another, either they can fuse (karyogamy, strictly speaking) to produce a single zygote nucleus (sea-urchin type of fertilization) or, alternatively, chromatin transforms into mitotic chromosomes, pronuclear membranes break down and the chromosomal groups derived from each parent intermingle to form the metaphase plate of the first mitotic division (*Ascaris* type of fertilization). In species exhibiting the *Ascaris* type of fertilization, pronuclei do not fuse to form a zygote nucleus, although the intermingling of chromosomes can be considered as karyogamy in the broad sense of the word. In the dipteran *Drosophila*, lepidopterans *Bombyx* and *Pieris*, and the linden bug *Pyrrhocoris*, each pronucleus produces a separated spindle that fuses to form a single spindle at metaphase. The maternal and paternal sets of chromosomes remain separate (gonomery) and meet only when they approach the spindle poles at the end of the first cleavage mitosis (for review, see Sander 1985). Although there may be intermediate forms, the sea-urchin type of fertilization is usually observed in oocytes fertilized at interphase after having completed meiosis, whereas the *Ascaris* type of fertilization is observed in oocytes fertilized prior to the completion of the meiosis, i.e. at prophase I, MI or MII (for review, see Longo 1985).

Acknowledgements. I would like to express my sincere gratitude to Drs. Michael Bedford, Antonio Cano, Takashi Iwamatsu, Yasuhiro Iwao, Kazuyuki Mikami, Celso Rodríguez, Heide Schatten, Stephen A. Stricker, Michael Whitaker, Graham J. Wishart and Ryuzo Yanagimachi who provided valuable comments on the original manuscript. I am also in debt to many outstanding colleagues working in different areas of fertilization who sent me reprints of their work. This study was supported by a grant from Conselleria de Cultura, Educació i Ciència, Generalitat Valenciana (GV99-138-1-04).

References

Ahnonkitpanit V, White D, Suwajanakorn S, Kan FW, Namking M, Wells G, Kamolvarin N, Tanphaichitr N (1999) Role of egg sulfolipid immobilizing protein 1 on mouse sperm–egg plasma membrane binding. Biol Reprod 61:749–756

Albrieux M, Lee HC, Villaz M (1998) Calcium signaling by cyclic ADP-ribose, NAADP, and inositol triphosphate are involved in distinct functions in ascidians' oocytes. J Biol Chem 273:14566–14574

Ausio J, Van Veghel MLJ, Gomez R, Barreda D (1997) The sperm nuclear basic proteins (SNBPs) of the sponge *Neofibularia nolitangere*: implications for the molecular evolution of SNBPs. Mol Evol 45:91–96

Bedford JM (1991) The coevolution of mammalian gametes. In: Dunbar BS, O'Rand MG (eds) A comparative overview of mammalian fertilization. Plenum Press, New York, pp 3–35

Bedford JM (1998) Mammalian fertilization misread? Sperm penetration of the eutherian zona pellucida is unlikely to be a lytic event. Biol Reprod 59:1275–1287

Bedford JM (1999) The implications of unusual sperm/female relationships in mammals. In: Gagnon C (ed) The mañe gamete. Cache River Press, Vienna, Illinois, pp 81–92

Ben-Yosef D, Shalgi R (1998) Early ionic events of the mammalian egg. Rev Reprod 3:96–103

Brewis IA, Wong CH (1999) Gamete recognition: sperm proteins that interact with the egg zona pellucida. Rev Reprod 4:135–142

Carré D, Sardet C, Rouviere C (1990) Fertilization in ctenophores. In: Dale B (ed) Mechanism of fertilization: plants to humans. Springer, Berlin Heidelberg New York, pp 625–636

Carré D, Rouviere C, Sardet C (1991) In vitro fertilization in ctenophores: sperm entry, mitosis, and the establishment of bilateral symmetry in *Beroe ovata*. Dev Biol 147:381–391

Chambers EL, McCulloh DH (1990) Excitation, activation and sperm entry in voltage-clamped sea-urchin eggs. J Reprod Fertil (Suppl) 42:117–132

Colas P, Dubé F (1998) Meiotic maturation in mollusc oocytes. Semin Cell Dev Biol 9:539–548

Correa LM, Carroll EJ Jr (1997) Identification of a new sea urchin vitelline envelope sperm binding glycoprotein. Dev Growth Differ 39:773–786

Dale B (1994) Oocyte activation in invertebrates and humans. Zygote 2:373–377

Dale B, Marino M, Wilding M (1999) Sperm-induced calcium oscillations. Soluble factor, factor or receptors? Mol Hum Reprod 5:1–4

David C, Halliwell J, Whitaker MJ (1988) Some properties of the membrane currents underlying the fertilization potential in sea-urchin eggs. J Physiol (Lond) 402:139–154

DiMaggio AJ Jr, Lonergan TA, Stewart-Savage J (1997) Cortical granule exocytosis in hamster eggs requires microfilaments. Mol Reprod Dev 47:334–340

Ducibella T (1996) The cortical reaction and development of activation competence in mammalian oocytes. Hum Reprod Update 2:29–42

Eckberg WR, Miller AL (1995) Propagated and nonpropagated calcium transients during egg activation in the annelid, *Chaetopterus*. Dev Biol 172:654–664

Edwards RG, Beard HK (1997) Oocyte polarity and cell determination in early mammalian embryos. Mol Hum Reprod 3:863–905

Eisen A, Kiehart DP, Wieland SJ, Reynolds GT (1984) Temporal sequence and spatial distribution of early events of fertilization in single sea urchin eggs. J Cell Biol 99:1647–1654

Elinson RP (1975) Site of sperm entry and a cortical contraction associated with egg activation in the frog *Rana pipiens*. Dev Biol 47:257–268

Fénichel P, Durand-Clément M (1998) Role of integrins during fertilization in mammals. Hum Reprod 13 (Suppl 4):31–46

Foltz KR (1995) Sperm-binding proteins. Int Rev Cytol 163:249–303

Freeman G (1996) The role of localized cell surface-associated glycoproteins during fertilization in the hydrozoan *Aequorea*. Dev Biol 179:17–26

Gilkey JC (1983) Roles of calcium and pH in activation of eggs of the medaka fish, *Oryzias latipes*. J Cell Biol 97:669–678

Goudeau H, Goudeau M (1989) A long-lasting electrically mediated block, due to the egg membrane hyperpolarization at fertilization, ensures physiological monospermy in eggs of the crab *Maia squinado*. Dev Biol 133:348–360

Guraya SS (1982) Recent progress in the structure, origin, composition, and function of cortical granules in animal egg. Int Rev Cytol 78:257–360

Hoodbhoy T, Talbot P (1994) Mammalian cortical granules: contents, fate, and function. Mol Reprod Dev 39:439–448

Hughes J, Ward CJ, Aspinwall R, Butler R, Harris PC (1999) Identification of a human homologue of the sea urchin receptor for egg jelly: a polycystic kidney disease-like protein. Hum Mol Genet 8:543–549

Iwamatsu T (1992) Egg activation. Fish Biol J Medaka 4:1–9

Jaffe LA, Gould M (1985) Polyspermy-preventing mechanisms. In: Metz CB, Monroy A (eds) Biology of fertilization, vol III. Academic Press, New York, pp 223–250

Jaffe LA, Sharp AP, Wolf DP (1983) Absence of electrical polyspermy block in the mouse. Dev Biol 96:317–323

Jeffery WR, Bates WR (1989) Ooplasmic segregation in the ascidian *Styela*. In: Shatten H, Shatten G (eds) The molecular biology of fertilization. Academic Press, San Diego, pp 340–367

Jones KT (1998) Protein kinase C action at fertilization: overstated or undervalued? Rev Reprod 3:7–12

Kishimoto T (1998) Cell cycle arrest and release in starfish oocytes and eggs. Semin Cell Dev Biol 9:549–557

Kobayashi W, Baba Y, Shimozawa T, Yamamoto TS (1994) The fertilization potential provides a fast block to polyspermy in lamprey eggs. Dev Biol 161:552–562

Koehler JK (1985) Gamete surface and their interactions. Liss, New York

Koyota S, Wimalasiri KM, Hoshi M (1997) Structure of the main saccharide chain in the acrosome reaction-inducing substance of the starfish, *Asterias amurensis*. J Biol Chem 272:10372–10376

Longo FJ (1985) Pronuclear events during fertilization. In: Metz CB, Monroy A (eds) Biology of fertilization, vol III. Academic Press, New York, pp 251–298

Longo FJ, Kunkle M (1977) Synthesis of RNA by male pronuclei of fertilized sea urchin eggs. J Exp Zool 201:431–438

Longo FJ, Scarpa J (1991) Expansion of the sperm nucleus and association of the maternal and paternal genomes in fertilized *Mulinia lateralis* eggs. Biol Bull 180:56–64

Lyons CE, Payette KL, Price JL, Huang RCC (1993) Expression and structural analysis of a teleost homolog of a mammalian zona pellucida gene. J Biol Chem 268:21351–21358

Maleszewski M, Kimura Y, Yanagimachi R (1996) Sperm membrane incorporation into oolemma contributes to the oolemma block to sperm penetration: evidence based on intracytoplasmic sperm injection experiments in the mouse. Mol Reprod Dev 44:256–259

Masui Y (1985) Meiotic arrest in animal oocytes. In: Metz CB, Monroy A (eds) Biology of fertilization, vol I. Academic Press, New York, pp 189–219

McCulloh DH, Rexroad CE Jr, Levitan H (1983) Insemination of rabbit eggs is associated with slow depolarization and repetitive diphasic membrane potentials. Dev Biol 95:372–377

McDougall A, Levasseur M (1998) Sperm-triggered calcium oscillations during meiosis in ascidian oocytes first pause, restart, then stop: correlations with cell cycle kinase activity. Development 125:4451–4459

McDougall A, Sardet C (1995) Function and characteristics of repetitive calcium waves associated with meiosis. Curr Biol 5:318–328

McDougall A, Sardet C, Lambert CC (1995) Different calcium-dependent pathways control fertilization-triggered glycoside release and the cortical contraction in ascidian eggs. Zygote 3:251–258

McLesky SB, Dowds C, Carballada R, White RR, Saling PM (1998) Molecules involved in mammalian sperm–egg interaction. Int Rev Cytol 177:57–113

Nishigaki T, Chiba K, Miki W, Hoshi M (1996) Structure and function of asterosaps, sperm-activating peptides from the jelly coat of starfish eggs. Zygote 4:237–245

Nuccitelli R (1980) The fertilization potential is not necessary for the block to polyspermy or the activation of development in the medaka egg. Dev Biol 76:499–504
Ohlendieck K, Lennarz WJ (1995) Role of the sea urchin egg receptor for sperm in gamete interactions. Trends Biochem Sci 20:29–33
Parekh AB, Penner R (1997) Store depletion and calcium influx. Physiol Rev 77:901–930
Parrington J, Jones KT, Lai A, Swann K (1999) The soluble sperm factor that causes Ca^{2+} release from sea-urchin (*Lytechinus pictus*) egg homogenates also triggers Ca^{2+} oscillations after injection into mouse eggs. Biochem J 341:1–4
Payne D, Flaherty SP, Barry MF, Matthews CD (1997) Preliminary observations on polar body extrusion and pronuclear formation in human oocytes using time-lapse video cinematography. Hum Reprod 12:532–541
Perez-Terzic CM, Chini EN, Shen SS, Dousa TP, Clapham DE (1995) Ca^{2+} release triggered by nicotinate adenine dinucleotide phosphate in intact sea urchin eggs. Biochem J 312(3):955–959
Perry ACF, Wakayama T, Yanagimachi R (1999) A novel trans-complementation assay suggests full mammalian oocyte activation is coordinately initiated by multiple, submembrane sperm components. Biol Reprod 60:747–755
Philipova R, Whitaker M (1998) MAP kinase activity increases during mitosis in early sea urchin embryos. J Cell Sci 111(17):2497–2505
Poccia D (1989) Reactivation and remodeling of the sperm nucleus following fertilization. In: Schatten H, Schatten G (eds) The molecular biology of fertilization. Academic Press, San Diego, pp 115–135
Poccia D, Collas P (1997) Nuclear envelope dynamics during male pronuclear development. Dev Growth Differ 39:541–550
Retief JD, Rees JS, Westerman M, Dixon GH (1995) Convergent evolution of cysteine residues in sperm protamines of one genus of marsupials, the Planigales. Mol Biol Evol 12:708–712
Sander K (1985) Fertilization and egg cell activation in insects. In: Metz CB, Monroy A (eds) Biology of fertilization, vol II. Academic Press, New York, pp 409–430
Sase I, Okinaga T, Hoshi M, Feigenson GW, Kinosita K Jr (1995) Regulatory mechanisms of the acrosome reaction revealed by multiview microscopy of single starfish sperm. J Cell Biol 131:963–973
Schatten G (1994) The centrosome and its mode of inheritance: the reduction of the centrosome during gametogenesis and its restoration during fertilization. Dev Biol 165:299–335
Shibuya EK, Boulton TG, Cobb MH, Ruderman JV (1992) Activation of p42 MAP kinase and the release of oocytes from cell cycle arrest. EMBO J 11:3963–3975
Shimizu T (1999) Cytoskeletal mechanisms of ooplasmic segregation in annelid eggs. Int J Dev Biol 43:11–18
Stricker SA (1997) Intracellular injections of a soluble sperm factor trigger calcium oscillations and meiotic maturation in unfertilized oocytes of a marine worm. Dev Biol 186:185–201
Stricker SA (1999) Comparative biology of calcium signaling during fertilization and egg activation in animals. Dev Biol 211:157–176
Strome S, Wood WB (1983) Generation of asymmetry and segregation of germ-line granules in early *C. elegans* embryos. Cell 35:15–25
Suzuki N (1995) Structure, function and biosynthesis of sperm-activating peptides and fucose sulfate glycoconjugate in the extracellular coat of sea urchin eggs. Zool Sci 12:13–27
Swann K, Lai FA (1997) A novel signalling mechanism for generating Ca^{2+} oscillations at fertilization in mammals. Bioessays 19:371–378
Swann K, Parrington J, Jones KT (1999) On the search for the sperm oscillogen. Mol Hum Reprod 4:1010–1012
Swanson WJ, Vacquier VD (1997) The abalone egg vitelline envelope receptor for sperm lysin is a giant multivalent molecule. Proc Natl Acad Sci USA 94:6724–6729
Tarín JJ, Cano A (1998) Distribution of 5-chloromethylfluorescein diacetate staining during meiotic maturation and fertilization in vitro of mouse oocytes. J Reprod Fertil 114:211–218

Tulsiani DRP, Yoshida-Komiya H, Araki Y (1997) Mammalian fertilization: a carbohydrate-mediated event. Biol Reprod 57:487–494
Waclawek M, Foisner R, Nimpf J, Schneider WJ (1998) The chicken homologue of zona pellucida protein-3 is synthesized by granulosa cells. Biol Reprod 59:1230–1239
Wassarman PM (1999) Mammalian fertilization: molecular aspects of gamete adhesion, exocytosis, and fusion. Cell 96:175–183
Whitaker M (1996) Control of meiotic arrest. Rev Reprod 1:127–135
Xu Z, Lefevre L, Ducibella T, Schultz RM, Kopf GS (1996) Effects of calcium-BAPTA buffers and the calmodulin antagonist W-7 on mouse egg activation. Dev Biol 180:594–604
Yanagimachi R (1988) Sperm–egg fusion. In: Bronner F, Duzgunes N (eds) Membrane fusion in fertilization, cellular transport and viral infection. Academic Press, Orlando, pp 3–43
Yanagimachi R (1994) Mammalian fertilization. In: Knobil E, Neill JD (eds) The physiology of reproduction, 2nd edn. Raven Press, New York, pp 189–317
Yoshimoto Y, Hiramoto Y (1991) Observation of intracellular Ca^{2+} with aequorin luminescence. Int Rev Cytol 129:45–73
Young GP, Young JD, Deshpande AK, Goldstein M, Koide SS, Cohn ZA (1984) A Ca^{2+}-activated channel from *Xenopus laevis* oocyte membranes reconstituted into planar bilayers. Proc Natl Acad Sci USA 81:5155–5159

Subject Index

MIX
Papier aus verantwortungsvollen Quellen
Paper from responsible sources
FSC® C105338

If you have any concerns about our products,
you can contact us on
ProductSafety@springernature.com

In case Publisher is established outside the EU,
the EU authorized representative is:
Springer Nature Customer Service Center GmbH
Europaplatz 3, 69115 Heidelberg, Germany

Printed by Libri Plureos GmbH
in Hamburg, Germany